Structural Optimization

Volume 2

Mathematical Programming

MATHEMATICAL CONCEPTS AND METHODS IN SCIENCE AND ENGINEERING

Series Editor: Angelo Miele
Mechanical Engineering and Mathematical Sciences
Rice University

Recent volumes in this series:

31 **NUMERICAL DERIVATIVES AND NONLINEAR ANALYSIS**
• *Harriet Kagiwada, Robert Kalaba, Nima Rasakhoo, and Karl Spingarn*

32 **PRINCIPLES OF ENGINEERING MECHANICS**
Volume 1: Kinematics—The Geometry of Motion • *Millard F. Beatty, Jr.*

33 **PRINCIPLES OF ENGINEERING MECHANICS**
Volume 2: Dynamics—The Analysis of Motion • *Millard F. Beatty, Jr.*

34 **STRUCTURAL OPTIMIZATION**
Volume 1: Optimality Criteria • *Edited by M. Save and W. Prager*

35 **OPTIMAL CONTROL APPLICATIONS IN ELECTRIC POWER SYSTEMS**
G. S. Christensen, M. E. El-Hawary, and S. A. Soliman

36 **GENERALIZED CONCAVITY** • *Mordecai Avriel, Walter E. Diewert, Siegfried Schaible, and Israel Zang*

37 **MULTICRITERIA OPTIMIZATION IN ENGINEERING AND IN THE SCIENCES**
• *Edited by Wolfram Stadler*

38 **OPTIMAL LONG-TERM OPERATION OF ELECTRIC POWER SYSTEMS**
• *G. S. Christensen and S.A. Soliman*

39 **INTRODUCTION TO CONTINUUM MECHANICS FOR ENGINEERS**
• *Ray M. Bowen*

40 **STRUCTURAL OPTIMIZATION**
Volume 2: Mathematical Programming • *Edited by M. Save and W. Prager*

41 **OPTIMAL CONTROL OF DISTRIBUTED NUCLEAR REACTORS**
• *G. S. Christensen, S. A. Soliman, and R. Nieva*

42 **NUMERICAL SOLUTION OF INTEGRAL EQUATIONS**
• *Edited by Michael A. Golberg*

Structural Optimization

Edited by

M. Save

Polytechnic Institute of Mons
Mons, Belgium

and

W. Prager

Volume 2
Mathematical Programming

A. Borkowski and S. Jendo

Institute of Fundamental Technological Research
Polish Academy of Sciences
Warsaw, Poland

With a contribution by

M. I. Reitman

Boston University
Boston, Massachusetts

Technical Editor

W. H. Warner

University of Minnesota
Minneapolis, Minnesota

PLENUM PRESS • NEW YORK AND LONDON

Library of Congress Cataloging in Publication Data

(Revised for vol. 2)
Structural optimization.

(Mathematical concepts and methods in science and engineering; 40)
Vol. 2: By A. Borkowski and S. Jendo, with a contribution by M. I. Reitman.
Includes bibliographical references and indexes.
Contents: v. 1. Optimality criteria – v. 2. Mathematical programming.
1. Structural optimization. I. Save, M. II. Prager, William, 1903–1980. III. Sacchi, G. (Gianantonnio) IV. Warner, William H. (William Hamer), 1929– . V. Series: Mathematical concepts and methods in science and engineering; 40, etc.
TA658.2.S757 1985 624.1′771 85-22590
ISBN 0-306-41862-2

A Division of Plenum Publishing Corporation
233 Spring Street, New York, N.Y. 10013

Printed in the United States of America

Preface

Following the general scheme that the late Professor William Prager and I agreed upon, this second volume of the treatise on structural optimization is devoted to numerical methods with a strong emphasis on mathematical programming, as indicated by its title. Though it is intended to be used as a textbook, like the first volume, which presents optimality conditions and solutions based on them, this volume, through its examples, brings us closer to the many practical applications of optimization techniques. The methods developed will be used in the forthcoming third volume together with any necessary modifications to account for technical constraints, rules, and codes to treat problems of practical optimal design of metal and concrete structures.

Most of the present volume is the joint work of Professors Adam Borkowski and Stefan Jendo of the Institute for Fundamental Technological Research of the Polish Academy of Sciences in Warsaw. Professor Borkowski's expertise in linear and nonlinear mathematical programming began with his collaboration with Professor A. A. Čyras of Vilniu University in the Soviet Union; he has continued his interests throughout his career at the Academy of Sciences in Warsaw. He has visited a number of important centers outside Poland, such as the Institute for Statics and Dynamics of Aerospace Structures at Stuttgart, directed by Professor J. Argyris. His name was suggested to me by the late Professor Antoni Sawczuk at the very beginning of this project as one that we should have on our team of authors.

Professor Borkowski himself wished to share the work with his colleague Professor Jendo, who has been involved for years in structural optimization based on mathematical programming methods. Besides his work in Poland, Professor Jendo has been a visiting professor at the University of Michigan, Ann Arbor, and Essen University, West Germany.

There remained the task of writing the very difficult chapter on optimization of plates and shells. The first two authors preferred to leave this to someone with more specialized knowledge in these areas. It took some time before this problem was resolved satisfactorily. Thanks to a suggestion by Professor M. Z. Cohn, Professor Mark Reitman, recently arrived in the United States and on the faculty of Boston University, was approached, and he agreed to accept this task.

Clearly, coordinating and improving texts by three authors whose native languages are Polish and Russian was not easy. I am pleased to emphasize here the success achieved in his task by the technical editor, Professor William Warner, after a great deal of work, during which his patience and perseverance were remarkable and for which he deserves many thanks.

M. Save

Contents

10

Elements of Mathematical Programming

10.1. Basic Concepts

Mathematical programming (MP) is a branch of mathematics dealing with the constrained extrema of functions. A complete and rigorous presentation of that theory can be found in numerous textbooks and monographs [e.g., Avriel (1976), Dantzig (1963), Mangasarian (1969), and Zangwill (1969)]. Here we confine ourselves to a concise statement of primary notions, assumptions, and results of MP theory.

Whenever possible we adopt matrix notation. Lowercase boldface letters denote single-column matrices (vectors), and boldface capital letters refer to rectangular matrices. Two kinds of brackets are used to distinguish single-column and rectangular matrices written explicitly in terms of their elements: e.g., $\mathbf{a} = \{a_i\}$ and $\mathbf{A} = [A_{ij}]$. The superscript "T" means transpose; hence $\mathbf{a}^\mathrm{T}$ is the single-row matrix.

Let $\mathbf{x}$ be a point in n-dimensional Euclidean space E^n and let the function f map that point into the scalar value $f(\mathbf{x})$. We assume in the sequel that f is continuous together with its first and second derivatives. The gradient of such a function is the n-dimensional matrix $\nabla f = \{\partial f/\partial x_i\}$, and its Hessian is the $n \times n$ matrix $\mathbf{F} = [\partial^2 f/\partial x_i\, \partial x_j]$.

As we shall see, the notion of convexity is the most crucial in MP theory. The set Ω is *convex* if every point on the line segment connecting two arbitrary points selected in that set also belongs to Ω (Fig. 10.1a). Figure 10.1b shows an example of a nonconvex set.

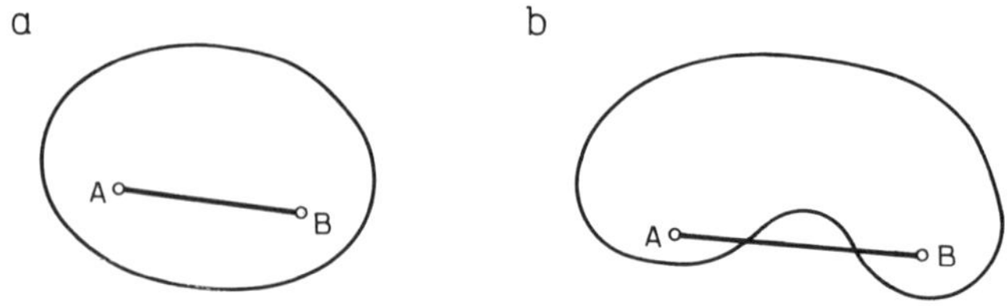

Figure 10.1. Convexity of Sets: (a) Convex Set. (b) Nonconvex Set.

The function f defined on the convex set Ω is convex if the line segment connecting two arbitrarily chosen function values never lies below the intermediate values of f (Fig. 10.2a). Analytically this requirement can be expressed as

$$f(\lambda x_1 + (1-\lambda)x_2) \leqslant \lambda f(x_1) + (1-\lambda)f(x_2) \tag{10.1}$$

for any $0 \leqslant \lambda \leqslant 1$ and any $x_1, x_2 \in \Omega$. Changing the sign of inequality (10.1) to "$\geqslant$," we obtain the definition of a concave function (Fig. 10.2b). An example of a function that is neither convex nor concave is shown in Fig. 10.2c. The function f is called strictly convex if strong inequality holds in (10.1). Then $-f$ is strictly concave. For example, the linear function $f = \mathbf{a}^T\mathbf{x}$ is simultaneously convex and concave, though obviously not strictly. The quadratic function $f = \frac{1}{2}\mathbf{x}^T\mathbf{F}\mathbf{x} + \mathbf{a}^T\mathbf{x}$ is convex if its Hessian is positive semidefinite—i.e., if $\mathbf{x}^T\mathbf{F}\mathbf{x} \geqslant 0$ for any $\mathbf{x} \neq \mathbf{0}$. If $\mathbf{x}^T\mathbf{F}\mathbf{x} > 0$ for any $\mathbf{x} \neq \mathbf{0}$, then $\mathbf{F}$ is a positive-definite matrix and f is strictly convex. Similarly the negative-definiteness of $\mathbf{F}$ leads to the concavity of f.

If the function $f = f(\mathbf{x}, \mathbf{y})$ is convex with respect to $\mathbf{x}$ and concave with respect to $\mathbf{y}$ (Fig. 10.3), it is called a *saddle-shaped function* or simply a *saddle* function. Performing a Legendre transformation with $\mathbf{y}$ active, we replace f by its jointly convex counterpart

$$f' = f - \mathbf{y}^T\nabla_y f \tag{10.2}$$

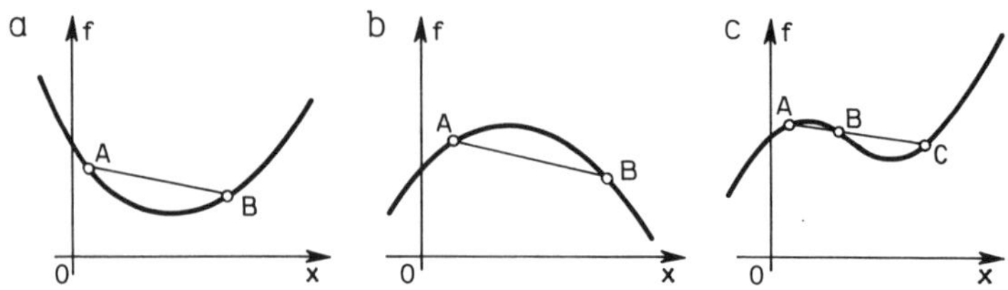

Figure 10.2. Convexity of Functions: (a) Convex Function. (b) Concave Function. (c) Nonconvex Function.

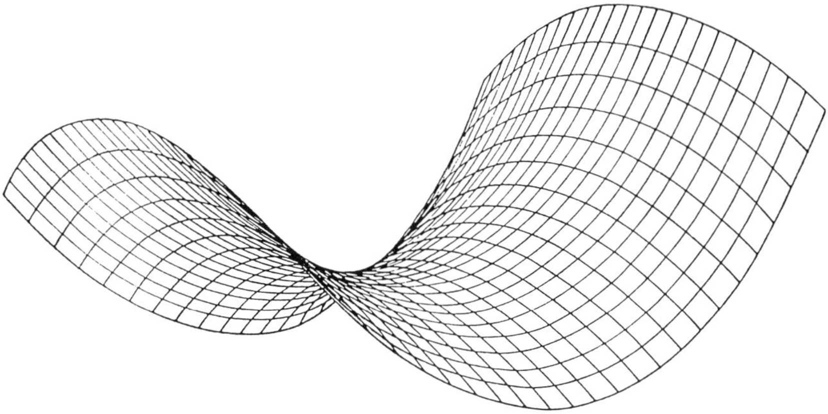

Figure 10.3. Saddle-Shaped Function.

whereas a Legendre transformation with $\mathbf{x}$ active yields the jointly concave function

$$f'' = f - \mathbf{x}^{\mathrm{T}}\nabla_x f. \tag{10.3}$$

Here $\nabla_x f$ and $\nabla_y f$ are the gradients of f with respect to $\mathbf{x}$ and $\mathbf{y}$.

Exercise 10.1

Check graphically that the set

$$\Omega_1 = \{\mathbf{x}\colon x_1^2 + x_2^2 - 4 \leqslant 0,\ x_1^2 - x_2 \leqslant 0,\ x_1 \geqslant 0\}$$

is convex and that the set

$$\Omega_2 = \{\mathbf{x}\colon x_1^2 + x_2^2 - 4 \leqslant 0,\ x_1^2 - x_2 \geqslant 0,\ x_1 \geqslant 0\}$$

is not.

Exercise 10.2

Check analytically that the function $f = 2x_1^2 + 6x_2^2 - x_1 - x_2 - 4$ is strictly convex over $\mathbf{x} \geqslant \mathbf{0}$.

Exercise 10.3

Find Legendre transforms of the saddle function $f = x^2 - y^2$.

10.2. Classification of MP Problems

The MP problem is generally formulated as follows: find an extremum of a given function $f(\mathbf{x})$ subject to a given set of equality and/or inequality constraints on $\mathbf{x}$. Since maximization of f is equivalent to minimization of $-f$ and the equality $g_i(\mathbf{x}) = 0$ can be replaced by the two inequalities $-g_i(\mathbf{x}) \leq 0$, $g_i(\mathbf{x}) \leq 0$, the MP problem can be formulated without loss of generality as

$$\min\{f(\mathbf{x}) \mid g_i(\mathbf{x}) \leq 0, i = 1, 2, \ldots, m, \mathbf{x} \in E^n\}. \tag{10.4}$$

The function f is called the *objective* function. The constraints define the *admissible domain* $\Omega = \{\mathbf{x}\colon g_i(\mathbf{x}) \leq 0,\ i = 1, 2, \ldots, m,\ \mathbf{x} \in E^n\}$. If f and g_i are convex functions, then Ω is a convex set and the problem (10.4) is called the *convex MP problem.* Any $\mathbf{x}_0 \in \Omega$ is an admissible solution of (10.4). The point $\mathbf{x}_*$ minimizing f over Ω is called the *optimal solution* or simply the solution of (10.4).

If the objective function f or any of the functions g_i is nonlinear, then (10.4) is called a *nonlinear programming problem* (NLP problem). The geometrical interpretation of such a problem is given in Fig. 10.4. The surfaces $g_i(\mathbf{x}) = 0$ bound the admissible region Ω, which can be closed (Fig. 10.4b) or open (Fig. 10.4a, c). The level lines $f =$ const reproduce the shape of the objective function. Figure 10.4a shows the rare case of the solution $\mathbf{x}_*$ situated inside the admissible region. Usually the unconstrained minimum of f falls outside Ω. Then $\mathbf{x}_*$ lies either on a single constraining surface (Fig. 10.4b) or at a vertex, i.e., at an intersection of several surfaces (Fig. 10.4c).

The *quadratic programming problem* (QP problem)

$$\min\{f = \tfrac{1}{2}\mathbf{x}^{\mathrm{T}}\mathbf{F}\mathbf{x} + \mathbf{c}^{\mathrm{T}}\mathbf{x} \mid \mathbf{A}\mathbf{x} \geq \mathbf{b}\} \tag{10.5}$$

has a quadratic objective function and linear constraints. Such constraints bound a convex polyhedron Ω in E^n (Fig. 10.5). Depending upon the objective function, the solution $\mathbf{x}_*$ can be situated inside the admissible region (Fig. 10.5a), on a constraining plane (Fig. 10.5b), or at a vertex (Fig. 10.5c).

In the *linear programming problem* (LP problem)

$$\min\{f = \mathbf{c}^{\mathrm{T}}\mathbf{x} \mid \mathbf{A}\mathbf{x} \geq \mathbf{b}\}, \tag{10.6}$$

both the objective function and the constraints are linear. As shown in Fig. 10.6, the level lines are straight and parallel to each other. The linearity of

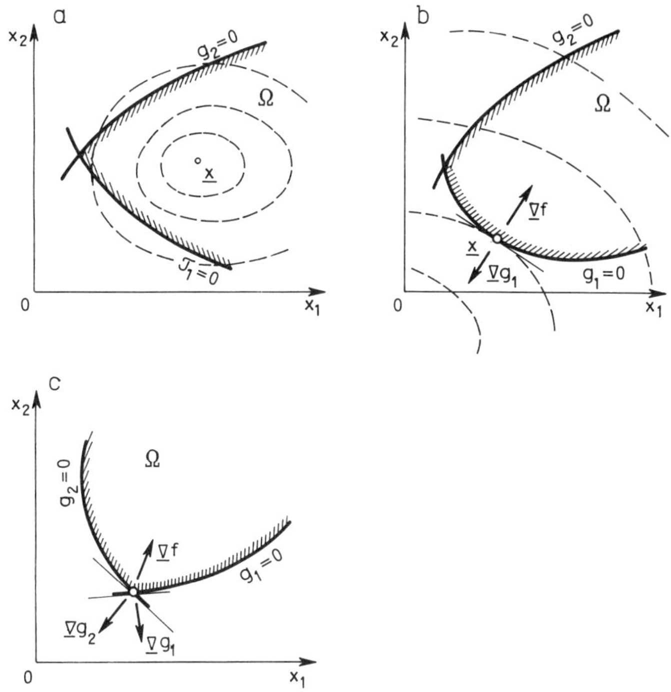

Figure 10.4. Geometrical Interpretation of the NLP Problem: (a) Solution Inside Ω. (b) Solution on the Boundary of Ω. (c) Solution at a Vertex of Ω.

the objective function excludes the possibility of $\mathbf{x}_*$ being inside Ω. The solution, if it exists, is located at a vertex of Ω (Fig. 10.6a) or corresponds to each point of an edge linking two vertices (Fig. 10.6b). In the second case, $\mathbf{x}_*$ is not unique.

An LP problem is correctly formulated if the extreme value of the objective function is finite. Such is the case shown in Fig. 10.6b: the objective function decreases with decreasing values of the variables and attains its finite minimum value at the edge *AB*. If we try to maximize that function over the same region, it would grow infinitely.

Occasionally the MP problem has no solution because its constraints are contradictory. See Fig. 10.7: two linear constraints describe the admissible set Ω_1, and the third one generates the set Ω_2. The joint admissible region Ω does not exist because $\Omega_1 \cap \Omega_2 = \varnothing$.

It is often convenient to treat separately the equality constraints and the constraints restricting the sign of the variables. Let the first m' constraints be equalities, and let the first n' variables be free with respect to their signs.

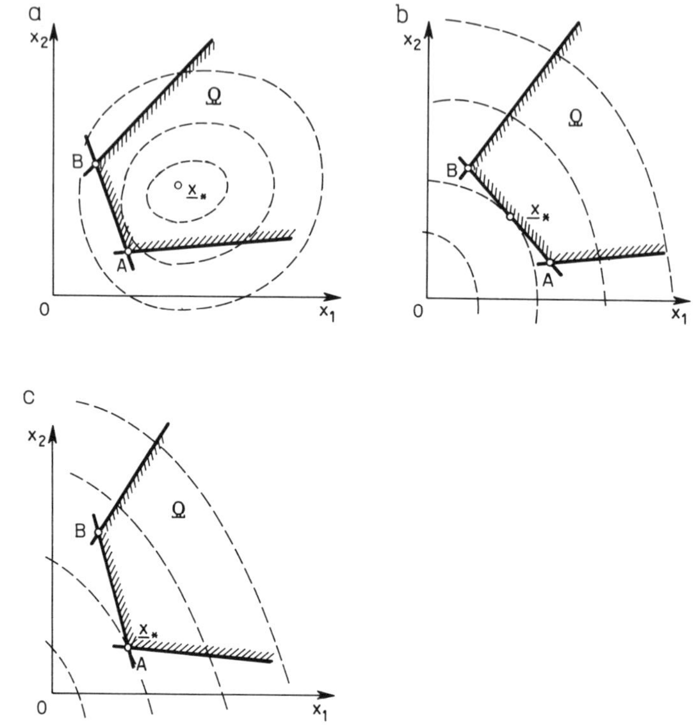

Figure 10.5. Geometrical Interpretation of QP Problem: (a) Solution Inside Ω. (b) Solution on the Boundary of Ω. (c) Solution at a Vertex of Ω.

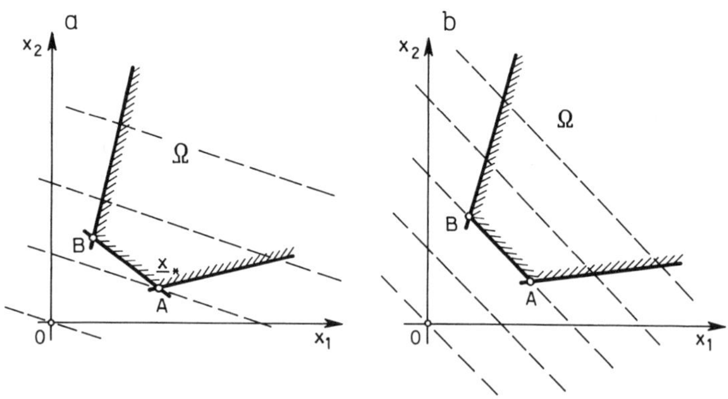

Figure 10.6. Geometrical Interpretation of the LP Problem: (a) Unique Solution. (b) Non-unique Solution.

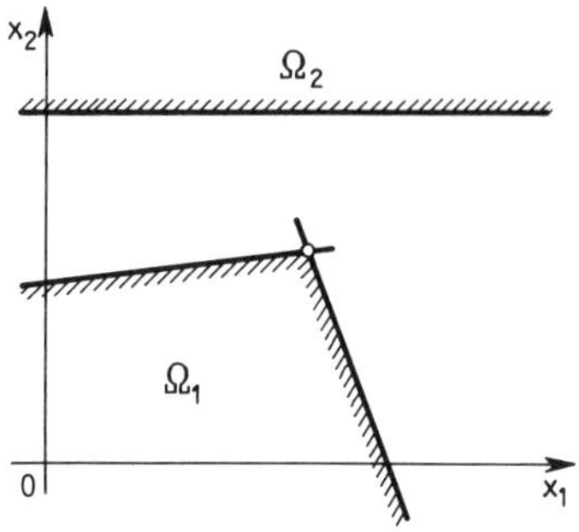

Figure 10.7. Contradictory Constraints.

The remaining $m - m'$ constraints are inequalities, and the remaining $n - n'$ unknowns are nonnegative. Thus, we obtain the NLP problem with mixed constraints and mixed variables:

$$\begin{aligned}
&\text{minimize } f(\mathbf{x}) \\
&\text{subject to} \\
&\qquad g_i(\mathbf{x}) = 0, \qquad i = 1, 2, \ldots, m', \\
&\qquad g_i(\mathbf{x}) \leq 0, \qquad i = m' + 1, m' + 2, \ldots, m, \\
&\qquad x_j \geq 0, \qquad j = n' + 1, n' + 2, \ldots, n.
\end{aligned} \tag{10.7}$$

In particular, we can have $n' = 0$, which means that all variables are nonnegative, or $m' = m$, which means that all constraints are equalities.

Exercise 10.4

Solve graphically the LP problem

$$\begin{aligned}
&\text{minimize } f = x_1 + x_2 \\
&\text{subject to} \\
&\qquad x_1 + 2x_2 \geq 4, \\
&\qquad -x_1 + 2x_2 \geq 0, \\
&\qquad x_1 \geq 1.
\end{aligned} \tag{10.8}$$

Exercise 10.5

Solve graphically the QP problem

$$\text{minimize } f = x_1^2 + x_2^2$$

subject to

$$x_1 + x_2 \geq 3, \qquad (10.9)$$
$$1 \leq x_1 \leq 4,$$
$$1 \leq x_2 \leq 4.$$

Exercise 10.6

Find a graphical interpretation of the NLP problem

minimize $f = \frac{1}{9}x_1^2 + \frac{1}{4}x_2^2$

subject to

$$x_1^2 - x_2 \leq 0, \qquad (10.10)$$
$$x_1^2 - 8x_1 - x_2 + 16 \leq 0,$$

and show that $\mathbf{x}_* = \{2, 4\}$ is the solution.

10.3. Karush–Kuhn–Tucker Conditions

The point $\mathbf{x}_*$ is called a *stationary* point of the function $f(\mathbf{x})$ if the gradient of that function vanishes at $\mathbf{x}_*$, i.e., if $\mathbf{x}_*$ satisfies the set of equations

$$\nabla f = \mathbf{0}. \qquad (10.11)$$

Depending upon the properties of f, the stationary point may or may not correspond to the extremum of the function (Fig. 10.8a, b). If it does correspond to the extremum, then the extremum can be local or global (Fig. 10.8c, a).

More definite statements hold for convex and concave functions: if f is convex (concave), then the stationary point $\mathbf{x}_*$ corresponds to its global minimum (maximum). The uniqueness of $\mathbf{x}_*$ is guaranteed only for strictly convex (concave) functions (Fig. 10.9).

Hence, the system of equations (10.11) are the necessary and sufficient conditions for a convex function to attain its minimum over E^n. In mathematical programming, however, we are looking for the minimum that is attained over a subset Ω of E^n. MP theory can be considered as a generalization of the classical approach developed by Lagrange for constrained minimization. According to that approach, the problem with equality constraints

$$\min\{f(\mathbf{x}) \,|\, g_i(\mathbf{x}) = 0, \ i = 1, 2, \ldots, m\} \qquad (10.12)$$

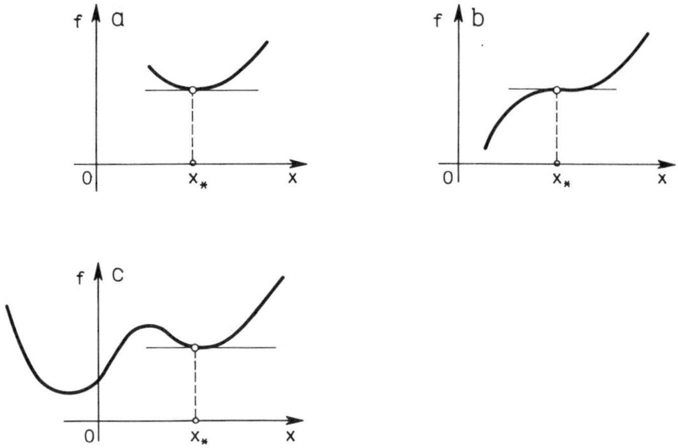

Figure 10.8. Stationary Points Corresponding to (a) Extremum. (b) Nonextremum. (c) Local Extremum.

is replaced by the unconstrained minimization of the Lagrangian function

$$L(\mathbf{x}, \mathbf{y}) = f(\mathbf{x}) + \sum_{i=1}^{m} y_i g_i(\mathbf{x}), \qquad (10.13)$$

where the y_i are Lagrange multipliers.

Applying Eq. (10.11) to the function L, we obtain necessary and sufficient conditions for $\mathbf{x}_*$ to be the solution of the set (10.12): there must exist $\mathbf{y}_* \in E^m$ such that $(\mathbf{x}_*, \mathbf{y}_*)$ satisfies the system of equations

$$\begin{aligned} \boldsymbol{\nabla}_x L &= \boldsymbol{\nabla} f + \mathbf{G}^{\mathrm{T}}\mathbf{y} = \mathbf{0}, \\ \boldsymbol{\nabla}_y L &= \mathbf{g} = \mathbf{0}, \end{aligned} \qquad (10.14)$$

where $\mathbf{G} = [\partial g_i/\partial x_j]$ and $\mathbf{g} = \{g_i(\mathbf{x})\}$ for $i = 1, 2, \ldots, m$, $j = 1, 2, \ldots, n$.

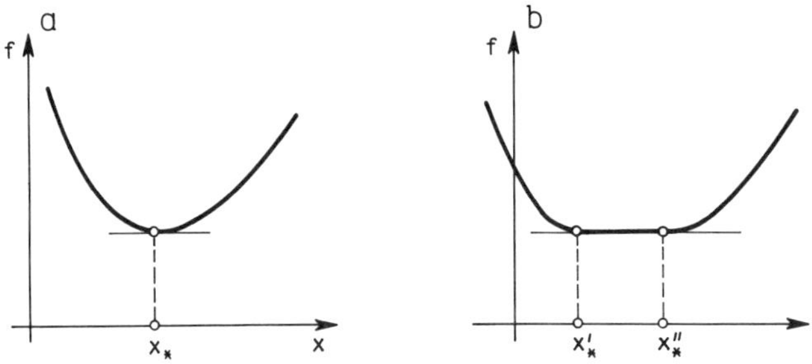

Figure 10.9. Minima of Convex Functions: (a) Uniquely Defined for a Strictly Convex Function. (b) Nonuniquely Defined for a Convex Function.

Now consider the NLP problem (10.4) instead of (10.12). Conditions (10.14) now must be replaced by the Karush–Kuhn–Tucker (KKT) conditions [Kuhn and Tucker (1951)]

$$
\begin{aligned}
\boldsymbol{\nabla}_x L &= \mathbf{0}, \\
\boldsymbol{\nabla}_y L &\leqslant \mathbf{0}, \\
\mathbf{y} &\geqslant \mathbf{0}, \\
\mathbf{y}^{\mathrm{T}} \boldsymbol{\nabla}_y L &= 0.
\end{aligned}
\tag{10.15}
$$

Now the Lagrange multipliers are nonnegative, the entries of the gradient $\boldsymbol{\nabla}_y L$ corresponding to them are nonpositive, and both have to satisfy the additional nonlinear condition $(10.15)_4$, called the *condition of complementary slackness.*

To explain why the solution $\mathbf{x}_*$ of (10.4) should satisfy the relations (10.15), we exploit the geometrical interpretation of (10.4). The point $\mathbf{x}_*$ can be situated inside the admissible region Ω (Fig. 10.4a), at the boundary of that region (Fig. 10.4b), or at a vertex of that region (Fig. 10.4c).

In the first case the constraints are irrelevant, and $\mathbf{x}_*$ should satisfy the classical condition (10.11). It is not difficult to see that in this case (10.15) reduces to (10.11). Since the derivatives $\partial L/\partial y_i$ are identical to the functions g_i by definition (10.13), all entries of the gradient $\boldsymbol{\nabla}_y L$ are negative at $\mathbf{x}_*$. Then the only possibility for the scalar product $\mathbf{y}^{\mathrm{T}}\boldsymbol{\nabla}_y L$ to vanish at $\mathbf{y}_*$ is $\mathbf{y}_* = \mathbf{0}$. Hence, the only nontrivial condition to be satisfied by $\mathbf{x}_*$ is $(10.15)_1$.

Now consider the situation in Fig. 10.4b: the solution $\mathbf{x}_*$ lies on the curve (surface in the n-dimensional case) $g_1 = 0$. Constraint 1 is active while constraints 2 and 3 are passive. Since $\partial L/\partial y_2 = g_2$ and $\partial L/\partial y_3 = g_3$ are negative at $\mathbf{x}_*$, the complementary slackness rule $(10.15)_4$ forces y_{2*} and y_{3*} to vanish. Hence, from (10.14) and (10.15), it follows that

$$-\boldsymbol{\nabla} f = y_1 \boldsymbol{\nabla} g_1;$$

i.e., the gradient of the objective function and the gradient of the active constraint are collinear at $\mathbf{x}_*$, and the steepest descent direction $-\boldsymbol{\nabla} f$ is perpendicular to the boundary of Ω.

Figure 10.4c depicts the situation when $\mathbf{x}_*$ is situated at the intersection of boundary curves $g_1 = 0$ and $g_2 = 0$. Using the previous argument, we can show that $y_{3*} = 0$ and that

$$-\boldsymbol{\nabla} f = y_1 \boldsymbol{\nabla} g_1 + y_2 \boldsymbol{\nabla} g_2$$

holds at $\mathbf{x}_*$. Geometrically this means that the steepest descent direction at that point lies inside the cone spanned by the gradients of the active constraints.

The KKT conditions (10.15) constitute the necessary conditions of optimality: if $\mathbf{x}_*$ is a solution of (10.4), it satisfies (10.15). Strictly speaking, the constraints of the MP problem (10.4) must fulfill certain regularity conditions for the preceding statement to be true. However, this restriction has more theoretical than practical importance, since most real-world problems satisfy such conditions.

If problem (10.4) is convex, then the KKT conditions are also sufficient: if $\mathbf{x}_*$ satisfies (10.15), then it is the solution of (10.4). As mentioned before, the minimum of a convex function is global. If the objective function is strictly convex, then $\mathbf{x}_*$ is unique.

Since we have already derived optimality conditions for both pure equality and pure inequality constraints, it is easy to combine them and to write the KKT conditions for the mixed NLP problem (10.7):

$$
\begin{aligned}
\partial L/\partial x_j &= 0, \qquad j = 1, 2, \ldots, n'; \\
\left.\begin{aligned}
\partial L/\partial x_j &\geqslant 0, \\
x_j &\geqslant 0, \\
x_j(\partial L/\partial x_j) &= 0,
\end{aligned}\right\} &\quad j = n'+1, n'+2, \ldots, n; \\
\partial L/\partial y_i &= 0, \qquad i = 1, 2, \ldots, m'; \\
\left.\begin{aligned}
\partial L/\partial y_i &\leqslant 0, \\
y_i &\geqslant 0, \\
y_i(\partial L/\partial y_i) &= 0,
\end{aligned}\right\} &\quad i = m'+1, m'+2, \ldots, m.
\end{aligned}
\tag{10.16}
$$

Finally, we formulate KKT conditions for linear and quadratic problems. For the QP problem (10.5) the Lagrange function is

$$L(\mathbf{x}, \mathbf{y}) = \tfrac{1}{2}\mathbf{x}^{\mathrm{T}}\mathbf{F}\mathbf{x} + \mathbf{c}^{\mathrm{T}}\mathbf{x} + \mathbf{y}^{\mathrm{T}}(\mathbf{b} - \mathbf{A}\mathbf{x}), \tag{10.17}$$

and the KKT conditions read

$$
\begin{aligned}
\nabla_x L &= \mathbf{F}\mathbf{x} + \mathbf{c} - \mathbf{A}^{\mathrm{T}}\mathbf{y} = \mathbf{0}, \\
\nabla_y L &= \mathbf{b} - \mathbf{A}\mathbf{x} \leqslant \mathbf{0}, \\
\mathbf{y} &\geqslant \mathbf{0}, \\
\mathbf{y}^{\mathrm{T}}(\mathbf{b} - \mathbf{A}\mathbf{x}) &= 0.
\end{aligned}
\tag{10.18}
$$

The Lagrangian for the LP problem (10.6) has the form

$$L(\mathbf{x}, \mathbf{y}) = \mathbf{c}^{\mathrm{T}}\mathbf{x} + \mathbf{y}^{\mathrm{T}}(\mathbf{b} - \mathbf{A}\mathbf{x}), \tag{10.19}$$

and (10.15) reduces to

$$\begin{aligned} \nabla_x L &= \mathbf{c} - \mathbf{A}^{\mathrm{T}}\mathbf{y} = \mathbf{0}, \\ \nabla_y L &= \mathbf{b} - \mathbf{A}\mathbf{x} \leqslant \mathbf{0}, \\ \mathbf{y} &\geqslant \mathbf{0}, \\ \mathbf{y}^{\mathrm{T}}(\mathbf{b} - \mathbf{A}\mathbf{x}) &= 0. \end{aligned} \tag{10.20}$$

If the unknowns x_j were nonnegative, then the equations $\nabla_x L = 0$ in (10.18) and (10.20) should be replaced by the inequalities $\nabla_x L \geqslant \mathbf{0}$.

Exercise 10.7

Write the KKT conditions for the LP problem (10.8) and check that the solution satisfies them.

Exercise 10.8

Write the KKT conditions for the QP problem (10.9) and check that the solution satisfies them.

Exercise 10.9

Write the KKT conditions for the NLP problem (10.10) and find the solution.

10.4. Dual Problems

An important feature of MP problems is that they possess alternative or dual formulations. For example, instead of solving problem (10.4), we can solve its dual

$$\max\{L(\mathbf{x}, \mathbf{y}) \mid \nabla_x L = \mathbf{0}, \mathbf{y} \geqslant \mathbf{0}\}, \tag{10.21}$$

where L is the Lagrangian function defined by (10.13). If the primal problem (10.4) is convex, then the *strong duality theorem* holds: if (10.4) has a solution $\mathbf{x}_*$, there exists $\mathbf{y}_*$ such that $(\mathbf{x}_*, \mathbf{y}_*)$ is the solution of (10.21) and $f(\mathbf{x}_*) = L(\mathbf{x}_*, \mathbf{y}_*)$.

The notion of duality and the KKT conditions are closely related. We obtain the latter by taking simultaneously into account all constraints of primal and dual problems supplemented by the complementary slackness conditions.

Duality of a general NLP problem is not a symmetric property: the dual of the dual is not the primal problem. Moreover, convexity of the primal problem does not imply concavity of L and the convexity of the dual admissible region. Before we come to the duality of quadratic and linear problems, where much stronger properties hold, we treat the case of mixed constraints and mixed variables. The following simple rules apply when the dual problem is formulated:

a. A free dual variable y_i is introduced for each equality constraint $g_i(\mathbf{x}) = 0$, whereas a nonnegative y_i corresponds to each inequality $g_i(\mathbf{x}) \leqslant 0$.
b. An equality constraint $\partial L/\partial x_j = 0$ is generated by each free variable x_j, whereas an inequality $\partial L/\partial x_j \geqslant 0$ corresponds to each nonnegative x_j.

Hence, the dual of problem (10.7) reads

$$\text{maximize } L = f(\mathbf{x}) + \sum_{i=1}^{m} y_i g_i(\mathbf{x})$$

subject to

$$\frac{\partial f}{\partial x_j} + \sum_{i=1}^{m} y_i \left(\frac{\partial g_i}{\partial x_j}\right) = 0, \qquad j = 1, 2, \ldots, n', \tag{10.22}$$

$$\frac{\partial f}{\partial x_j} + \sum_{i=1}^{m} y_i \left(\frac{\partial g_i}{\partial x_j}\right) \geqslant 0, \qquad j = n' + 1, n' + 2, \ldots, n,$$

$$y_i \geqslant 0, \qquad i = m' + 1, m' + 2, \ldots, m.$$

Applying definition (10.21) to the QP problem (10.5), we obtain its dual

$$\max\{\tfrac{1}{2}\mathbf{x}^{\mathrm{T}}\mathbf{F}\mathbf{x} + \mathbf{c}^{\mathrm{T}}\mathbf{x} + \mathbf{y}^{\mathrm{T}}(\mathbf{b} - \mathbf{A}\mathbf{x}) \mid \mathbf{F}\mathbf{x} + \mathbf{c} - \mathbf{A}^{\mathrm{T}}\mathbf{y} = \mathbf{0}, \mathbf{y} \geqslant \mathbf{0}\}.$$

Since equality constraints must be satisfied by each admissible solution, we must also have $\mathbf{y}^{\mathrm{T}}\mathbf{A}\mathbf{x} = \mathbf{x}^{\mathrm{T}}\mathbf{F}\mathbf{x} + \mathbf{c}^{\mathrm{T}}\mathbf{x}$. This allows us to simplify the objective function and to write the dual of (10.5) as follows:

$$\max\{-\tfrac{1}{2}\mathbf{x}^{\mathrm{T}}\mathbf{F}\mathbf{x} + \mathbf{b}^{\mathrm{T}}\mathbf{y} \mid -\mathbf{F}\mathbf{x} + \mathbf{A}^{\mathrm{T}}\mathbf{y} = \mathbf{c}, \mathbf{y} \geqslant \mathbf{0}\}. \tag{10.23}$$

We can obtain the dual of the linear problem (10.6) by using definition (10.21) or by taking $\mathbf{F} = \mathbf{0}$ in (10.23). The result is

$$\max\{\mathbf{b}^{\mathrm{T}}\mathbf{y} \mid \mathbf{A}^{\mathrm{T}}\mathbf{y} = \mathbf{c}, \mathbf{y} \geqslant \mathbf{0}\}. \tag{10.24}$$

Note that the dual of (10.24) is again the primal (10.6). The dual LP problems (10.6) and (10.24) are also uncoupled with respect to $\mathbf{x}$ and $\mathbf{y}$.

In general, dual LP problems can have mixed (i.e., equality and inequality) constraints and mixed (i.e., free and nonnegative) variables. The internal structure of such duals is given in Table 10.1. We shall use such tables to represent initial data for numerical examples.

Exercise 10.10

Write the dual of (10.8) and check the validity of the strong duality theorem.

Exercise 10.11

Write the dual of type (10.23) for the QP problem (10.9).

Exercise 10.12

Formulate the dual of (10.10) and solve it.

Table 10.1. Initial Simplex Tables for Dual LP Problems with Mixed Variables and Constraints

a. Primal problem

x_1^T	$\mathbf{x}_2^T$			
$\mathbf{A}_{11}$	$\mathbf{A}_{12}$	$=$	$\mathbf{b}_1$	(m')
$\mathbf{A}_{21}$	$\mathbf{A}_{22}$	$\geqslant$	$\mathbf{b}_2$	$(m - m')$
$\mathbf{c}_1^T$	$\mathbf{c}_2^T$	min		
(n')	$(n - n')$			

b. Dual problem

$\mathbf{y}_1^T$	$\mathbf{y}_2^T$			
$\mathbf{A}_{11}^T$	$\mathbf{A}_{21}^T$	$=$	$\mathbf{c}_1$	(n')
$\mathbf{A}_{12}^T$	$\mathbf{A}_{22}^T$	$\leqslant$	$\mathbf{c}_2$	$(n - n')$
$\mathbf{b}_1^T$	$\mathbf{b}_2^T$	max		
(m')	$(m - m')$			

10.5. Saddle Point Problems and Symmetric Dual QP Problems

Consider a set of linear algebraic equations

$$\mathbf{A}\mathbf{x} + \mathbf{b} = \mathbf{0} \tag{10.25}$$

with a symmetric matrix of coefficients $\mathbf{A}$ or order n. The quadratic function $f(\mathbf{x}) = \frac{1}{2}\mathbf{x}^{\mathrm{T}}\mathbf{A}\mathbf{x} + \mathbf{b}^{\mathrm{T}}\mathbf{x}$ is called the *potential* of this set because $\nabla f = \mathbf{A}\mathbf{x} + \mathbf{b}$. Hence, the solution $\mathbf{x}_*$ of (10.25) is simultaneously the stationary point of f according to definition (10.11). If $\mathbf{A}$ is not only symmetric but also positive (negative) semidefinite, then $\mathbf{x}_*$ corresponds to the minimum (maximum) of the potential f.

Suppose that the set of equations has a slightly more complicated structure:

$$\begin{aligned} \mathbf{A}_{xx}\mathbf{x} + \mathbf{A}_{xy}\mathbf{y} + \mathbf{b}_x &= \mathbf{0}, \\ \mathbf{A}_{yx}\mathbf{x} + \mathbf{A}_{yy}\mathbf{y} + \mathbf{b}_y &= \mathbf{0}, \end{aligned} \tag{10.26}$$

where the global matrix of coefficients is again symmetric ($\mathbf{A}_{xx}^{\mathrm{T}} = \mathbf{A}_{xx}$, $\mathbf{A}_{yy}^{\mathrm{T}} = \mathbf{A}_{yy}$, $\mathbf{A}_{xy}^{\mathrm{T}} = \mathbf{A}_{yx}$), the submatrix $\mathbf{A}_{xx}$ is positive semidefinite, and $\mathbf{A}_{yy}$ is negative semidefinite. Then the potential of (10.26) is the saddle function

$$f(\mathbf{x}, \mathbf{y}) = \tfrac{1}{2}\mathbf{x}^{\mathrm{T}}\mathbf{A}_{xx}\mathbf{x} + \tfrac{1}{2}\mathbf{y}^{\mathrm{T}}\mathbf{A}_{yy}\mathbf{y} + \mathbf{x}^{\mathrm{T}}\mathbf{A}_{xy}\mathbf{y} + \mathbf{b}_x^{\mathrm{T}}\mathbf{x} + \mathbf{b}_y^{\mathrm{T}}\mathbf{y}, \tag{10.27}$$

which is convex with respect to $\mathbf{x}$ and concave with respect to $\mathbf{y}$. The solution $(\mathbf{x}_*, \mathbf{y}_*)$ of the saddle point problem (SP problem)

$$\min_{\mathbf{x}} \max_{\mathbf{y}} f(\mathbf{x}, \mathbf{y}) \tag{10.28}$$

has to satisfy the equations

$$\nabla_x f = \mathbf{0}, \qquad \nabla_y f = \mathbf{0}. \tag{10.29}$$

Hence, it is simultaneously the solution of (10.26).

The SP problem (10.28) is equivalent to the following dual pair of QP problems:

$$\begin{aligned} &\min\{f' \mid \nabla_y f = \mathbf{0}\}, \\ &\max\{f'' \mid \nabla_x f = \mathbf{0}\}, \end{aligned} \tag{10.30}$$

where

$$f' = \tfrac{1}{2}\mathbf{x}^{\mathrm{T}}\mathbf{A}_{xx}\mathbf{x} - \tfrac{1}{2}\mathbf{y}^{\mathrm{T}}\mathbf{A}_{yy}\mathbf{y} + \mathbf{b}_x^{\mathrm{T}}\mathbf{x},$$
$$f'' = -\tfrac{1}{2}\mathbf{x}^{\mathrm{T}}\mathbf{A}_{xx}\mathbf{x} + \tfrac{1}{2}\mathbf{y}^{\mathrm{T}}\mathbf{A}_{yy}\mathbf{y} + \mathbf{b}_y^{\mathrm{T}}\mathbf{y} \tag{10.31}$$

are the Legendre transforms and

$$\nabla_x f = \mathbf{A}_{xx}\mathbf{x} + \mathbf{A}_{xy}\mathbf{y} + \mathbf{b}_x,$$
$$\nabla_y f = \mathbf{A}_{yx}\mathbf{x} + \mathbf{A}_{yy}\mathbf{y} + \mathbf{b}_y \tag{10.32}$$

are the gradients of the saddle potential f.

Since no inequalities have been involved so far, the equivalence among the set of equations of a certain structure, the SP problem, and the pair of dual QP problems can be established in the frame of the classical Lagrangian approach. The situation changes if the saddle point is to be looked for among nonnegative variables:

$$\min_{\mathbf{x}\geqslant\mathbf{0}} \max_{\mathbf{y}\geqslant\mathbf{0}} f(\mathbf{x}, \mathbf{y}). \tag{10.33}$$

The solution $(\mathbf{x}_*, \mathbf{y}_*)$ of such an SP problem must satisfy the relations

$$\begin{aligned} &\nabla_x f \geqslant \mathbf{0}, && \nabla_y f \leqslant \mathbf{0}, \\ &\mathbf{x} \geqslant \mathbf{0}, && \mathbf{y} \geqslant \mathbf{0}, \\ &\mathbf{x}^{\mathrm{T}}\nabla_x f = 0, && \mathbf{y}^{\mathrm{T}}\nabla_y f = 0, \end{aligned} \tag{10.34}$$

which can be easily recognized as the KKT conditions related to the following pair of dual QP problems:

$$\min\{f' \mid \nabla_y f \leqslant \mathbf{0}, \mathbf{x} \geqslant \mathbf{0}\},$$
$$\max\{f'' \mid \nabla_x f \geqslant \mathbf{0}, \mathbf{y} \geqslant \mathbf{0}\}. \tag{10.35}$$

For the sake of later reference we write them down explicitly:

$$\begin{aligned} &\text{minimize } \tfrac{1}{2}\mathbf{x}^{\mathrm{T}}\mathbf{A}_{xx}\mathbf{x} - \tfrac{1}{2}\mathbf{y}^{\mathrm{T}}\mathbf{A}_{yy}\mathbf{y} + \mathbf{b}_x^{\mathrm{T}}\mathbf{x} \\ &\text{subject to} \\ &\qquad \mathbf{A}_{yx}\mathbf{x} + \mathbf{A}_{yy}\mathbf{y} + \mathbf{b}_y \leqslant \mathbf{0}, \qquad \mathbf{x} \geqslant \mathbf{0}; \\ &\text{maximize } -\tfrac{1}{2}\mathbf{x}^{\mathrm{T}}\mathbf{A}_{xx}\mathbf{x} + \tfrac{1}{2}\mathbf{y}^{\mathrm{T}}\mathbf{A}_{yy}\mathbf{y} + \mathbf{b}_y^{\mathrm{T}}\mathbf{y} \\ &\text{subject to} \\ &\qquad \mathbf{A}_{xx}\mathbf{x} + \mathbf{A}_{xy}\mathbf{y} + \mathbf{b}_x \geqslant \mathbf{0}, \qquad \mathbf{y} \geqslant \mathbf{0}. \end{aligned} \tag{10.36}$$

These are symmetric duals introduced by Cottle (1963): the dual of $(10.36)_2$ is $(10.36)_1$.

Taking $\mathbf{A}_{xx} = \mathbf{A}_{yy} = \mathbf{0}$, $\mathbf{A}_{xy} = -\mathbf{A}$, $\mathbf{b}_x = \mathbf{c}$, $\mathbf{b}_y = \mathbf{b}$, we obtain a special case of (10.36), the symmetric dual LP problems

$$\begin{aligned} &\min\{\mathbf{c}^{\mathrm{T}}\mathbf{x} \mid \mathbf{A}\mathbf{x} \geqslant \mathbf{b}, \mathbf{x} \geqslant \mathbf{0}\}, \\ &\max\{\mathbf{b}^{\mathrm{T}}\mathbf{y} \mid \mathbf{A}^{\mathrm{T}}\mathbf{y} \leqslant \mathbf{c}, \mathbf{y} \geqslant \mathbf{0}\}. \end{aligned} \tag{10.37}$$

Mixed variables and mixed constraints are treated in symmetric duality as they are in conventional duality. Namely, each free primal variable generates a dual constraint of the equality type, and each nonnegative primal variable yields an inequality. Conversely, each primal equality constraint generates a free dual variable, and each inequality constraint of the primal problem corresponds to a nonnegative variable of the dual problem.

Exercise 10.13

Find the potential and the equivalent pair of dual QP problems for the following set of equations:

$$\begin{aligned} 4x_1 + x_2 - x_3 + 2x_4 &= 4, \\ x_1 + 4x_2 - x_3 \qquad &= 2, \\ -x_1 - x_2 - 2x_3 - x_4 &= 1, \\ 2x_1 \qquad - x_3 - x_4 &= 3. \end{aligned}$$

Exercise 10.14

Modify the QP problems obtained in the previous exercise to account for the nonnegative variables x_2 and x_3.

10.6. Methods of Solution and Available Computer Programs

Numerous computational methods are used in mathematical programming. Therefore we merely briefly describe several typical algorithms and encourage the reader to examine the references by Avriel (1976), Dantzig (1963), Mangasarian (1969), and Zangwill (1969) for an exhaustive presentation.

10.6.1. The Simplex Method

We begin with linear and quadratic programming. We usually solve LP problems by the *simplex method*. This method exploits the fact that the solution of a linear problem, if it exists, is located at a vertex of Ω or it belongs to an edge connecting two adjacent vertices (Fig. 10.6). Hence, the procedure is accomplished in two stages. During the first stage an admissible solution $\mathbf{x}_0$ corresponding to an arbitrary vertex of the admissible region is looked for. Since at least n constraints of the problem (10.6) are active at such a vertex, it is theoretically sufficient to pick n independent constraints, construct a system of equations $\hat{\mathbf{A}}\mathbf{x} = \hat{\mathbf{b}}$, and solve that system to obtain the desired point $\mathbf{x}_0 = \hat{\mathbf{A}}^{-1}\hat{\mathbf{b}}$. However, practically it is cumbersome to select simultaneously n linearly independent rows of $\mathbf{A}$ as candidates for the square submatrix $\hat{\mathbf{A}}$. Fortunately we can obtain the same result through a finite number of Jordan transformations performed with the matrices $\mathbf{A}$, $\mathbf{b}$, and $\mathbf{c}$. A special pivoting rule can be devised that leads to the vertex $\mathbf{x}_0$ or reveals the incompatibility of constraints.

Suppose we have found $\mathbf{x}_0$. Then the second stage of the simplex method—the search for the optimal solution—is activated. It consists of the following steps:

a. Take $\mathbf{x}_0$ as a current trial solution $\mathbf{x}_k$.
b. Check whether $f(\mathbf{x}_k) \leq f(\mathbf{x}_j)$ for all adjacent vertices $\mathbf{x}_j$ (i.e., vertices having a common edge with $\mathbf{x}_k$).
c. If "yes," then $\mathbf{x}_* = \mathbf{x}_k$ and calculations are finished; otherwise choose that adjacent vertex $\hat{\mathbf{x}}_j$ that corresponds to the smallest value of the objective function.
d. Take $\mathbf{x}_k = \hat{\mathbf{x}}_j$ and repeat (b).

It can be proved that this procedure terminates in a finite number of steps: either the solution $\mathbf{x}_*$ is found, or the unboundedness of f is detected. Moreover, it is not necessary to solve the dual problem (10.24) separately: its solution $\mathbf{y}_*$ is obtained simultaneously with $\mathbf{x}_*$.

Since QP problems have a constraint structure similar to linear problems, we would naturally expect the simplex method to be usable for solving quadratic problems as well. Various modifications have been proposed in the literature. It seems that the methods of Beale (1955), Lemke (1965), and Wolfe (1959) have been most frequently used. These algorithms solve the set of KKT conditions (10.18) in one way or another. They require a larger amount of computer memory than the standard simplex needs for the same values of m and n.

10.6.2. Sequential Methods

We turn now to the solution of a general NLP problem (10.4). From calculus we know that an arbitrary nonlinear function $f(\mathbf{x})$ can be approximated in the vicinity of a given point $\mathbf{x}_k$ either linearly

$$f(\mathbf{x}) \approx f(\mathbf{x}_k) + \nabla f(\mathbf{x}_k)(\mathbf{x} - \mathbf{x}_k) \tag{10.38}$$

or quadratically

$$f(\mathbf{x}) \approx f(\mathbf{x}_k) + \nabla f(\mathbf{x}_k)(\mathbf{x} - \mathbf{x}_k) + \tfrac{1}{2}(\mathbf{x} - \mathbf{x}_k)^{\mathrm{T}}\mathbf{F}(\mathbf{x}_k)(\mathbf{x} - \mathbf{x}_k). \tag{10.39}$$

Therefore instead of solving the NLP problem, we can solve a sequence of approximating LP or QP problems. If the linear approximation is used, then the relevant procedure is called the *sequential linear programming method* (SLP method). For the second-order approximation (10.39), the *sequential quadratic programming* (SQP) *method* is used. The obvious advantage of both is that we can use rapid and robust LP or QP solvers. However, the total amount of time needed to solve a specific NLP problem via sequential approximations also depends on the number of steps leading to the solution. If the problem is only slightly nonlinear, the linear approximation (10.38) will work. In many practical cases, however, it turns out to be too crude and drifts from the exact solution. Many nonlinear examples in this book were solved by the SQP-type computer program NLPQP developed by Schittkowski (1981, 1984).

The NLPQP program implements the method of Wilson, Han, and Powell with some modifications. This method can be briefly described as follows. Given a current admissible point $\mathbf{x}_k$, one calculates a direction $\mathbf{z}_k$ and a step length α_k such that the next solution is $\mathbf{x}_{k+1} = \mathbf{x}_k + \alpha_k \mathbf{z}_k$. The direction $\mathbf{z}_k$ is obtained from the solution of the subproblem

$$\begin{aligned} &\text{minimize } (\tfrac{1}{2}\mathbf{z}_k^{\mathrm{T}}\mathbf{F}(\mathbf{x}_k)\mathbf{z}_k + \mathbf{z}_k^{\mathrm{T}}\nabla f(\mathbf{x}_k)) \\ &\text{subject to} \\ &\qquad g_i(\mathbf{x}_k) + \mathbf{z}_k^{\mathrm{T}}\nabla g_i(\mathbf{x}_k) \leq 0, \qquad i = 1, 2, \ldots, m, \\ &\qquad\qquad\qquad \mathbf{x}_k, \mathbf{z}_k \in E^n. \end{aligned} \tag{10.40}$$

We recognize the quadratic approximation (10.39) of the objective function and the linearized constraints of (10.4). An accurate evaluation of the Hessian matrix $\mathbf{F}$ at each step would be too time-consuming. Hence, we construct $\mathbf{F}$ approximately by using information from the previous step (the BFGS formula). Omitting technical details that can be found in Schittkowski (1981), we indicate only that the step length α_k is found by a line search.

10.6.3. Penalty Function Methods

Problem (10.4) can be converted into one of unconstrained minimization by adding a penalty term to the objective function for any violation of the constraints. In general, the new cost function can be written as

$$\Phi(\mathbf{x}, r) = f(\mathbf{x}) + r \sum_{j=1}^{m} G_j(g_j(\mathbf{x})), \tag{10.41}$$

where G_j is an operator on the jth constraint and r is a positive constant called the *penalty parameter.* The unconstrained minimum of Φ can be sought by any method. It is essential to solve the unconstrained problem for a sequence r_k $(k = 1, 2, \ldots)$ of the values of penalty parameter such that the resulting minima of Φ converge to the solution of the original problem.

The penalty function formulation for inequality-constrained problems can be divided into two categories: *interior* and *exterior* (Fig. 10.10). In the interior formulation the unconstrained minima of $\Phi(\mathbf{x}, r_k)$ are all located inside the admissible domain, and they converge to the solution with $r_k \to 0$. A typical form of the operator G_j in the interior method is

$$G_j(g_j) = -1/g_j. \tag{10.42}$$

The interior method, first proposed by Carroll (1961), is also called the *barrier method,* because the penalty term in Φ is chosen so that it

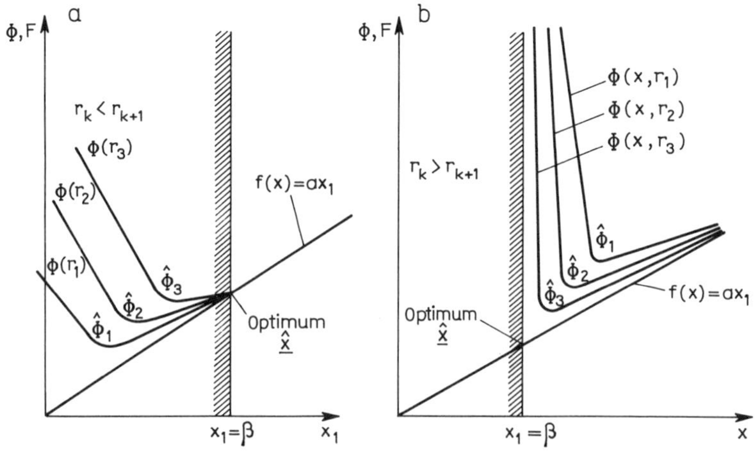

Figure 10.10. Penalty Function Methods: (a) Interior. (b) Exterior.

increases unboundedly when the constraints are approached. Thus, it sets a barrier against the minima leaving the admissible domain. With G defined by Eq. (10.42), the auxiliary function becomes

$$\Phi(\mathbf{x}, r) = f(\mathbf{x}) - r \sum_{j=1}^{m} \frac{1}{g_j(\mathbf{x})}. \tag{10.43}$$

In the exterior penalty function method, the unconstrained minima converge to the solution from outside the admissible region as $r_k \to \infty$. The most common forms of G_j in the exterior method are

$$G_j(g_j) = \max[g_j, 0] \tag{10.44}$$

and

$$G_j(g_j) = (\max[g_j, 0])^2. \tag{10.45}$$

The usual form of the auxilary function is then

$$\Phi(\mathbf{x}, r) = f(\mathbf{x}) + r \sum_{j=1}^{m} \langle g_j(\mathbf{x})\rangle^{\beta} \tag{10.46}$$

where $\langle\cdot\rangle$ denotes $\max[\cdot, 0]$ and β is a nonnegative exponent. If the jth constraint is violated, then a penalty proportional to the βth power of the amount of violation is added to the cost function. The individual minima of Φ can fall outside the admissible domain, but their sequence converges to the solution of the original problem as $r_k \to \infty$. The value of β influences the behavior of the auxiliary function. For $\beta = 0$, Φ is discontinuous at the boundary of the admissible region, and it is difficult to minimize it. For $0 < \beta < 1$, Φ is continuous, but it can fail to penalize strongly enough the violations of admissibility. In addition, such Φ has discontinuous first derivatives along the boundary, which is again an unpleasant feature for the minimization procedure. For $\beta = 1$, as shown by Zangwill (1967a), there always exists a large enough $\hat{r}$ that makes the minimum of Φ coincide with the constrained minimum of f for all $r_k \geqslant \hat{r}$. The most convenient choice, however, is $\beta > 1$, because then Φ has continuous first derivatives. Usually $\beta = 2$ is taken.

We shall use penalty function methods in Chapter 15 for the optimum design of skeletal structures.

10.6.4. The Tangent Search Method

The tangent search method (TSEARCH) developed by Hilleary (1966) is a special case of the feasible direction methods. These methods rely on conventional hill-climbing algorithms inside the admissible domain, whereas in the vicinity of the boundary they orient themselves according to the gradients of active constraints. TSEARCH, for example, uses a direct search according to the Hook and Jeeves (1962) scheme until some constraint gets violated. Then it backs off to the last admissible point and performs a unidirectional optimization along the hyperplane tangent to the active constraint. As a result, the move direction becomes orthogonal to the gradient of that constraint and the next point maintains a certain distance to the boundary. Evidently this method requires the calculation of the gradient vectors of active constraints when the current point arrives at the boundary. At the expense of this slight extra effort, TSEARCH combines efficient operation inside the admissible region with satisfactory performance along the boundary of that region. We apply this method in Chapter 15 to optimize the transverse web of the tanker ship.

10.6.5. Computer Codes

Because linear programming is widely applied in economics, management, and related areas, computer codes solving LP problems are almost as easily accessible as the solvers of linear algebraic equations. One should be aware, however, that practical applications of linear programming in structural optimization lead to problems with numerous variables and constraints. The proper computational tools are therefore highly developed commercial packages, such as the MPSX code supplied by IBM (1977) or the XDLA system of ICL (1971). The most powerful LP solver is the MPSIII system distributed for IBM mainframes by KENTRON, Inc. This system is reported capable of solving LP problems with up to 16,000 constraints and up to 100,000 unknowns [Schittkowski (1985)].

The supply of QP solvers on the software market is not numerous, but several reliable codes are available: e.g., QPSOL developed at Stanford University [Gill *et al.* (1983)] and the ZQPCVX procedure elaborated by Powell (1983).

The NLPQL code has proven to be very reliable and can be recommended for structural optimization. This does not mean, however, that programs based on other MP algorithms cannot be successfully used. Many interesting results in structural optimization have been obtained by means of the sequential unconstrained minimization technique (SUMT) [Fiacco and McCormick (1968)] and the gradient projection method [Rosen (1960,

1961)]. Among current NLP solvers, the best known seems to be the MINOS system developed at Stanford University [Murtagh and Saunders (1983)]. The code OPT2 by Wang and Ragsdell (1983) and the program IDESIGN by Belsare and Arora (1983) have proved to be very useful in structural optimization. Very instructive information on the comparative effectiveness of computational algorithms in NLP can be found in Schittkowski (1980), and a review of available software is given in Schittkowski (1985).

11

Matrix Modeling of a Structure

11.1. Introductory Remarks

In the first chapter of Volume 1, the state of a structure was described by means of generalized loads, displacements, stresses, and strains. These quantities were assumed to be functions of position measured in a properly chosen reference system (e.g., along the axis of a beam). A mathematician would say, therefore, that the mechanical state of the structure was described by elements of an infinite-dimensional function space.

We have already shown in Chapter 10 that mathematical programming (MP) problems are defined in the finite-dimensional space E^n. Hence, application of that theory to structural optimization requires prior discretization of the structure. The aim of such a discretization is to replace the functions $P(x)$, $p(x)$, $Q(x)$, and $q(x)$ by a finite set of parameters. Since the displacements, stresses, and strains subject to discretization are unknown a priori, we must make certain assumptions about their possible distributions. Hence, a finite-dimensional or discrete model of a structure is usually an approximate one, except for simple skeletal systems (e.g., trusses) where it is exact.

Several standard procedures of discretization are known, such as finite differences or finite series, but the most frequently used procedure is the finite element method (FEM). If the reader is not familiar with the FEM, he should see, e.g., Davies (1980), Gallagher (1975), or Zienkiewicz (1977).

In the following sections we introduce a matrix model of a structure that results from discretization, omitting the details of that derivation. [A complete derivation of the matrix model is given by Borkowski (1987).] Examples of the resulting matrices are given for skeletal structures only, though the model itself is applicable for an arbitrary structure. A detailed discussion of the consistency requirements for the discrete model is given in Anderhegen and Knöpfel (1972) and Corradi and Poggi (1984).

11.2. Adjoint Equations of Kinematics and Statics

Let the deformed configuration of the structure be completely described by n parameters w_j arranged as a matrix of generalized displacements $\mathbf{w}$. We call them generalized because the quantities w_j can include rotations as well as translations. We assume that the displacements required to be zero by the support conditions are not included in $\mathbf{w}$. (Such displacement components will hereafter be called "suppressed.") Hence, n defines the number of degrees of freedom of the discretized structure.

Any external loading can be reduced to a statically equivalent system of forces p_j acting in the directions of the displacements w_j. Then the dimension of the resulting load matrix $\mathbf{p}$ is n. The entries p_j of that matrix are generalized forces in the sense that they can include moments. The scalar product $\mathbf{w}^T\mathbf{p}$ defines the external work acting on the discrete system.

Let the strain state of the discretized structure be completely described by m parameters q_i arranged into the strain matrix $\mathbf{q}$. Each kinematic variable has its static counterpart—a stress parameter s_i. Hence, the stress state of the discretized system is described by a stress matrix $\mathbf{s}$ of dimension m. Both q_i and s_i are generalized quantities: the former can be elongations of bars, end rotations of beams, etc., and the latter are stress resultants such as axial forces, bending moments, etc. The pairs q_i, s_i are chosen in such a way that the scalar product $\mathbf{q}^T\mathbf{s}$ represents the internal work produced in the discretized structure.

The kinematic matrices $\mathbf{w}$, $\mathbf{q}$ and the static matrices $\mathbf{p}$, $\mathbf{s}$ are related by the pair of adjoint equations

$$\mathbf{q} = \mathbf{C}\mathbf{w}, \tag{11.1}$$

$$\mathbf{p} = \mathbf{C}^T\mathbf{s}. \tag{11.2}$$

This property holds for arbitrary structures subjected to small displacements and can be derived from the virtual work principle.

The $m \times n$ matrix $\mathbf{C}$ appearing in Eqs. (11.1) and (11.2) is called the *compatibility matrix.* In a properly constructed discrete model of the structure, that matrix has rank n; i.e., all its columns are linearly independent. Then an index $k = m - n$ can serve for the following classification:

(a) $k > 0$: hyperstatic system,
(b) $k = 0$: isostatic system,
(c) $k < 0$: substatic system (mechanism).

Hyperstatic systems are our primary concern here because they offer the most interesting possibilities of optimization. If k is positive, then the set of kinematic equations (11.1) is overdetermined. It can be solved with respect to displacements

$$\mathbf{w} = \mathbf{C}^{-}\mathbf{q}, \tag{11.3}$$

provided the strains are compatible:

$$\mathbf{C}^{0}\mathbf{q} = \mathbf{0}. \tag{11.4}$$

Here $\mathbf{C}^{-}$ is a generalized inverse of $\mathbf{C}$ satisfying the condition

$$\mathbf{C}^{-}\mathbf{C} = \mathbf{I}, \tag{11.5}$$

and $\mathbf{C}^{0}$ is a kernel matrix of $\mathbf{C}$ defined so that

$$\mathbf{C}^{0}\mathbf{C} = \mathbf{0}. \tag{11.6}$$

In Eq. (11.5), $\mathbf{I}$ is the identity matrix of order n and, in Eq. (11.6), $\mathbf{0}$ is the $k \times n$ matrix with zero entries.

The set of equilibrium equations (11.2) is underdetermined for $k > 0$. A general solution of that set with respect to stresses is

$$\mathbf{s} = \mathbf{C}^{-\mathrm{T}}\mathbf{p} + \mathbf{C}^{0\mathrm{T}}\mathbf{z}, \tag{11.7}$$

where $\mathbf{C}^{-\mathrm{T}}$ and $\mathbf{C}^{0\mathrm{T}}$ are the transposed matrices of $\mathbf{C}^{-}$ and $\mathbf{C}^{0}$, respectively. The matrix $\mathbf{z}$ of dimension k contains free (redundant) stress parameters.

For *isostatic* systems the compatibility matrix $\mathbf{C}$ is square, and Eqs. (11.1) and (11.2) can be given unique inverse form:

$$\mathbf{w} = \mathbf{C}^{-1}\mathbf{q}, \tag{11.8}$$

$$\mathbf{s} = \mathbf{C}^{-\mathrm{T}}\mathbf{p}, \tag{11.9}$$

where $\mathbf{C}^{-1}$ is the ordinary inverse of $\mathbf{C}$, and $\mathbf{C}^{-T}$ is its transpose. *Mechanisms* or *substatic* systems fall outside the scope of our considerations.

11.2.1. Example of a Truss

As an example, consider the simple truss in Fig. 11.1a. It consists of four nodes connected by three elements (bars). Elements of the truss can undergo elongation or shortening when the truss is subjected to an external load. When such a load is applied, the truss is deformed as shown in Fig.

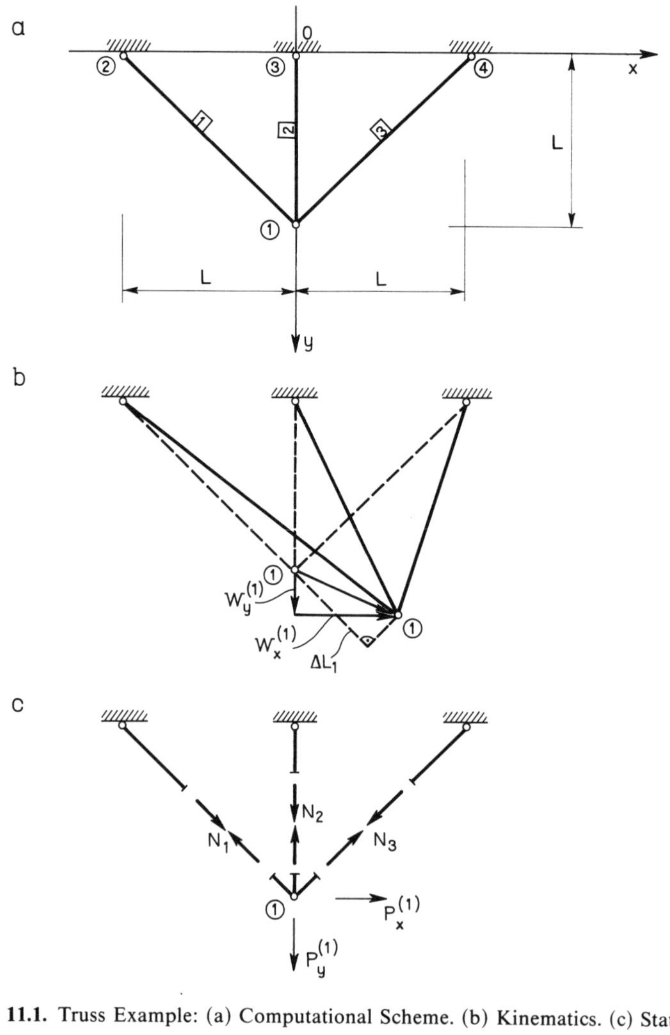

Figure 11.1. Truss Example: (a) Computational Scheme. (b) Kinematics. (c) Statics.

11.1b. Since nodes 2, 3, 4 are fixed, the deformed configuration is completely described by the matrix $\mathbf{w} = \{W_x^{(1)}, W_y^{(1)}\}$ containing the components of the displacement of node 1 in a fixed reference frame Oxy.

Let the increments of length introduced in elements 1, 2, 3 by the transition from the initial to the deformed configuration be arranged into the strain matrix $\mathbf{q} = \{\Delta L_1, \Delta L_2, \Delta L_3\}$. The entries of $\mathbf{q}$ can be found, provided that $\mathbf{w}$ is known. In general, the relation $\mathbf{q} = \mathbf{q}(\mathbf{w})$ is nonlinear. We restrict ourselves, however, to situations where the deformed configuration is close to the initial one. In other words, we assume that displacements are small and that the strain-displacement relations are linearized.

For the truss in Fig. 11.1, it is easy to check that these relations are

$$\begin{aligned} \Delta L_1 &= 0.707\, W_x^{(1)} + 0.707\, W_y^{(1)}, \\ \Delta L_2 &= W_y^{(1)}, \\ \Delta L_3 &= -0.707\, W_x^{(1)} + 0.707\, W_y^{(1)}. \end{aligned}$$

These are the kinematic equations (11.1) with

$$\mathbf{C} = \begin{bmatrix} 0.707 & 0.707 \\ 0.000 & 1.000 \\ -0.707 & 0.707 \end{bmatrix}.$$

Having described the kinematics of the truss, we now examine the relations between external and internal forces. Since a pin-jointed truss is considered, the stress state of the ith element is characterized by its axial force N_i. Hence, the matrix $\mathbf{s} = \{N_1, N_2, N_3\}$ completely describes the distribution of internal forces. We assume, furthermore, that the external force is applied to the free node of the truss. This force can be represented by the load matrix $\mathbf{p} = \{P_x^{(1)}, P_y^{(1)}\}$ containing the load components in the coordinate system Oxy.

Writing down the equilibrium conditions for node 1 (Fig. 11.1c), we obtain the equations

$$\begin{aligned} P_x^{(1)} &= 0.707\, N_1 - 0.707\, N_3, \\ P_y^{(1)} &= 0.707\, N_1 + N_2 + 0.707\, N_3. \end{aligned}$$

These are the equations of equilibrium (11.2).

The truss in Fig. 11.1 is hyperstatic, since $m = 3$ and $n = 2$, so $k = 1$. In structural mechanics the value of k is called the *degree of statical indeterminancy.* If we remove element 3, the resulting truss (Fig. 11.2a) is isostatic ($m = n = 2, k = 0$). Deleting element 2 converts the truss to the mechanism in Fig. 11.2b ($m = 1, n = 2, k = -1$).

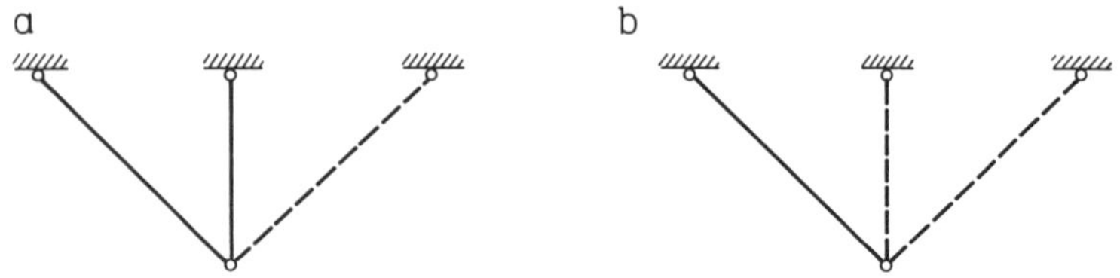

Figure 11.2. Modifications of Truss from Fig. 11.1: (a) Isostatic Structure. (b) Mechanism.

This simple example will help us to clarify the physical meaning of matrices $\mathbf{C}^{-}$ and $\mathbf{C}^{0}$. It follows from (11.7) that the columns of $\mathbf{C}^{-T}$ represent stresses induced in a primary isostatic system by the loads $p_j = 1$, $j = 1, 2, \ldots, n$. Similarly, the columns of $\mathbf{C}^{0T}$ would be the stresses caused in the same system by $z_i = 1$, $i = 1, 2, \ldots, k$. Since there is usually some freedom in choosing the primary system, $\mathbf{C}^{-}$ and $\mathbf{C}^{0}$ are not unique. Nevertheless, they can be computed automatically by means of Jordan transformations and even optimized with respect to the accuracy of numerical calculations [Robinson (1973)].

If the isostatic truss in Fig. 11.2a serves as the primary system, then

$$\mathbf{C}^{-} = \begin{bmatrix} 1.414 & -1.000 & 0.000 \\ 0.000 & 1.000 & 0.000 \end{bmatrix}$$

and

$$\mathbf{C}^{0} = [1.000 \quad -1.414 \quad 1.000].$$

It is easy to check that these matrices fulfill definitions (11.5) and (11.6). Since $k = 1$, the matrix $\mathbf{z}$ contains a single entry corresponding to the redundant axial force in member 3.

If the truss in Fig. 11.2a were analyzed independently, we would have $\mathbf{s} = \{N_1, N_2\}$ and

$$\mathbf{C} = \begin{bmatrix} 0.707 & 0.707 \\ 1.000 & 0.000 \end{bmatrix}.$$

The stress state induced by a given load $\mathbf{p}$ would be then uniquely defined by Eq. (11.9), with

$$\mathbf{C}^{-1} = \begin{bmatrix} 0.000 & 1.000 \\ 1.414 & -1.000 \end{bmatrix}.$$

Finally, if bars 2 and 3 were removed simultaneously from the initial truss, it would be converted to the mechanism. The mechanism would have a single degree of freedom since $k = -1$.

11.2.2. Example of a Frame

Trusses are the most convenient object of discretization, since the stress or strain state of each of their generic elements can be described uniquely and exactly by a single parameter. For flexural systems the situation is more complicated. The stress state of a cross section of a beam can be described with three generalized stresses: the bending moment M, the shear force T, and the axial force N. If distributed loads are absent, then T and N remain constant along the beam and M is a linear function of the position. Hence, the stress state can be represented uniquely by four quantities: the axial force, the shear force, and the two bending moments (one at each end of the beam). The model can be simplified further if we neglect the generalized strains associated with shear and axial elongations. Then T and N become internal reactions and can be excluded from the matrix $\mathbf{s}$. The result is a discrete model of a structure subject to bending only, which we explain by using the example of the simple frame shown in Fig. 11.3a.

This frame consists of two elements connected to three nodes. We label the end points of each element by a and b. Since the right end of element 2 is simply supported, the bending moment $M_b^{(2)}$ must vanish. Therefore, three moments ($M_a^{(1)}$, $M_b^{(1)}$, $M_a^{(2)}$) completely describe the stress state of the frame. The corresponding generalized strains are the angles $\theta_a^{(1)}$, $\theta_b^{(1)}$, and $\theta_a^{(2)}$ between the chord of each element and the tangents to its deformed axis at points a and b (Fig. 11.3b). Note that $\theta_a^{(1)}$ and $\theta_b^{(1)}$ are measured with respect to the chord of the deformed element, not with respect to the centerline of the element in the initial configuration. This allows us to exclude the influence of rigid-body motion upon strains. Imagine, for example, that after the frame has been deformed, as shown in Fig. 11.3.b, we introduce hinges at the end points of element 1. Then that element would straighten up, and its deformed axis would coincide with the chord. Hence, the strains $\theta_a^{(1)}$ and $\theta_b^{(1)}$ would vanish, as expected. On the other hand, the angle between the chord and the initial centerline would remain constant, since the rigid-body part of the motion would be the same as it was before the structure was modified.

The integral of the product "bending moment" times "curvature" taken along the axis of beam, and representing the internal work, can be replaced by the sum of products "bending moment" times "rotation," evaluated at points a and b. Thus we take

$$\mathbf{s} = \{M_a^{(1)}, M_b^{(1)}, M_a^{(2)}\}, \quad \mathbf{q} = \{\theta_a^{(1)}, \theta_b^{(1)}, \theta_a^{(2)}\}, \quad m = 3.$$

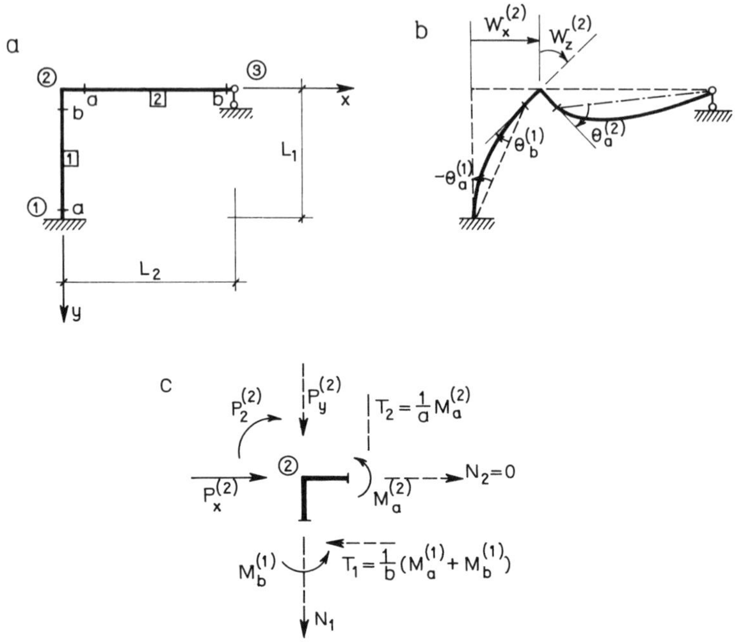

Figure 11.3. Frame Example: (a) Computational Scheme. (b) Kinematics. (c) Equilibrium of Node 2.

The absence of axial strain implies that node 2 can undergo only horizontal shift $W_x^{(2)}$ and rotation $W_z^{(2)}$, which are the generalized displacements. Their static counterparts are the horizontal force $P_x^{(2)}$ and the moment $P_z^{(2)}$ both acting on node 2. The vertical force $P_y^{(2)}$, if present, cannot be included in **p** because it does not contribute to the external work. Thus,

$$\mathbf{p} = \{P_x^{(2)}, P_z^{(2)}\}, \quad \boldsymbol{w} = \{W_x^{(2)}, W_z^{(2)}\}, \quad \textit{and} \quad n = 2.$$

The following kinematic relations can be deduced from Fig. 11.3b:

$$\begin{aligned} \theta_a^{(1)} &= -W_x^{(2)}/L_1, \\ \theta_b^{(1)} &= -W_x^{(2)}/L_1 + W_z^{(2)}, \\ \theta_a^{(2)} &= W_z^{(2)}. \end{aligned}$$

Therefore the compatibility matrix is

$$\mathbf{C} = \begin{bmatrix} -1/L_1 & 0 \\ -1/L_1 & 1 \\ 0 & 1 \end{bmatrix}.$$

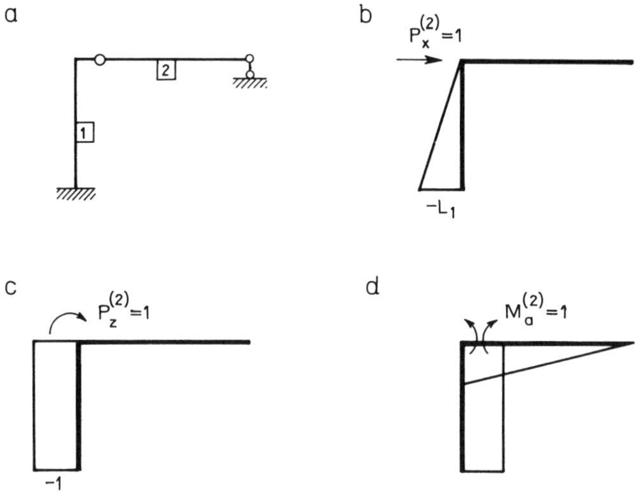

Figure 11.4. Physical Meaning of $\mathbf{C}^-$ and $\mathbf{C}^0$: (a) Primary System. Stress State Corresponding to the (b) First Row of $\mathbf{C}^-$. (c) Second Row of $\mathbf{C}^-$. (d) Single Row of $\mathbf{C}^0$.

The equilibrium equations for node 2 follow from Fig. 11.3c:

$$P_x^{(2)} = -(M_a^{(1)} + M_b^{(1)})/L_1,$$
$$P_z^{(2)} = M_b^{(1)} + M_a^{(2)}.$$

Their matrix form is that of adjoint equations (11.2).

For the frame under consideration, $k = 1$. One possible primary system is shown in Fig. 11.4a. The single component of $\mathbf{z}$ is then the redundant bending moment $M_a^{(2)}$, and the matrices entering Eq. (11.7) are

$$\mathbf{C}^- = \begin{bmatrix} -L & 0 & 0 \\ -1 & 1 & 0 \end{bmatrix},$$
$$\mathbf{C}^0 = [1 \quad -1 \quad 1].$$

The rows of $\mathbf{C}^-$ are the internal forces caused by the two unit-load systems (Fig. 11.4b,c); $\mathbf{C}^0$ represents the stress state induced by the unit redundant moment (Fig. 11.4d).

11.3. Automatic Generation of Adjoint Equations

In the previous section we showed how to write Eqs. (11.1) and (11.2) by considering the kinematics and equilibrium of a structure. This approach is suitable for analyzing simple structures because it uses our ability to interpret graphical data.

However, for more complicated structural systems an automatic procedure that would allow us to generate the compatibility matrix **C** by computer is preferable. We can obtain such a procedure similar to the way we obtained the stiffness matrix in the FEM: we first generate the compatibility matrices for individual elements referred to their local coordinate systems. We then transform these matrices to global coordinates and construct the global matrix **C**.

As an example, consider the beam element in Fig. 11.5. The overbar is used to indicate quantities referred to local coordinates $(\bar{x}, \bar{y})$. The element is rigidly attached to nodes a and b, the end points of the element.

We assume that the shear strain is negligible. The resulting kinematics is shown in Fig. 11.5b. The nodes can undergo rotations $W_z^{(a)}$, $W_z^{(b)}$ and displacements $W_x^{(a)}$, $W_y^{(a)}$, $W_x^{(b)}$, $W_y^{(b)}$. The generalized strains are the

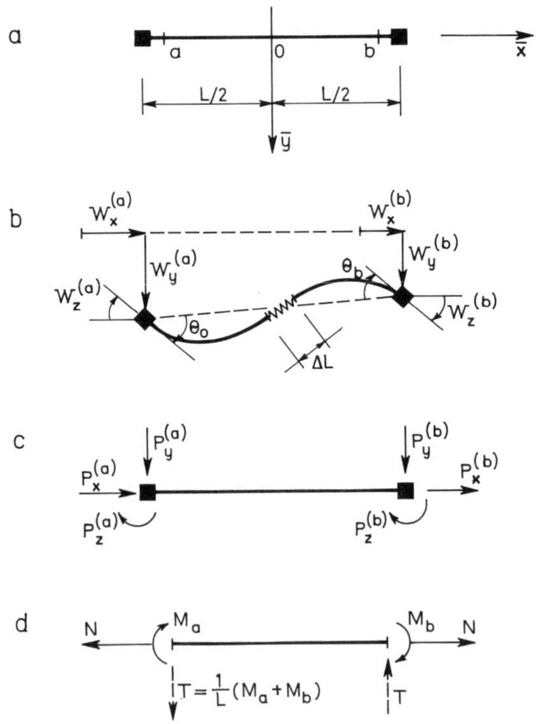

Figure 11.5. Element Subjected to Bending and Tension/Compression: (a) Computational Scheme. (b) Kinematics. (c) Generalized Loads. (d) Generalized Stresses.

angles of rotation θ_a, θ_b of the tangents to the beam axis measured clockwise from the chord at points a and b as well as the elongation ΔL [the generalized strains in our approach coincide with the "natural strains" proposed by Argyris and Mlejnek (1982)].

Static counterparts of the generalized displacements are the nodal moments $P_z^{(a)}$, $P_z^{(b)}$ and the nodal forces $P_x^{(a)}$, $P_y^{(a)}$, $P_x^{(b)}$, $P_y^{(b)}$ (Fig. 11.5c). The shear force T should be treated as an internal reaction because it does not contribute to the internal work. The bending moment is linear along the beam and can be uniquely represented by the values M_a, M_b at the end points. The axial force N is constant throughout the element.

The generalized variables for this element have different dimensions: the kinematic variables include rotations and linear displacements, whereas the static quantities include moments and direct forces. If we accept this situation, we end up with entries of $\bar{\mathbf{C}}$ having large differences in their absolute values (compare $1/L_1$ and unity in the previous example). Such a discrepancy could have a negative influence on the accuracy of the numerical calculations, especially if the problem is solved on a microcomputer. Therefore it is better to work with nondimensional variables:

$$\begin{aligned}
m &= M/P_0L_0, & n &= N/P_0, & & \\
p_x &= P_x/P_0, & p_y &= P_y/P_0, & p_z &= P_z/P_0L_0, \\
w_x &= W_x/L_0, & w_y &= W_y/L_0, & w_z &= W_z, \\
l &= L/L_0, & \varepsilon &= \Delta L/L_0, & &
\end{aligned} \tag{11.10}$$

where the reference length L_0 and the reference force P_0, though arbitrary, should be chosen so that the important nondimensional variables have the same order. In particular, the length scaling parameter l should be 1 for all comparable spans.

The kinematic equations of the beam,

$$\begin{aligned}
\theta_a &= (W_y^{(a)} - W_y^{(b)})/L + W_z^{(a)}, \\
\theta_b &= (W_y^{(a)} - W_y^{(b)})/L + W_z^{(b)}, \\
\Delta L &= -W_x^{(a)} + W_x^{(b)},
\end{aligned}$$

follow from Fig. 11.5b. Transforming them into nondimensional form, we obtain

$$\begin{aligned}
\theta_a &= (w_y^{(a)} - w_y^{(b)})/l + w_z^{(a)}, \\
\theta_b &= (w_y^{(a)} - w_y^{(b)})/l + w_z^{(b)}, \\
\varepsilon &= -w_x^{(a)} + w_x^{(b)}.
\end{aligned} \tag{11.11}$$

These are the local kinematic equations

$$\bar{\mathbf{q}} = \bar{\mathbf{C}}\bar{\mathbf{w}}, \tag{11.12}$$

with

$$\bar{\mathbf{q}} = \{\theta_a, \theta_b, \varepsilon\},$$
$$\bar{\mathbf{w}} = \{w_x^{(a)}, w_y^{(a)}, w_z^{(a)} \mid w_x^{(b)}, w_y^{(b)}, w_z^{(b)}\},$$

and

$$\bar{\mathbf{C}} = \left[\begin{array}{ccc|ccc} 0 & 1/l & 1 & 0 & -1/l & 0 \\ 0 & 1/l & 0 & 0 & -1/l & 1 \\ -1 & 0 & 0 & 1 & 0 & 0 \end{array}\right]. \tag{11.13}$$

Thus the characteristic dimensions are $\bar{m} = 3$ and $\bar{n} = 6$.

Note that the displacement matrix consists of two subvectors related to nodes a and b: $\bar{\mathbf{w}} = \{\bar{\mathbf{w}}^{(a)}, \bar{\mathbf{w}}^{(b)}\}$. Similarly, the compatibility matrix can be decomposed into two submatrices, as indicated by the dashed line in (11.13): $\bar{\mathbf{C}} = [\bar{\mathbf{C}}^{(a)}, \bar{\mathbf{C}}^{(b)}]$.

The equilibrium equations for nodes a and b are (compare Fig. 11.3c)

$$P_x^{(a)} = -N, \quad P_y^{(a)} = T = (M_a + M_b)/L, \quad P_z^{(a)} = M_a,$$
$$P_x^{(b)} = N, \quad P_y^{(b)} = -T = -(M_a + M_b)/L, \quad P_z^{(b)} = M_b.$$

Transforming to nondimensional quantities, we obtain the set of equations

$$\begin{aligned} p_x^{(a)} &= -n, \\ p_y^{(a)} &= (m_a + m_b)/l, \\ p_z^{(a)} &= m_a, \\ p_x^{(b)} &= n, \\ p_y^{(b)} &= -(m_a + m_b)/l, \\ p_z^{(b)} &= m_b. \end{aligned} \tag{11.14}$$

That set corresponds to the matrix equation

$$\bar{\mathbf{p}} = \bar{\mathbf{C}}^{\mathrm{T}}\bar{\mathbf{s}}, \tag{11.15}$$

adjoint to (11.12).

In agreement with the general philosophy of the FEM, each structure is represented as an assembly of n_E elements connected to n_N nodes. After a global reference system Oxy is chosen, the layout of the structure is described through given coordinates of the nodes supplemented by information about the way in which elements are connected to nodes. The latter information can be stored in the $n_E \times n_N$ matrix of connectivity, **B**, which relates the local labels a, b of nodes to their global indices:

$$B_{ij} = \begin{cases} a & \text{if } j\text{th node is node } a \text{ of } i\text{th element,} \\ b & \text{if } j\text{th node is node } b \text{ of } i\text{th element,} \\ 0 & \text{otherwise.} \end{cases} \tag{11.16}$$

For example, the connectivity matrix of the frame in Fig. 11.3a is

$$\mathbf{B} = \begin{bmatrix} a & b & 0 \\ 0 & a & b \end{bmatrix}. \tag{11.17}$$

Before they are assembled, all matrices $\bar{\mathbf{C}}_i$ generated for individual elements must be transformed to the global system of coordinates. Let α_i be the angle between the local axis $\bar{x}_i$ and the global axis x (Fig. 11.6). Then the transformation matrix

$$\mathbf{A}_i = \begin{bmatrix} \cos\alpha_i & -\sin\alpha_i & 0 \\ \sin\alpha_i & \cos\alpha_i & 0 \\ 0 & 0 & 1 \end{bmatrix} \tag{11.18}$$

allows us to replace Eq. (11.12) by its equivalent in the reference system Oxy:

$$\mathbf{q}_i = \mathbf{C}_i \mathbf{w}_i, \tag{11.19}$$

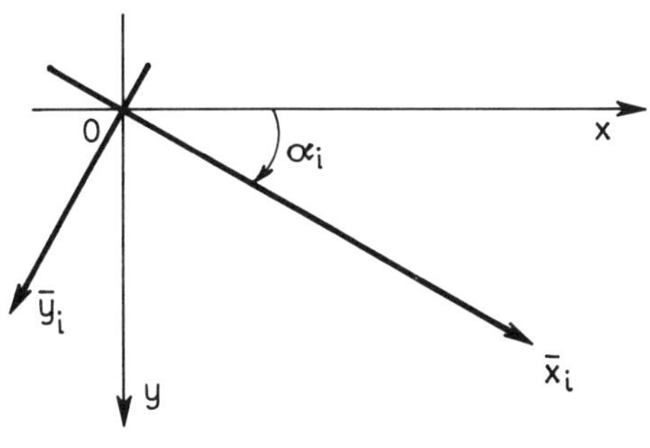

Figure 11.6. Local and Global Coordinate Systems.

with $\mathbf{C}_i = [\mathbf{C}_i^{(a)}, \mathbf{C}_i^{(b)}]$ and

$$\mathbf{C}_i^{(a)} = \bar{\mathbf{C}}_i^{(a)}\mathbf{A}_i, \qquad \mathbf{C}_i^{(b)} = \bar{\mathbf{C}}_i^{(b)}\mathbf{A}_i. \tag{11.20}$$

The local equilibrium equation transforms similarly:

$$\mathbf{p}_i = \mathbf{C}_i^{\mathrm{T}}\mathbf{s}_i. \tag{11.21}$$

Let n_s be the number of suppressed degrees of freedom. Then the dimensions of the global compatibility matrix $\mathbf{C}$ can be computed as

$$\begin{aligned} m &= \bar{m} \times n_E = m', \\ n &= \bar{n} \times n_N - n_s = n' - n_s. \end{aligned} \tag{11.22}$$

The assembly of $\mathbf{C}$ can be accomplished in many ways, and the quality of the resulting computer code depends substantially upon special techniques used by the programmer in order to save memory and to reduce the number of required operations. Ignoring the technical details, which can be found, e.g., in Lawo and Thierauf (1980), we demonstrate a crude but simple solution of the problem.

Let us temporarily forget about supports. Then the matrices $\mathbf{p}'$, $\mathbf{w}'$, $\mathbf{s}'$, and $\mathbf{q}'$ could be constructed from submatrices of constant dimension:

$$\begin{aligned} \mathbf{p}' &= \{\mathbf{p}^{(1)}, \mathbf{p}^{(2)}, \ldots, \mathbf{p}^{(n_N)}\}, \\ \mathbf{w}' &= \{\mathbf{w}^{(1)}, \mathbf{w}^{(2)}, \ldots, \mathbf{w}^{(n_N)}\}, \\ \mathbf{s}' &= \{\mathbf{s}^{(1)}, \mathbf{s}^{(2)}, \ldots, \mathbf{s}^{(n_E)}\}, \\ \mathbf{q}' &= \{\mathbf{q}^{(1)}, \mathbf{q}^{(2)}, \ldots, \mathbf{q}^{(n_E)}\}. \end{aligned}$$

The corresponding compatibility matrix $\mathbf{C}'$ would then have dimensions $m' \times n'$ and a very regular internal structure. As Fig. 11.7 shows, that structure is identical to that of matrix $\mathbf{B}$. To generate $\mathbf{C}'$, we simply insert the submatrices $\mathbf{C}_i^{(a)}$, $\mathbf{C}_i^{(b)}$ into positions indicated by the entries $B_{ij} = a$ and $B_{ij} = b$, respectively. The remaining blocks of $\mathbf{C}'$ are filled with zero entries. After the scratch matrix $\mathbf{C}'$ has been generated, we can take care of the supports. It is clear that each suppressed degree of freedom allows us to eliminate a single column of $\mathbf{C}'$.

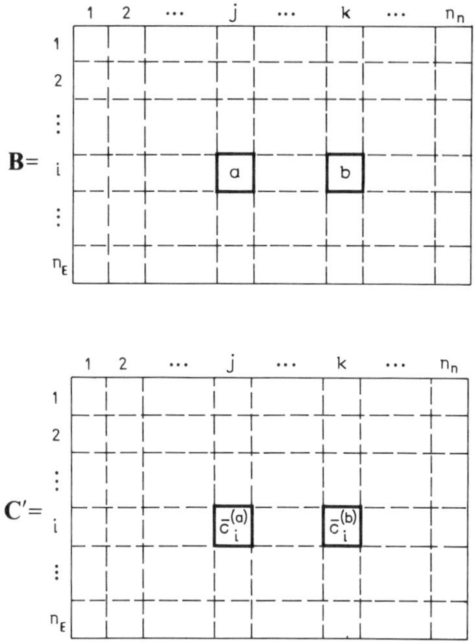

Figure 11.7. Similarity between Internal Structure of Matrices **B** and **C**′.

The procedure we have introduced can be best explained with an example. Recall the frame depicted in Fig. 11.3a. First we find that the characteristic dimensions of that structure are

Number of elements $n_E = 2$;
Number of nodes $n_N = 3$;
Number of suppressed degrees of freedom $n_s = 4$.

Taking the beam element from Fig. 11.5 for discretization purposes, we have $\bar{m} = 3$ and $\bar{n} = 6$. Therefore, $\mathbf{C}'$ must have dimensions $m' = 3 \times 2 = 6$ and $n' = 3 \times 3 = 9$. After reduction, the final dimensions of $\mathbf{C}$ are $m = 6$ and $n = 9 - 4 = 5$.

Next we have to compute transformation matrices. Since $\alpha_1 = 3\pi/2$ and $\alpha_2 = 0$, it follows from (11.18) that

$$\mathbf{A}_1 = \begin{bmatrix} 0 & -1 & 0 \\ 1 & 0 & 0 \\ 0 & 0 & 1 \end{bmatrix}, \qquad \mathbf{A}_2 = \begin{bmatrix} 1 & 0 & 0 \\ 0 & 1 & 0 \\ 0 & 0 & 1 \end{bmatrix}. \tag{11.23}$$

Applying transformation rule (11.20) to matrices (11.13) and (11.23),

Table 11.1. The Matrix $\mathbf{C}'$

		$w_x^{(1)}$	$w_y^{(1)}$	$w_z^{(1)}$	$w_x^{(2)}$	$w_y^{(2)}$	$w_z^{(2)}$	$w_x^{(3)}$	$w_y^{(3)}$	$w_z^{(3)}$
		1	2	3	4	5	6	7	8	9
$\theta_a^{(1)}$	1	$\frac{1}{l_1}$	0	1	$-\frac{1}{l_1}$	0	0	0	0	0
$\theta_b^{(1)}$	2	$\frac{1}{l_1}$	0	0	$-\frac{1}{l_1}$	0	1	0	0	0
ε_1	3	0	1	0	0	−1	0	0	0	0
$\theta_a^{(2)}$	4	0	0	0	0	$\frac{1}{l_2}$	1	0	$-\frac{1}{l_2}$	0
$\theta_b^{(2)}$	5	0	0	0	0	$\frac{1}{l_2}$	0	0	$-\frac{1}{l_2}$	1
ε_2	6	0	0	0	−1	0	0	1	0	0

we have

$$\mathbf{C}_1^{(a)} = \begin{bmatrix} 1/l_1 & 0 & 1 \\ 1/l_1 & 0 & 0 \\ 0 & 1 & 0 \end{bmatrix}, \qquad \mathbf{C}_1^{(b)} = \begin{bmatrix} -1/l_1 & 0 & 0 \\ -1/l_1 & 0 & 1 \\ 0 & -1 & 0 \end{bmatrix},$$

$$\mathbf{C}_2^{(a)} = \begin{bmatrix} 0 & 1/l_2 & 1 \\ 0 & 1/l_2 & 0 \\ -1 & 0 & 0 \end{bmatrix}, \qquad \mathbf{C}_2^{(b)} = \begin{bmatrix} 0 & -1/l_2 & 0 \\ 0 & -1/l_2 & 1 \\ 1 & 0 & 0 \end{bmatrix}.$$

Inserting these submatrices into the position indicated by the connectivity matrix (11.17), we obtain $\mathbf{C}'$, as shown in Table 11.1.

Now the supports must be accounted for (Fig. 11.8). Node 1 is fixed, i.e., $w_x^{(1)} = w_y^{(1)} = w_z^{(1)} = 0$. Node 2 is free, and the vertical displacement of

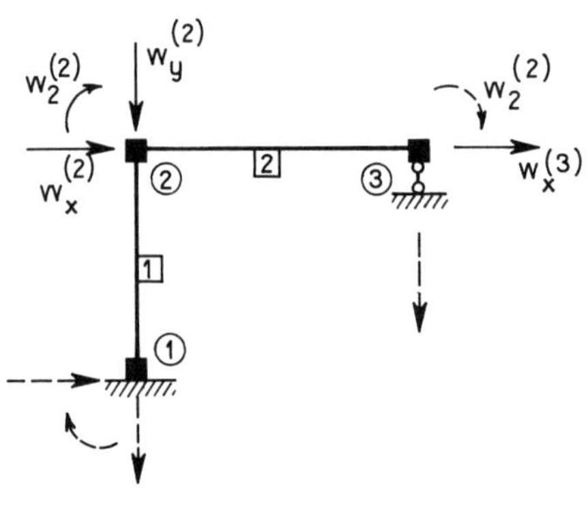

Figure 11.8. Free and Suppressed Degrees of Freedom of Frame Shown in Fig. 11.3.

node 3 is suppressed: $w_y^{(3)} = 0$. Therefore we have to delete columns 1, 2, 3, and 8 of $\mathbf{C}'$. The result is the global compatibility matrix:

$$\mathbf{C} = \begin{bmatrix} -1/l_1 & 0 & 0 & 0 & 0 \\ -1/l_1 & 0 & 1 & 0 & 0 \\ 0 & -1 & 0 & 0 & 0 \\ 0 & 1/l_2 & 1 & 0 & 0 \\ 0 & 1/l_2 & 0 & 0 & 1 \\ -1 & 0 & 0 & 1 & 0 \end{bmatrix} \tag{11.24}$$

The corresponding matrices of state variables are

$$\mathbf{w} = \{w_x^{(2)}, w_y^{(2)}, w_z^{(2)}, w_x^{(3)}, w_z^{(3)}\},$$

$$\mathbf{p} = \{p_x^{(2)}, p_y^{(2)}, p_z^{(2)}, p_x^{(3)}, p_z^{(3)}\},$$

$$\mathbf{q} = \{\theta_a^{(1)}, \theta_b^{(1)}, \varepsilon_1, \theta_a^{(2)}, \theta_b^{(2)}, \varepsilon_2\},$$

$$\mathbf{s} = \{m_a^{(1)}, m_b^{(1)}, n_1, m_a^{(2)}, m_b^{(2)}, n_2\}.$$

Note that the simple support of the right end of element 2 is correctly modeled by taking $p_x^{(3)} = p_z^{(3)} = 0$. If there were an internal hinge in the frame, we would introduce two nodes adjacent to that hinge, as shown in Fig. 11.9. These nodes would have common linear displacements but possibly different rotations.

Such a methodology allows us, in principle, to discretize any skeletal structure by means of finite elements like that shown in Fig. 11.5. However, that would lead to unnecessarily large matrices for simpler structures such as trusses or continuous beams. Therefore, it is reasonable to use a broader catalog of elements such as those given in Section 11.5.

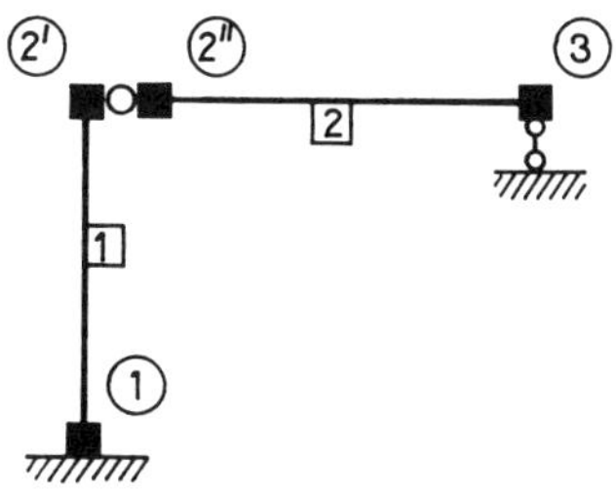

Figure 11.9. Computational Scheme of Frame with Internal Hinge.

Exercise 11.1

Find the matrix $\mathbf{C}^T$ for the truss in Fig. 11.10.

Exercise 11.2

Find the matrices $\mathbf{C}^T$, $\mathbf{C}^-$, and $\mathbf{C}^0$ for the beam in Fig. 11.11.

Exercise 11.3

Find the matrices $\mathbf{C}^T$, $\mathbf{C}^-$, and $\mathbf{C}^0$ for the frame in Fig. 11.12.

Exercise 11.4

Derive the matrix $\mathbf{C}^T$ for the frame in Fig. 11.13, taking into account (a) bending moments only, (b) bending moments and axial forces.

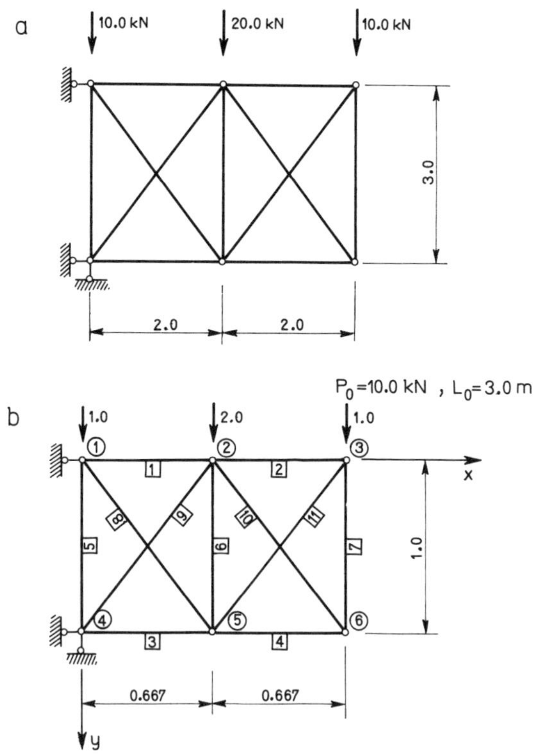

Figure 11.10. Example of Cantilever Truss.

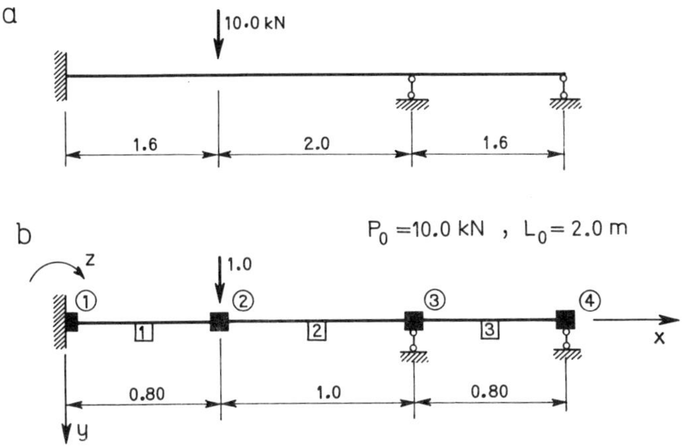

Figure 11.11. Example of Continuous Beam.

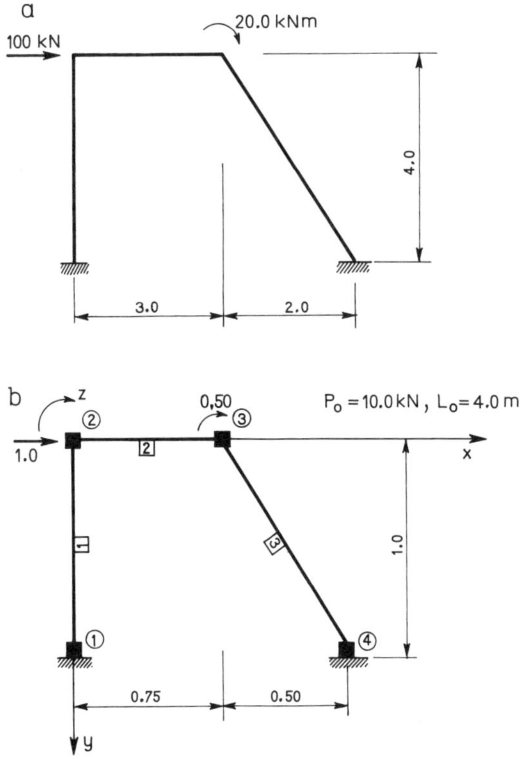

Figure 11.12. Example of Frame with Nonorthogonal Layout.

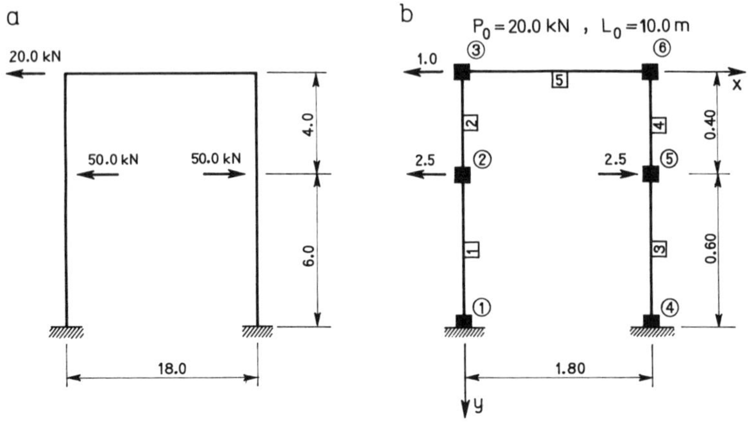

Figure 11.13. Example of Frame with Orthogonal Layout.

11.4. Constitutive Relations

Since most structures remain elastic under service loads, it is natural to begin with the linear elastic physical law in our review. However, the requirement that stresses everywhere in the structure always remain under the yield level is too conservative, at least in civil engineering. The modern tendency in design is to ensure a sufficient safety margin against such undesirable global effects as plastic failure or nonadaptation to cyclic load. This leads to structural optimization in the frame of plasticity theory.

Taking into account the above considerations, we introduce in the sequel the matrix form of constitutive laws most frequently used in engineering practice.

11.4.1. Linear Elastic Behavior

If the material of a generic element remains linearly elastic (Fig. 11.14a), then the following relations hold:

$$\bar{\mathbf{s}} = \bar{\mathbf{E}}\bar{\mathbf{q}}, \qquad \bar{\mathbf{q}} = \bar{\mathbf{E}}^{-1}\bar{\mathbf{s}}. \tag{11.25}$$

Here $\bar{\mathbf{E}}$ is a positive-definite matrix of elasticity of order $\bar{m}$.

The entries of $\mathbf{E}$ depend, first, upon the type of element, second, upon physical properties of the material (e.g., Young's modulus E) and, finally, upon such geometrical characteristics as cross-sectional area A, moment of inertia I of the cross section, or length L of the element.

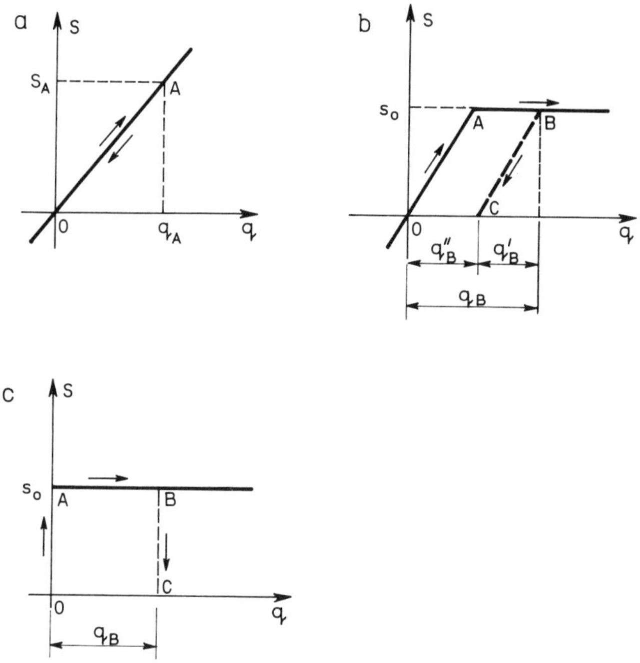

Figure 11.14. Constitutive Models of Axially Loaded Element: (a) Linear Elastic. (b) Elastic, Perfectly Plastic. (c) Rigid, Perfectly Plastic.

For example, for an axially loaded strut we have

$$N = (EA/L)\,\Delta L,$$

or, in nondimensional quantities,

$$n = e_N \varepsilon, \tag{11.26}$$

where

$$e_N = EAL_0/P_0L \tag{11.27}$$

is the longitudinal stiffness of the element. Since the only entries in $\bar{\mathbf{s}}$ and $\bar{\mathbf{q}}$ are n and ε, respectively, the local elasticity matrix $\bar{\mathbf{E}}$ reduces to the scalar e_N.

For elements subject to bending, there is an interaction of strains at both ends of the beam:

$$M_a = 4(EI/L)\theta_a + 2(EI/L)\theta_b,$$
$$M_b = 2(EI/L)\theta_a + 4(EI/L)\theta_b.$$

Equivalent relations in nondimensional form are

$$
\begin{aligned}
m_a &= 4e_M\theta_a + 2e_M\theta_b, \\
m_b &= 2e_M\theta_a + 4e_M\theta_b,
\end{aligned}
\tag{11.28}
$$

where

$$e_M = EI/P_0L_0L \tag{11.29}$$

is the flexural stiffness of the element.

Combining (11.26) with (11.28), we obtain the elasticity matrix

$$\bar{\mathbf{E}} = \begin{bmatrix} 4e_M & 2e_M & 0 \\ 2e_M & 4e_M & 0 \\ 0 & 0 & e_N \end{bmatrix} \tag{11.30}$$

of the element in Fig. 11.5. Its inverse is

$$\bar{\mathbf{E}}^{-1} = \begin{bmatrix} 1/3e_M & 1/6e_M & 0 \\ 1/6e_M & 1/3e_M & 0 \\ 0 & 0 & 1/e_N \end{bmatrix}. \tag{11.31}$$

The elasticity matrices for typical finite elements are given in Section 11.5.

The constitutive equations for an assembled structure have the form

$$\mathbf{s} = \mathbf{E}\mathbf{q}, \qquad \mathbf{q} = \mathbf{E}^{-1}\mathbf{s}, \tag{11.32}$$

where $\mathbf{E}$ is the global elasticity matrix of rank m. Generation of that matrix is very simple, since it is sufficient to place the submatrices $\bar{\mathbf{E}}_i$ along the diagonal:

$$\mathbf{E} = \lceil \bar{\mathbf{E}}^{(1)}, \bar{\mathbf{E}}^{(2)}, \ldots, \bar{\mathbf{E}}^{(n_E)} \rfloor. \tag{11.33}$$

Exercise 11.5

Find the matrix $\mathbf{E}$ for the truss in Fig. 11.10, where Young's modulus $E = 200$ MPa and the cross-sectional area $A = 30.0\ \text{cm}^2$ for all bars.

Exercise 11.6

Find the matrix $\mathbf{E}$ for the frame in Fig. 11.12, where Young's modulus $E = 200$ MPa and the moment of inertia $I = 70.000\ \text{cm}^4$ for all bars.

11.4.2. Elastic, Perfectly Plastic Behavior

If the material is plastic and if we disregard strain hardening, then an axially loaded strut behaves as shown in Fig. 11.14b. It is elastic up to the yield stress s_0 and then exhibits elongation at constant stress level. A crucial feature is that the behavior at an intermediate point B depends upon the stress rate $\dot{s}$: for $\dot{s} = 0$ the strut continues to yield (loading process), whereas for $\dot{s} < 0$ it returns to elastic behavior (unloading process). A total strain q_B is a sum of an elastic part q'_B and a plastic part q''_B. The latter can be detected through unloading along the path BC.

In the general case of an m-dimensional stress state the scalar parameter s_0 has to be replaced by a yield function $f = f(\bar{\mathbf{s}})$. A surface $f = 0$ in stress space, called the *yield locus*, describes the boundary of a convex domain Ω_s surrounding the origin of stress space (Fig. 11.15a). Points such as N located inside Ω_s correspond to elastic behavior of the element, whereas

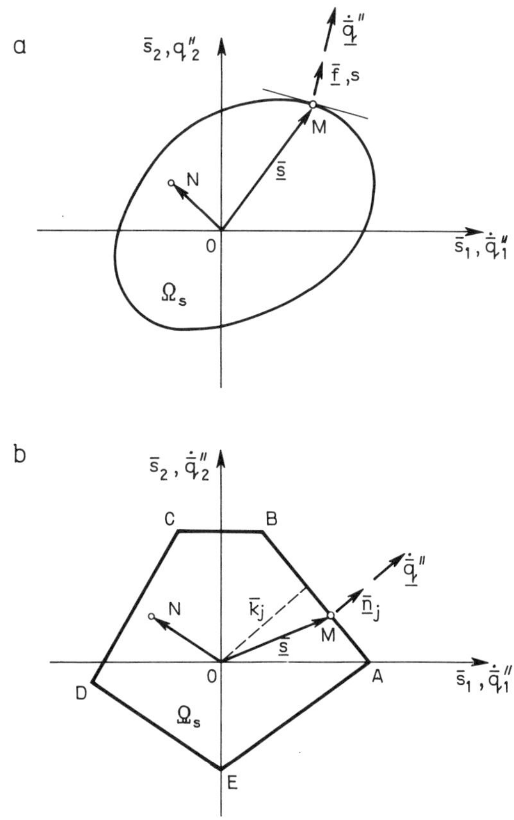

Figure 11.15. Yield Surfaces: (a) Smooth. (b) Piecewise Linear.

stress states located on the yield locus, such as M, may give rise to yielding. Stress states falling outside $f = 0$ are not possible.

Elastic unloading from yield occurs when $\dot{\mathbf{s}}$ is directed toward the interior of Ω_s, i.e., when the current stress point tends to leave the yield locus (Fig. 11.16a). Since the stress vector can not cross that surface, the stress rate $\dot{\mathbf{s}}$ can be at most tangent to $f = 0$ (Fig. 11.16b). When the latter is true, the rate of plastic strain follows from the associated flow rule

$$\dot{\mathbf{q}}'' = \dot{\lambda} \nabla f, \tag{11.34}$$

where ∇f is the gradient of f calculated at the current stress point and $\dot{\lambda}$ is a nonnegative factor called the *plastic multiplier.* Geometrically, relation (11.34) means that the vector $\dot{\mathbf{q}}''$ is directed along the outward normal to the yield locus.

Smooth yield loci such as shown in Fig. 11.15a are convenient to describe analytically, but lead inevitably to nonlinear constraints in the structure's description. As mentioned in Chapter 10, computational efficiency of NLP algorithms is considerably less than that of QP or LP algorithms. Hence, it is desirable to avoid nonlinearity where possible. One step toward that is a piecewise linear approximation of the yield surface.

Elastic-plastic behavior governed by piecewise linear yield functions has been extensively studied by Maier (1970). Following him, we describe the behavior of the element by the following set of matrix relations:

$$\begin{aligned}
&\bar{\mathbf{f}} = \bar{\mathbf{N}}^{\mathrm{T}}\bar{\mathbf{s}} - \bar{\mathbf{k}} \leqslant \mathbf{0},\\
&\dot{\mathbf{q}} = \dot{\mathbf{q}}' + \dot{\mathbf{q}}'',\\
&\dot{\mathbf{q}}' = \bar{\mathbf{E}}^{-1}\dot{\mathbf{s}},\\
&\dot{\mathbf{q}}'' = \bar{\mathbf{N}}\dot{\boldsymbol{\lambda}},\\
&\dot{\boldsymbol{\lambda}} \geqslant \mathbf{0},\\
&\bar{\mathbf{f}}^{\mathrm{T}}\dot{\boldsymbol{\lambda}} = 0,\\
&\dot{\mathbf{f}}_a \leqslant \mathbf{0},\\
&\dot{\mathbf{f}}_a^{\mathrm{T}}\dot{\boldsymbol{\lambda}}_a = 0.
\end{aligned} \tag{11.35}$$

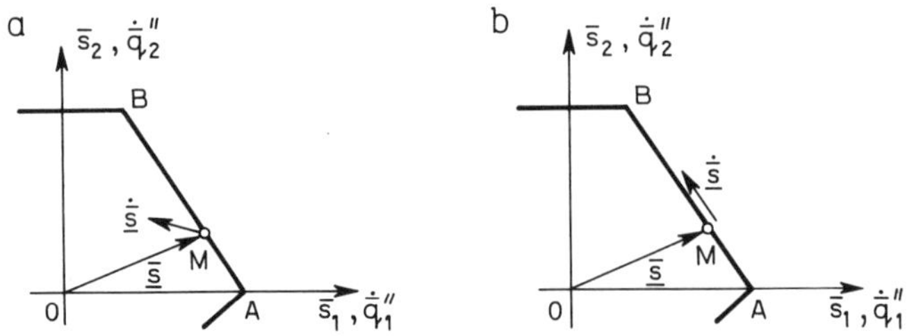

Figure 11.16. Types of Incremental Processes: (a) Unloading. (b) Loading.

The first relation defines Ω_s as a convex polyhedron bounded by $\bar{t}$ hyperplanes $\bar{f}_j = 0$ (Fig. 11.15b). Entries of the matrix $\bar{\mathbf{f}} \in E^{\bar{t}}$ are called *plastic potentials.* Elements of the matrix $\bar{\mathbf{k}} \in E^{\bar{t}}$ are called *plastic moduli.* They can be expressed through such physical characteristics of the cross section as the yield force N_0 or yield moment M_0. Geometrically speaking, $\bar{k}_j$ defines the distance of the jth yield plane from the origin of the stress space. The jth column of the $\bar{m} \times \bar{t}$ matrix $\bar{\mathbf{N}}$ contains a unit outward normal $\bar{\mathbf{n}}_j$ that defines the orientation of the jth plane. For a given stress point $\bar{\mathbf{s}}$ each hyperplane can be either potentially active ($\bar{f}_j = 0$) or passive ($\bar{f}_j < 0$). The potentials of active facets enter the matrix $\bar{\mathbf{f}}_a$, which appears in $(11.35)_7$ and $(11.35)_8$.

Equations $(11.35)_2$ to $(11.35)_4$ describe the decomposition of the total strain rate $\dot{\mathbf{q}}$ into its elastic and plastic parts. The latter part is defined by $(11.35)_4$, which is a special case of rule (11.34). Relations $(11.35)_5$ to $(11.35)_8$ account for the transition between elastic and plastic states. In particular, Eq. $(11.35)_6$ together with relations $(11.35)_1$ and $(11.35)_5$ excludes yielding on passive hyperplanes: $\bar{f}_j < 0$ implies $\dot{\lambda}_j = 0$. Inequality $(11.35)_7$ asserts that the stress point is not allowed to cross the active plane: the time derivative of $\dot{\bar{\mathbf{f}}}_a$ is equal to $\bar{\mathbf{N}}^{\mathrm{T}}\dot{\mathbf{s}}$ and must be nonpositive in order to keep $\dot{\mathbf{s}}$ directed tangent to the active plane or to the interior of Ω_s. If the latter case occurs, then the derivative $\dot{\bar{f}}_{aj}$ becomes negative and condition $(11.35)_8$ implies $\dot{\lambda}_{aj} = 0$.

Consider an axially loaded prismatic element of cross-sectional area A. If the yield stress of the material is equal to σ_0, then the yield force is

$$N^0 = \sigma_0 A. \tag{11.36}$$

For compressed struts we may need to use a different yield stress magnitude than for the same material in tension. We may also require a safe design against the possibility of buckling. For either reason, the magnitude $\tilde{N}^0$ of the limiting compressive force may differ from the tensile value N^0.

Hence, the admissible axial force N should be in the interval between $-\tilde{N}^0$ and N^0. That requirement is equivalent to the inequalities

$$\bar{f}^{\mathrm{I}} = N - N^0 \leqslant 0,$$
$$\bar{f}^{\mathrm{II}} = -N - \tilde{N}^0 \leqslant 0.$$

A nondimensional form of these inequalities is

$$\begin{aligned} \bar{f}^{\mathrm{I}} &= n - n^0 \leqslant 0, \\ \bar{f}^{\mathrm{II}} &= -n - \tilde{n}^0 \leqslant 0, \end{aligned} \tag{11.37}$$

where

$$n^0 = N^0/P_0, \qquad \tilde{n}^0 = \tilde{N}^0/P_0 \tag{11.38}$$

are the nondimensional yield forces. It is easy to convert conditions (11.37) into the general matrix form (11.35), taking $\bar{t} = 2$ and

$$\bar{\mathbf{f}} = \{\bar{f}^{\mathrm{I}}, \bar{f}^{\mathrm{II}}\}, \quad \bar{\mathbf{k}} = \{n^0, \tilde{n}^0\}, \quad \bar{\mathbf{N}} = [1, -1]. \tag{11.39}$$

If the element is subjected only to a bending moment, then the proper plastic modulus is the yield moment

$$M^0 = 2\sigma_0 S, \tag{11.40}$$

where S denotes the static moment of the shaded part of the cross-section with respect to the axis $\bar{z}$ (Fig. 11.17).

Similarly to axial forces, admissible bending moments M must remain between $-\tilde{M}^0$ and M^0, where $\tilde{M}^0$ accounts for negative bending. However, we have two critical cross sections in the beam. Therefore, in total, the admissibility condition is equivalent to the four inequalities

$$\begin{aligned}
\bar{f}_a^{\mathrm{I}} &= M_a - M^0 \leqslant 0,\\
\bar{f}_a^{\mathrm{II}} &= -M_a - \tilde{M}^0 \leqslant 0,\\
\bar{f}_b^{\mathrm{I}} &= M_b - M^0 \leqslant 0,\\
\bar{f}_b^{\mathrm{II}} &= -M_b - \tilde{M}^0 \leqslant 0,
\end{aligned}$$

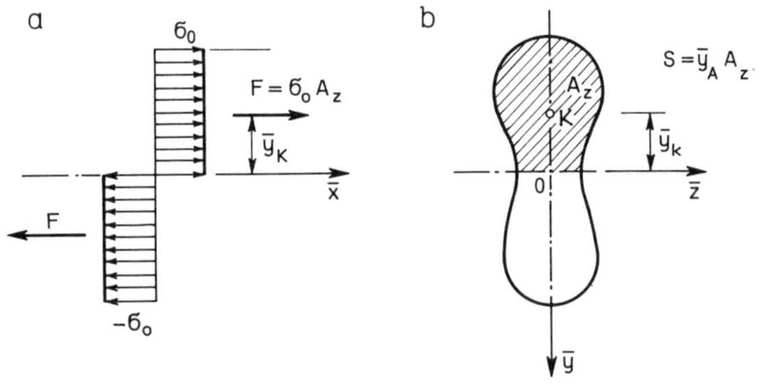

Figure 11.17. Yield Moment of Cross Section: (a) Distribution of Stresses. (b) Evaluation of Static Moment S.

or, in nondimensional form,

$$
\begin{aligned}
\bar{f}_a^{\mathrm{I}} &= m_a - m^0 \leqslant 0, \\
\bar{f}_a^{\mathrm{II}} &= -m_a - \tilde{m}^0 \leqslant 0, \\
\bar{f}_b^{\mathrm{I}} &= m_b - m^0 \leqslant 0, \\
\bar{f}_b^{\mathrm{II}} &= -m_b - \tilde{m}^0 \leqslant 0,
\end{aligned}
\tag{11.41}
$$

where

$$
m^0 = M^0/P_0L_0, \qquad \tilde{m}^0 = \tilde{M}^0/P_0L_0 \tag{11.42}
$$

are the nondimensional yield moments.

The inequalities (11.41) fit into the general form (11.35) if we take $\bar{t} = 4$ and

$$
\begin{aligned}
\bar{\mathbf{f}} &= \{\bar{f}_a^{\mathrm{I}}, \bar{f}_a^{\mathrm{II}} \,\vdots\, \bar{f}_b^{\mathrm{I}}, \bar{f}_b^{\mathrm{II}}\}, \\
\bar{\mathbf{k}} &= \{m^0, \tilde{m}^0 \,\vdots\, m^0, \tilde{m}^0\}, \\
\bar{\mathbf{N}} &= \begin{bmatrix} 1 & -1 & 0 & 0 \\ 0 & 0 & 1 & -1 \end{bmatrix}.
\end{aligned}
\tag{11.43}
$$

When yield is caused by the combined action of axial force and bending moment, the yield function $f = f(M, N)$ is generally nonlinear. To simplify things, we restrict ourselves to the yield curves symmetric with respect to the M and N axes. Such yield functions are applicable for cross sections of steel structures, whereas composite materials such as reinforced concrete behave differently when subjected to positive or negative stress.

For example, the domain Ω_s for the rectangular cross section is bounded by two parabolas, as shown in Fig. 11.18a. Replacing that domain by an

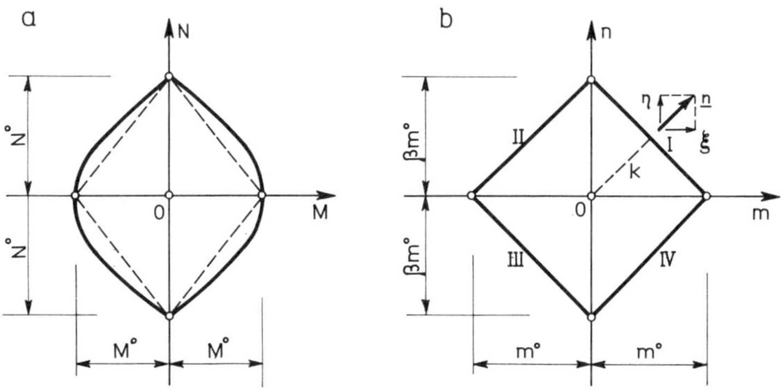

Figure 11.18. Yield Curves for Homogeneous Rectangular Cross Section: (a) Exact Parabolic and Approximate Rhombic Curves. (b) Normalized Nondimensional Form.

inscribed rhombus, we assume that M and N must satisfy the inequalities

$$\begin{aligned} \bar{f}^{\mathrm{I}} &= M/M^0 + N/N^0 \leqslant 1, \\ \bar{f}^{\mathrm{II}} &= -M/M^0 + N/N^0 \leqslant 1, \\ \bar{f}^{\mathrm{III}} &= -M/M^0 - N/N^0 \leqslant 1, \\ \bar{f}^{\mathrm{IV}} &= M/M^0 - N/N^0 \leqslant 1. \end{aligned} \tag{11.44}$$

Such an approximation can be interpreted physically as the replacement of the real cross section by an idealized model consisting of two thin outer layers that carry bending moment and axial forces and of a core that transmits shear force and maintains a constant distance between the layers. Usually such a model is called the *idealized sandwich cross section.*

The admissibility conditions (11.44) can be concisely written as

$$\pm M \pm (M^0/N^0)N \leqslant M^0. \tag{11.45}$$

Introducing nondimensional generalized stresses m, n defined by Eqs. (11.10), the nondimensional yield moment m^0 defined by Eq. (11.42), and the parameter

$$\beta = N^0 L_0 / M^0, \tag{11.46}$$

we transform inequalities (11.45) into the nondimensional form

$$\pm m \pm \frac{1}{\beta} n \leqslant m^0. \tag{11.47}$$

Geometrically speaking, this form is again a rhombus with the vertices located at the points $(\pm m^0, 0)$ and $(0, \pm\beta m^0)$. The first quadrant edge of that rhombus is at perpendicular distance k from the origin and has unit outward normal $\bar{\mathbf{n}} = \{\xi, \eta\}$. It follows from simple geometrical considerations that [Polizzotto and Mazzarella (1979)]

$$\begin{aligned} \xi &= \beta/\sqrt{1+\beta^2}, \\ \eta &= 1/\sqrt{1+\beta^2}, \\ k &= \xi m_0. \end{aligned} \tag{11.48}$$

Admissibility conditions of type (11.47) must be imposed at both ends of the beam. Taking into account that $n_a = n_b = n$, we obtain matrix inequality $(11.35)_1$ and $\bar{t} = 8$ and

$$\begin{aligned}
\bar{\mathbf{f}} &= \{f_a^{\mathrm{I}}, f_a^{\mathrm{II}}, f_a^{\mathrm{III}}, f_a^{\mathrm{IV}}, f_b^{\mathrm{I}}, f_b^{\mathrm{II}}, f_b^{\mathrm{III}}, f_b^{\mathrm{IV}}\},\\
\bar{\mathbf{k}} &= \{1, 1, 1, 1, 1, 1, 1, 1\}k\\
\bar{\boldsymbol{\lambda}} &= \{\lambda_a^{\mathrm{I}}, \lambda_a^{\mathrm{II}}, \lambda_a^{\mathrm{III}}, \lambda_a^{\mathrm{IV}}, \lambda_b^{\mathrm{I}}, \lambda_b^{\mathrm{II}}, \lambda_b^{\mathrm{III}}, \lambda_b^{\mathrm{IV}}\},\\
\bar{\mathbf{N}} &= \left[\begin{array}{cccc|cccc} \xi & -\xi & -\xi & \xi & 0 & 0 & 0 & 0\\ 0 & 0 & 0 & 0 & \xi & -\xi & -\xi & \xi\\ \eta & \eta & -\eta & -\eta & \eta & \eta & -\eta & -\eta \end{array}\right].
\end{aligned} \qquad (11.49)$$

The yield curves for rolled I sections and for welded box sections are also parabolic (Fig. 11.19a). They can be approximated either by the rhombic curve introduced earlier or, less conservatively, by the hexagonal curve shown in Fig. 11.19a by the dashed line. The latter results from the assumption that the influence of the axial force upon yielding can be neglected when the absolute value of that force is less than or equal to αN^0. Such a heuristic rule is included in many civil engineering codes. Massonet and Save (1965) recommend values of $\alpha = 0.15$ for wide-flange sections in bending about the stronger axis and $\alpha = 0.40$ for the same cross section working about the weaker axis. Intermediate values should be taken for box sections. The matrices $\bar{\mathbf{f}}$, $\bar{\boldsymbol{\lambda}}$, $\bar{\mathbf{k}}$, and $\bar{\mathbf{N}}$ for the idealized I sections and box sections are given in Section 11.5.

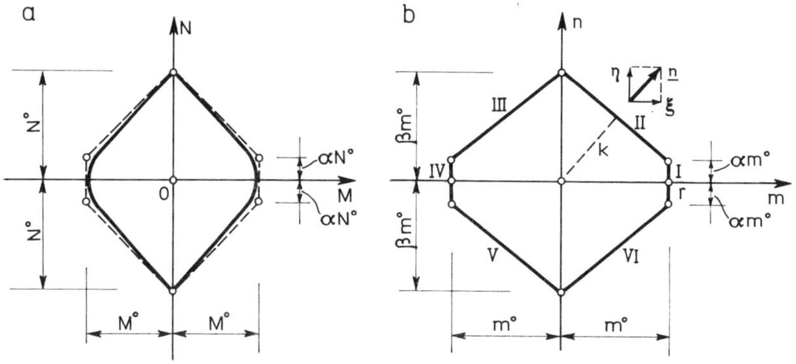

Figure 11.19. Yield Curves for *I* Section and Box Section: (a) Exact Parabolic and Approximate Hexagonal Curves. (b) Normalized Nondimensional Form.

Since the reference length L_0 can be taken arbitrarily, we can adjust it so that $\beta = 1$. According to Eq. (11.46), this requires that we take

$$L_0 = M^0/N^0. \tag{11.50}$$

This choice is advantageous from the stress admissibility condition point of view, but it can destroy the balance of the compatibility matrix **C**. Moreover, the ratio M^0/N^0 need not be the same for each element of the structure, and in optimization that ratio is not known in advance. Despite these difficulties, we should keep β as close to unity as possible.

After the properties of individual elements have been described, constructing the structure is straightforward. The constitutive relations are defined in terms of the local quantities (stresses and strains), which makes the transformation from local to global coordinates obsolete. Hence, the global matrices are simply formed as aggregates of submatrices for the elements:

$$\begin{aligned}
\mathbf{f} &= \{\bar{f}^{(1)}, \bar{f}^{(2)}, \ldots, \bar{f}^{(n_E)}\},\\
\mathbf{k} &= \{\bar{\mathbf{k}}^{(1)}, \bar{\mathbf{k}}^{(2)}, \ldots, \bar{\mathbf{k}}^{(n_E)}\},\\
\boldsymbol{\lambda} &= \{\bar{\boldsymbol{\lambda}}^{(1)}, \bar{\boldsymbol{\lambda}}^{(2)}, \ldots, \bar{\boldsymbol{\lambda}}^{(n_E)}\},\\
\mathbf{N} &= [\bar{\mathbf{N}}^{(1)}, \bar{\mathbf{N}}^{(2)}, \ldots, \bar{\mathbf{N}}^{(n_E)}].
\end{aligned} \tag{11.51}$$

The dimension of **f**, **k**, and $\boldsymbol{\lambda}$ is $t = \bar{t} \times n_E$, whereas **N** is an $m \times t$ matrix.

For the sake of later reference, we rewrite the constitutive relations (11.35) in their global form:

$$\begin{aligned}
&\mathbf{f} = \mathbf{N}^{\mathrm{T}}\mathbf{s} - \mathbf{k} \leqslant \mathbf{0},\\
&\dot{\mathbf{q}} = \dot{\mathbf{q}}' + \dot{\mathbf{q}}'',\\
&\dot{\mathbf{q}}' = \mathbf{E}^{-1}\dot{\mathbf{s}},\\
&\dot{\mathbf{q}}'' = \mathbf{N}\dot{\boldsymbol{\lambda}},\\
&\dot{\boldsymbol{\lambda}} \geqslant \mathbf{0},\\
&\mathbf{f}^{\mathrm{T}}\dot{\boldsymbol{\lambda}} = 0,\\
&\mathbf{f}_a \leqslant \mathbf{0}, \qquad \mathbf{f}_a^{\mathrm{T}}\dot{\boldsymbol{\lambda}}_a = 0.
\end{aligned} \tag{11.52}$$

Exercise 11.7

Find the matrix **N** for the frame in Fig. 11.12, taking into account (a) bending moments only, (b) interaction between bending moments and axial forces according to the rhombic yield curve. For the latter case, assume $\beta = 2.0$.

11.4.3. Rigid, Perfectly Plastic Behavior

Structures made of elastic, perfectly plastic material can carry loads up to a certain level only, called the *ultimate load.* When the loading reaches that level, the structure collapses due to unrestricted plastic deformation. In Chapter 13 we deal with optimization of structures having a prescribed safety margin against such a plastic collapse. It can be proved that the ultimate load does not depend upon the elastic properties of structural elements. This property allows us to use in calculations related to the ultimate load-carrying capacity a very simple model of rigid, perfectly plastic behavior.

This behavior for an axially loaded element is shown in Fig. 11.14c. There is no deformation while $s < s_0$, whereas unrestricted plastic elongation is possible at $s = s_0$. Obviously, the rigid-plastic model can be derived from the elastic-plastic one under the assumption of an infinitely large elastic stiffness. Indeed, taking $\bar{\mathbf{E}}^{-1} = \mathbf{0}$ in $(11.35)_3$, we obtain $\dot{\mathbf{q}}' = \mathbf{0}$, and the constitutive relations for a rigid, perfectly plastic element are

$$\begin{aligned} \bar{\mathbf{f}} &= \bar{\mathbf{N}}^{\mathrm{T}}\bar{\mathbf{s}} - \bar{\mathbf{k}} \leqslant \mathbf{0}, \\ \dot{\mathbf{q}} &= \bar{\mathbf{N}}\dot{\boldsymbol{\lambda}}, \\ \dot{\boldsymbol{\lambda}} &\geqslant \mathbf{0}, \\ \bar{\mathbf{f}}^{\mathrm{T}}\dot{\boldsymbol{\lambda}} &= 0. \end{aligned} \tag{11.53}$$

There is no need to include conditions on $\dot{\mathbf{f}}_a$ because the ultimate load does not depend upon possible unloading at a lower load level. We have also dropped the label distinguishing the plastic strain rate since there is no other kind of strain rate to be considered.

The global constitutive relations for the rigid, perfectly plastic structure have the following form:

$$\begin{aligned} &\mathbf{f} = \mathbf{N}^{\mathrm{T}}\mathbf{s} - \mathbf{k} \leqslant \mathbf{0}, \\ &\dot{\mathbf{q}} = \mathbf{N}\dot{\boldsymbol{\lambda}}, \\ &\dot{\boldsymbol{\lambda}} \geqslant \mathbf{0}, \\ &\mathbf{f}^{\mathrm{T}}\dot{\boldsymbol{\lambda}} = 0. \end{aligned} \tag{11.54}$$

The global matrices have the same structure (11.54) as for elastic–plastic behavior.

One concept remains to be introduced, which is characteristic for the rigid-plastic model of a structural element, namely, the concept of a plastic hinge. It amounts to the experimentally verified assumption that plastic strains tend to concentrate in narrow zones adjacent to the ends of the element. As a result, the element subject to bending remains practically straight except in the neighborhood of cross sections a and b, where it exhibits locally large curvatures. Idealizing such a situation, we assume that all plastic strains occur in plastic hinges that can develop at points a and b (Fig. 11.20).

The instantaneous rotations $\dot{\theta}_a$, $\dot{\theta}_b$ at those hinges together with the elongation rate $\dot{\varepsilon}$ enter the matrix $\dot{\bar{\mathbf{q}}}$. Strictly speaking, we should have introduced generalized plastic hinges that allow not only rotations but axial strains as well. Then $\dot{\varepsilon}_a$ and $\dot{\varepsilon}_b$ would have appeared separately. Nevertheless, it would be necessary to introduce $\dot{\varepsilon} = \dot{\varepsilon}_a + \dot{\varepsilon}_b$ as the kinematic counterpart of n, because the contribution of the axial force to the internal power (work rate) is

$$n_a\dot{\varepsilon}_a + n_b\dot{\varepsilon}_b = n(\dot{\varepsilon}_a + \dot{\varepsilon}_b) = n\dot{\varepsilon}.$$

Note that the kinematic variables introduced in the rigid-plastic model (Fig. 11.20) are measured in the same way as the relevant variables used in the elastic beam (Fig. 11.5). This reflects the known fact that the adjoint equations and, hence, the compatibility matrix $\bar{\mathbf{C}}$ do not depend on constitutive assumptions. In fact, we could imagine elastic hinges (springs) intro-

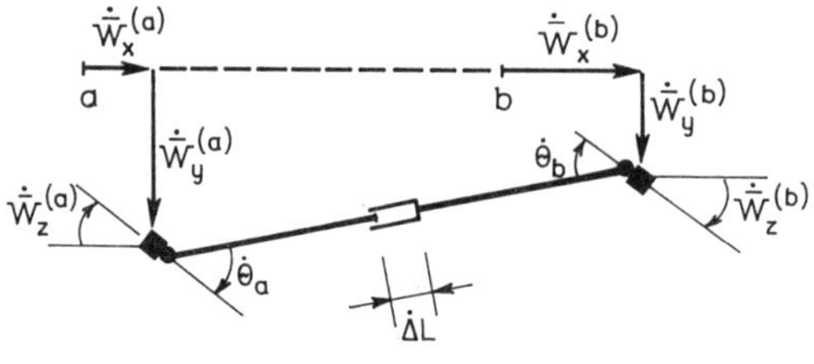

Figure 11.20. Kinematics of Rigid-Plastic Element with Plastic Hinges.

duced at cross sections a, b that would properly simulate the elastic behavior of the element.

11.5. Catalog of Finite Elements

11.5.1. General Notation

11.5.1.1. Computational Scheme

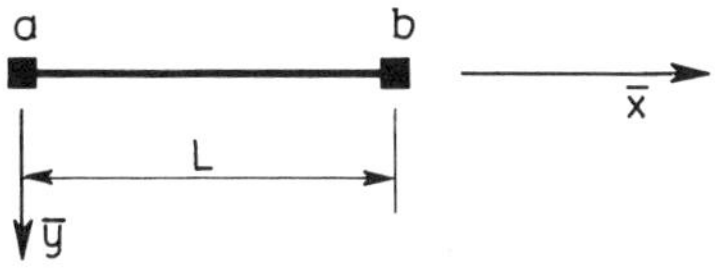

11.5.1.2. Generalized Displacements

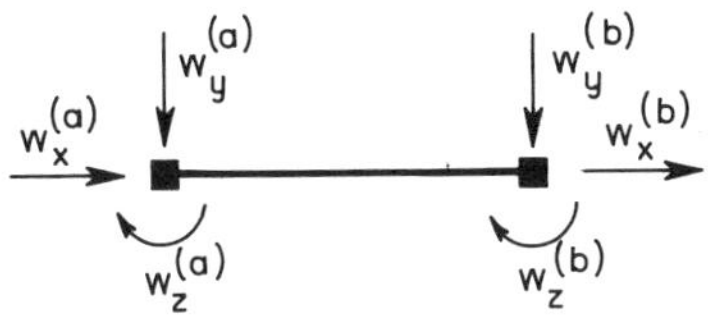

$$\bar{\mathbf{w}} = \{w_x^{(a)}, w_y^{(a)}, w_z^{(a)} \,\vdots\, w_x^{(b)}, w_y^{(b)}, w_z^{(b)}\}$$

11.5.1.3. Generalized Nodal Forces

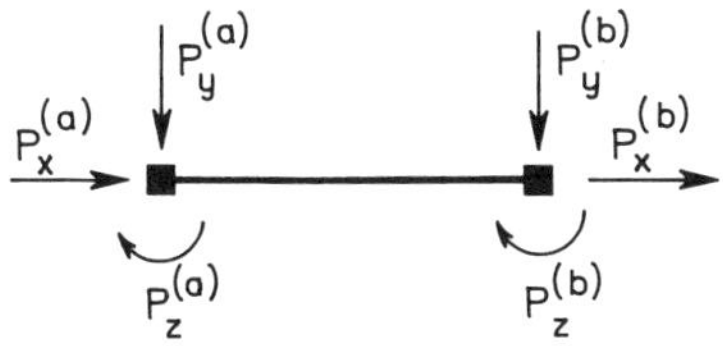

$$\bar{\mathbf{P}} = \{P_x^{(a)}, P_y^{(a)}, P_z^{(a)} \,\vdots\, P_x^{(b)}, P_y^{(b)}, P_z^{(b)}\}$$

11.5.2. Pin-ended Strut

11.5.2.1. Generalized Strains and Stresses

$$\bar{m} = 1 \qquad \bar{\mathbf{C}} = [-1 \quad 0 \quad 0 \,\vdots\, 1 \quad 0 \quad 0]$$

$$\bar{\mathbf{s}} = \{n\} \qquad \bar{\mathbf{E}} = [e^N]$$

$$\bar{\mathbf{q}} = \{\varepsilon\} \qquad e^N = EAL_0/P_0L$$

Yield locus

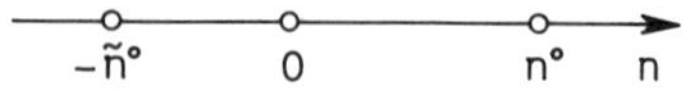

$\bar{l} = 2$

$\bar{\mathbf{f}} = \{\bar{f}^{\mathrm{I}}, \bar{f}^{\mathrm{II}}\}$

$\bar{\boldsymbol{\lambda}} = \{\bar{\lambda}^{\mathrm{I}}, \bar{\lambda}^{\mathrm{II}}\}$

$\bar{\mathbf{k}} = \{\bar{k}^{\mathrm{I}}, \bar{k}^{\mathrm{II}}\}$

$\bar{\mathbf{k}}_1 = n^0, \bar{k}_2 = \tilde{n}^0$

$\bar{\mathbf{N}} = [1 \quad -1]$

11.5.3. Beam Subjected to Bending

11.5.3.1. Generalized Strains and Stresses

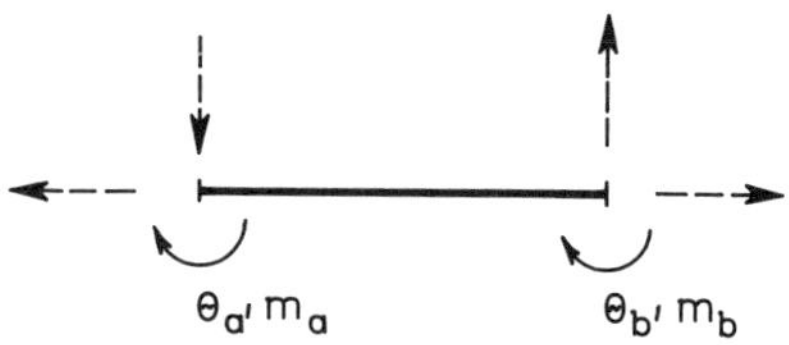

$\bar{m} = 2$

$$\bar{\mathbf{C}} = \begin{bmatrix} 0 & 1/l & 1 & 0 & -1/l & 0 \\ 0 & 1/l & 0 & 0 & -1/l & 1 \end{bmatrix}$$

$\bar{\mathbf{s}} = \{m_a, m_b\}$

$$\bar{\mathbf{E}} = \begin{bmatrix} 4e_M & 2e_M \\ 2e_M & 4e_M \end{bmatrix}$$

$\bar{\mathbf{q}} = \{\theta_a, \theta_b\}$

$e_M = EI/P_0L_0L$

Yield locus

$\bar{l} = 4$

$\bar{\mathbf{f}} = \{\bar{f}^{\mathrm{I}}_a, \bar{f}^{\mathrm{II}}_a \mid \bar{f}^{\mathrm{I}}_b, \bar{f}^{\mathrm{II}}_b\}$

$\bar{\boldsymbol{\lambda}} = \{\bar{\lambda}^{\mathrm{I}}_a, \bar{\lambda}^{\mathrm{II}}_a \mid \bar{\lambda}^{\mathrm{I}}_b, \bar{\lambda}^{\mathrm{II}}_b\}$

$\bar{\mathbf{k}} = \{\bar{k}^{\mathrm{I}}_a, \bar{k}^{\mathrm{II}}_a \mid \bar{k}^{\mathrm{I}}_b, \bar{k}^{\mathrm{II}}_b\}$

$\bar{k}^{\mathrm{I}}_i = m^0; \bar{k}^{\mathrm{II}}_i = \tilde{m}^0 \quad (i = a, b)$

$$\bar{\mathbf{N}} = \begin{bmatrix} 1 & -1 & 0 & 0 \\ 0 & 0 & 1 & -1 \end{bmatrix}$$

11.5.4. Beam Subjected to Bending and Tension/Compression

11.5.4.1. Generalized Strains and Stresses

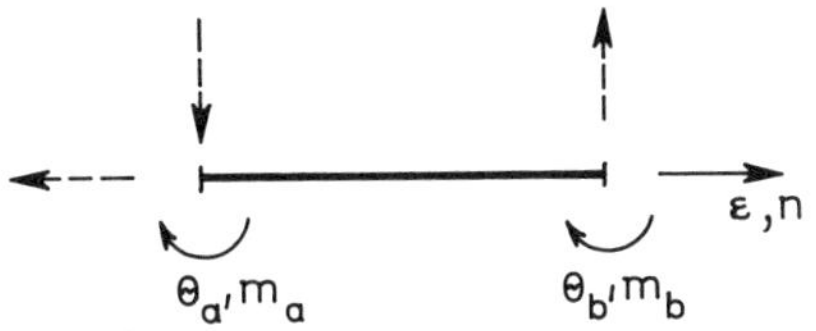

$$\bar{m} = 3$$

$$\bar{\mathbf{s}} = \{m_a, m_b, n\}$$

$$\bar{\mathbf{q}} = \{\theta_a, \theta_b, \varepsilon\}$$

$$\bar{\mathbf{C}} = \left[\begin{array}{ccc|ccc} 0 & 1/l & 1 & 0 & -1/l & 0 \\ 0 & 1/l & 0 & 0 & -1/l & 1 \\ -1 & 0 & 0 & 1 & 0 & 0 \end{array}\right]$$

$$\bar{\mathbf{E}} = \begin{bmatrix} 4e_M & 2e_M & 0 \\ 2e_M & 4e_M & 0 \\ 0 & 0 & e_N \end{bmatrix}$$

$$e_N = EAL_0/P_0L$$

$$e_M = EJ/P_0L_0L$$

a. Idealized sandwich cross-section

Yield locus

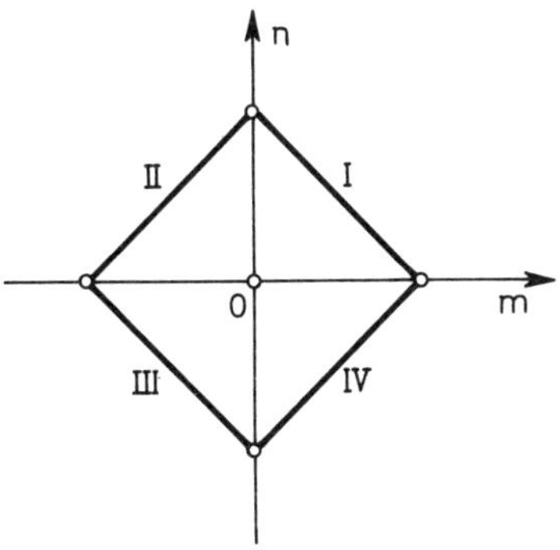

$$\bar{l} = 8$$

$$\bar{\mathbf{f}} = \{f_a^{\mathrm{I}}, f_a^{\mathrm{II}}, f_a^{\mathrm{III}}, f_a^{\mathrm{IV}} \mid f_b^{\mathrm{I}}, f_b^{\mathrm{II}}, f_b^{\mathrm{III}}, f_b^{\mathrm{IV}}\},$$

$$\bar{\boldsymbol{\lambda}} = \{\lambda_a^{\mathrm{I}}, \lambda_a^{\mathrm{II}}, \lambda_a^{\mathrm{III}}, \lambda_a^{\mathrm{IV}} \mid \lambda_b^{\mathrm{I}}, \lambda_b^{\mathrm{II}}, \lambda_b^{\mathrm{III}}, \lambda_b^{\mathrm{IV}}\},$$

$$\bar{\mathbf{k}} = \{\xi, \xi, \xi, \xi \mid \xi, \xi, \xi, \xi\}m^0$$

$$\bar{\mathbf{N}} = \left[\begin{array}{cccc|cccc} \xi & -\xi & -\xi & \xi & 0 & 0 & 0 & 0 \\ 0 & 0 & 0 & 0 & \xi & -\xi & -\xi & \xi \\ \eta & \eta & \eta & -\eta & \eta & \eta & -\eta & -\eta \end{array}\right]$$

b. Idealized I cross section

Yield locus

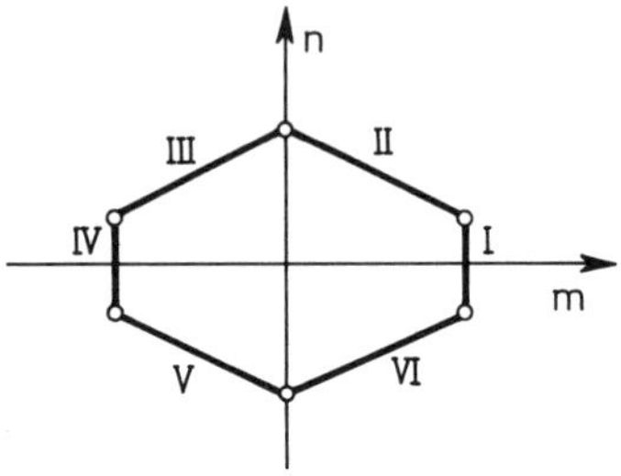

$$\bar{l} = 12$$

$$\eta = 1/[1 + (1-\alpha)^2\beta^2]^{1/2}$$

$$\xi = (1-\alpha)\beta\eta$$

$$\bar{\mathbf{f}} = \{f_a^{\mathrm{I}}, f_a^{\mathrm{II}}, f_a^{\mathrm{III}}, f_a^{\mathrm{IV}}, f_a^{\mathrm{V}}, f_a^{\mathrm{VI}} \;\vdots\; f_b^{\mathrm{I}}, f_b^{\mathrm{II}}, f_b^{\mathrm{III}}, f_b^{\mathrm{IV}}, f_b^{\mathrm{V}}, f_b^{\mathrm{VI}}\},$$

$$\bar{\boldsymbol{\lambda}} = \{\lambda_a^{\mathrm{I}}, \lambda_a^{\mathrm{II}}, \lambda_a^{\mathrm{III}}, \lambda_a^{\mathrm{IV}}, \lambda_a^{\mathrm{V}}, \lambda_a^{\mathrm{VI}} \;\vdots\; \lambda_b^{\mathrm{I}}, \lambda_b^{\mathrm{II}}, \lambda_b^{\mathrm{III}}, \lambda_b^{\mathrm{IV}}, \lambda_b^{\mathrm{V}}, \lambda_b^{\mathrm{VI}}\},$$

$$\bar{\mathbf{k}} = \{1, \beta\eta, \beta\eta, 1, \beta\eta, \beta\eta \;\vdots\; 1, \beta\eta, \beta\eta, 1, \beta\eta, \beta\eta\}m^0$$

$$\bar{\mathbf{N}} = \left[\begin{array}{ccccc|cccccc} 1 & \xi & -\xi & -1 & -\xi & 0 & 0 & 0 & 0 & 0 & 0 \\ 0 & 0 & 0 & 0 & 0 & 1 & \xi & -\xi & -1 & -\xi & -\xi \\ 0 & \eta & \eta & 0 & -\eta & 0 & \eta & \eta & 0 & -\eta & -\eta \end{array}\right]$$

12

Optimization of Elastic Structures

12.1. Introduction

It is natural to begin with the optimal design of elastic structures for two reasons: first, the linear elastic model of the material is the simplest one; second, most structures remain elastic under service loads.

In the following section we recall basic principles of elastic analysis. These principles constitute the background for the more ambitious task of optimal design. First we collect the complete set of governing equations; then we prove that set to be equivalent to the saddle point problem and to the dual pair of QP problems. In this way the discrete formulation of the complementary energy principles is derived. The canonical equations of the two basic methods of elastic analysis—the direct stiffness method and the flexibility method—follow directly from the energy principles.

Results presented in Section 12.2 are general, in the sense that they are valid for any discretized elastic structure. The rest of the chapter is devoted to optimal design of trusses. This is done deliberately: we take the first steps toward optimal design in the simple environment of pin-jointed structures. The presentation of the more complex optimization of flexural systems is postponed until Chapter 15.

12.2. Summary of Elastic Analysis

12.2.1. Formulation of the Problem and the Energy Principles

The problem of static analysis of an elastic structure is formulated as follows: find the displacement $\mathbf{w}_*$, the strain $\mathbf{q}_*$, and the stress $\mathbf{s}_*$ corresponding to the given load $\mathbf{p}_*$. The structure itself is given as well, which means that the matrices $\mathbf{C}$ and $\mathbf{E}$ are known.

The complete set of governing equations for such a problem includes the adjoint equations (12.1) and (12.2) and the constitutive equation $(11.25)_1$ or $(11.25)_2$. That set can be written in the following form:

$$
\begin{array}{rcc|cccc}
 & \mathbf{q}^{\mathrm{T}} & \mathbf{w}^{\mathrm{T}} & \mathbf{s}^{\mathrm{T}} & 1 & & \\
\hline
\nabla f_q = & \mathbf{E} & \mathbf{0} & -\mathbf{I} & \mathbf{0} & = \mathbf{0} & (m) \\
\nabla f_w = & \mathbf{0} & \mathbf{0} & \mathbf{C}^{\mathrm{T}} & -\mathbf{p}_* & = \mathbf{0} & (n) \\
\hdashline
\nabla f_s = & -\mathbf{I} & \mathbf{C} & \mathbf{0} & \mathbf{0} & = \mathbf{0} & (m) \\
\hline
 & (m) & (n) & (m) & (1) & &
\end{array}
\qquad (12.1)
$$

where the symbols in parentheses indicate the dimensions of relevant submatrices.

Clearly, the global square matrix of coefficients of the set (12.1) has order $2m + n$ and is symmetric. The internal structure of that matrix coincides with the pattern in (10.26). Let the kinematic variables $\mathbf{q}$, $\mathbf{w}$ play the role of $\mathbf{x}$, and assume the static variable $\mathbf{s}$ is of type $\mathbf{y}$. Then the global matrix splits into the four submatrices as indicated in (12.1) by the dashed lines. Let us investigate the crucial submatrices along the diagonal. The only nonzero submatrix that appears in $\mathbf{A}_{xx}$ is the positive-definite elasticity matrix $\mathbf{E}$. Hence the square matrix $\mathbf{A}_{xx}$ of order $m + n$ is positive semidefinite. The submatrix $\mathbf{A}_{yy} = \mathbf{0}$ can be formally treated as negative semidefinite. Thus, all the requirements of the scheme (10.26) are met by the complete set of governing equations (12.1).

According to formula (10.27), the potential for (12.1) is the following function of the state variables:

$$f(\mathbf{q}, \mathbf{w}, \mathbf{s}) = \tfrac{1}{2}\mathbf{q}^{\mathrm{T}}\mathbf{E}\mathbf{q} - \mathbf{q}^{\mathrm{T}}\mathbf{I}\mathbf{s} + \mathbf{w}^{\mathrm{T}}\mathbf{C}^{\mathrm{T}}\mathbf{s} - \mathbf{p}_*^{\mathrm{T}}\mathbf{w}. \qquad (12.2)$$

We see here a term quadratic in strains, a term linear in displacements, and two bilinear terms. Since linear functions can be classified as concave or convex, the potential (12.2) can be treated formally as being saddle-shaped—convex with respect to the strains and displacements but concave with respect to the stresses.

Then the solution of Eq. (12.1) corresponds to the saddle point of the potential (12.2):

$$f(\mathbf{q}_*, \mathbf{w}_*, \mathbf{s}_*) = \min_{\mathbf{q},\mathbf{w}} \max_{\mathbf{s}} f(\mathbf{q}, \mathbf{w}, \mathbf{s}). \tag{12.3}$$

Thus we obtain a discrete formulation of the free variational principle usually associated with Hu and Washizu [Lippmann (1972)].

The complementary energy principles that govern the elastic response of the structure follow immediately from the equivalence scheme in Chapter 10. Thus, applying the formulas (10.31), we find the Legendre transforms of the potential:

$$\begin{aligned} f' &= \tfrac{1}{2}\mathbf{q}^{\mathrm{T}}\mathbf{E}\mathbf{q} - \mathbf{p}_*^{\mathrm{T}}\mathbf{w}, \\ f'' &= -\tfrac{1}{2}\mathbf{q}^{\mathrm{T}}\mathbf{E}\mathbf{q}. \end{aligned} \tag{12.4}$$

Since the gradient ∇f_x includes the subvectors ∇f_q and ∇f_w, whereas the gradient ∇f_y corresponds to ∇f_s, the dual QP problems defined by formulas (10.30) assume the following form:

$$\begin{aligned} &\min\{\tfrac{1}{2}\mathbf{q}^{\mathrm{T}}\mathbf{E}\mathbf{q} - \mathbf{p}_*^{\mathrm{T}}\mathbf{w} \mid -\mathbf{q} + \mathbf{C}\mathbf{w} = \mathbf{0}\}, \\ &\max\{-\tfrac{1}{2}\mathbf{q}^{\mathrm{T}}\mathbf{E}\mathbf{q} \mid \mathbf{E}\mathbf{q} - \mathbf{s} = \mathbf{0}, \mathbf{C}^{\mathrm{T}}\mathbf{s} = \mathbf{p}_*\}. \end{aligned} \tag{12.5}$$

The primal problem $(12.5)_1$ expresses the well-known principle of minimum total potential energy. The dual problem $(12.5)_2$ can be transformed further: the first of its constraints allows us to eliminate strains from the unknowns. Having done that, we replace the maximization by a minimization by changing the sign of the cost function. The final result is the familiar principle of minimum complementary energy:

$$\min\{\tfrac{1}{2}\mathbf{s}^{\mathrm{T}}\mathbf{E}^{-1}\mathbf{s} \mid \mathbf{C}^{\mathrm{T}}\mathbf{s} = \mathbf{p}_*\}. \tag{12.6}$$

12.2.2. Isostatic Structures

We mentioned in Section 11.2 that for an isostatic structure $m = n$ and $\mathbf{C}$ is a nonsingular square matrix of order n. The analysis of such structures is trivial: first, stresses are determined from the solution of equilibrium equations

$$\mathbf{s}_* = \mathbf{C}^{-\mathrm{T}}\mathbf{p}_*; \tag{12.7}$$

then strains are found from the constitutive law

$$\mathbf{q}_* = \mathbf{E}^{-1}\mathbf{s}_*; \tag{12.8}$$

lastly, displacements are calculated from the inverse compatibility equation

$$\mathbf{w}_* = \mathbf{C}^{-1}\mathbf{q}_*. \tag{12.9}$$

Consider the cantilever truss in Fig. 12.1a. It is statically determinate because $m = 5$ and $n = 4 \times 2 - 3 = 5$. Let $P_0 = P$ and $L_0 = L$. We assume further that all elements have a common cross-sectional area A_0 and are made of identical material with Young's modulus E_0.

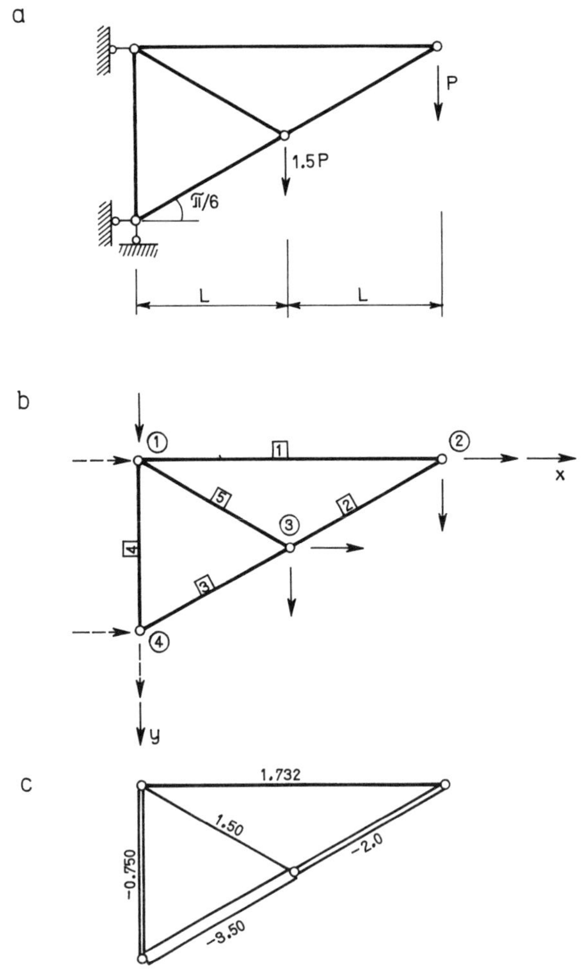

Figure 12.1. Isostatic truss: (a) Layout, Dimensions, and Loading. (b) Computational Scheme. (c) Axial Forces.

According to the computational scheme in Fig. 12.1b, we have the following matrices of state variables:

$$\mathbf{w} = \{w_y^{(1)}, w_x^{(2)}, w_y^{(2)}, w_x^{(3)}, w_y^{(3)}\},$$

$$\mathbf{p} = \{p_y^{(1)}, p_x^{(2)}, p_y^{(2)}, p_x^{(3)}, p_y^{(3)}\},$$

$$\mathbf{q} = \{\varepsilon_1, \varepsilon_2, \varepsilon_3, \varepsilon_4, \varepsilon_5\},$$

$$\mathbf{s} = \{n_1, n_2, n_3, n_4, n_5\}.$$

The equilibrium matrix $\mathbf{C}^{\mathrm{T}}$ can be generated by the computer procedure of Section 11.3 or by writing the equilibrium equations for the nodes of the truss. Both methods yield the same result:

$$\mathbf{C}^{\mathrm{T}} = \begin{bmatrix} 0 & 0 & 0 & -1.0 & -0.50 \\ 1.0 & 0.866 & 0 & 0 & 0 \\ 0 & -0.50 & 0 & 0 & 0 \\ 0 & -0.866 & 0.866 & 0 & 0.866 \\ 0 & 0.50 & -0.50 & 0 & 0.50 \end{bmatrix}.$$

Its inverse is

$$\mathbf{C}^{-\mathrm{T}} = \begin{bmatrix} 0 & 1.0 & 1.732 & 0 & 0 \\ 0 & 0 & -2.0 & 0 & 0 \\ 0 & 0 & -2.0 & 0.577 & -1.0 \\ -1.0 & 0 & 0 & -0.289 & -0.50 \\ 0 & 0 & 0 & 0.577 & 1.0 \end{bmatrix}.$$

Multiplying by the load matrix

$$\mathbf{p}_* = \{0, 0, 1.0, 0, 1.50\},$$

we obtain the nondimensional axial forces occurring in the elements of the truss:

$$\mathbf{s}_* = \{1.732, -2.000, -3.500, -0.750, 1.500\}.$$

Those forces are shown in Fig. 12.1c, where the width of the element has been taken proportional to the value of s_{j*}. We can obtain true axial forces, if needed, by multiplying the entries of $\mathbf{s}_*$ by the reference force P_0.

Though the stress state of the truss is independent of its material and cross-sectional areas, we must take them into account when calculating elastic displacements. Thus, according to formula (11.27), the reciprocal of the axial stiffness of the jth element is $e_{jN}^{-1} = P_0 L_j / E_0 A_0 L_0$. Remembering that the length of the first element is $2L_0$, whereas remaining elements have length $1.1547L_0$, we calculate the nondimensional axial flexibilities:

$$e_{1N}^{-1} = \frac{2.0P_0}{e_0 A_0}, \quad e_{jN}^{-1} = \frac{1.1547P_0}{E_0 A_0} \qquad \text{for } j = 2, 3, 4, 5.$$

Hence,

$$\mathbf{E}^{-1} = \frac{P_0}{E_0 A_0} \lceil 2.000, 1.1547, 1.1547, 1.1547, 1.1547 \rfloor.$$

Multiplying this matrix by $\mathbf{s}_*$, we obtain the nondimensional strains

$$\mathbf{q}_* = \frac{P_0}{E_0 A_0} \{3.464, -2.309, -4.041, -0.866, 1.732\}.$$

According to Eq. (11.8), the product $\mathbf{C}^{-1}\mathbf{q}_*$ gives the required displacements:

$$\mathbf{w}_* = \frac{P_0}{E_0 A_0} \{0.866, 3.464, 18.701, -1.083, 6.206\}.$$

To transform these displacements into dimensional quantities, we must multiply $\mathbf{w}_*$ by the reference length L_0.

Exercise 12.1

Find the axial forces induced in the truss shown in Fig. 12.5a (p. 91) by the load $P = 10.0$ kN.

12.2.3. Direct Stiffness Method

For the dual QP problems (12.5), all variables are free, and there are no inequalities among the constraints. We can therefore derive explicit conditions for the relevant extrema in the form of matrix equations. Thus, we can solve the single constraint of (12.6) with respect to $\mathbf{q}$ and substitute the result into the expression for potential energy. This converts (12.5) into the unconstrained minimization problem

$$\min\{\tfrac{1}{2}\mathbf{w}^{\mathrm{T}}\mathbf{K}\mathbf{w} - \mathbf{p}_*^{\mathrm{T}}\mathbf{w}\}, \tag{12.10}$$

where

$$\mathbf{K} = \mathbf{C}^{\mathrm{T}}\mathbf{E}\mathbf{C} \tag{12.11}$$

is the stiffness matrix of the structure. Obviously, the minimum (12.10) is attained at that $\mathbf{w}_*$ that corresponds to the vanishing gradient of the minimized function. Hence, $\mathbf{w}_*$ has to satisfy the equation

$$\mathbf{K}\mathbf{w} = \mathbf{p}_*, \tag{12.12}$$

which is the canonical equation of the direct stiffness method.

After $\mathbf{w}_*$ has been obtained from the solution of Eq. (12.12), we can calculate the generalized strains from Eq. (11.1):

$$\mathbf{q}_* = \mathbf{C}\mathbf{w}_*. \tag{12.13}$$

The generalized stresses follow from the physical law (11.32)

$$\mathbf{s}_* = \mathbf{E}\mathbf{q}_* = \mathbf{E}\mathbf{C}\mathbf{w}_*. \tag{12.14}$$

As an example, we apply the direct stiffness method to the three-bar truss in Fig. 11.1a. Let A_1, A_2, A_3 be the cross-sectional areas of the struts forming that truss. The material of the struts has a common Young's modulus E. Then the global elasticity matrix of the truss is

$$\mathbf{E} = \frac{E}{L}\left[\frac{A_1}{\sqrt{2}}, A_2, \frac{A_3}{\sqrt{2}}\right].$$

The compatibility matrix $\mathbf{C}$ has been already derived in Section 11.2.1. Hence, we can calculate the stiffness matrix according to the definition (12.11):

$$\mathbf{K} = \begin{bmatrix} \sqrt{2}/2 & 0 & -\sqrt{2}/2 \\ \sqrt{2}/2 & 1 & \sqrt{2}/2 \end{bmatrix} \frac{E}{L} \begin{bmatrix} A_1/\sqrt{2} & 0 & 0 \\ 0 & A_2 & 0 \\ 0 & 0 & A_3/\sqrt{2} \end{bmatrix} \begin{bmatrix} \sqrt{2}/2 & \sqrt{2}/2 \\ 0 & 1 \\ -\sqrt{2}/2 & \sqrt{2}/2 \end{bmatrix}$$

$$= \frac{E\sqrt{2}}{4L}\begin{bmatrix} A_1 + A_3 & A_1 - A_3 \\ A_1 - A_3 & A_1 + 2\sqrt{2}A_2 + A_3 \end{bmatrix}.$$

For the symmetric truss (i.e., when $A_1 = A_3$), this result simplifies to

$$\mathbf{K} = \frac{E\sqrt{2}}{2L}\begin{bmatrix} A_1 & 0 \\ 0 & A_1 + A_2\sqrt{2} \end{bmatrix}. \tag{12.15}$$

The inverse matrix is

$$\mathbf{K}^{-1} = \frac{L\sqrt{2}}{E}\begin{bmatrix} 1/A_1 & 0 \\ 0 & 1/(A_1 + A_2\sqrt{2}) \end{bmatrix}.$$

According to the solution $\mathbf{w}_* = \mathbf{K}^{-1}\mathbf{p}_*$ of Eq. (12.9), the entries of $\mathbf{K}^{-1}$ reflect the influence of load upon the displacements. Thus, the first column of that matrix collects the influence coefficients related to the horizontal force $P_x^{(1)}$, whereas the second column corresponds to the vertical component $P_y^{(1)}$.

Due to the simplicity of the problem, it is possible to find analytical expressions for the elements of the matrix $\mathbf{K}^{-1}$. Hence, the displacements $W_x^{(1)}$ and $W_y^{(1)}$ can be expressed analytically as functions of the geometry, the material properties, and the load:

$$W_x^{(1)} = \frac{1}{A_1}\frac{L\sqrt{2}}{E}P_x^{(1)}, \qquad W_y^{(1)} = \frac{1}{A_1 + A_2\sqrt{2}}\frac{L\sqrt{2}}{E}P_y^{(1)}.$$

The same applies for the generalized strains and stresses that follow from Eqs. (11.1) and (11.25), respectively.

The complete solution for the displacements and generalized stresses of the symmetric truss is given in Table 12.1. Note that even for such a simple structure the entries of **w** and **s** are nonlinear functions of the cross-sectional areas. In most practical cases these functions cannot be obtained explicitly.

Table 12.1. Influence Coefficients for Three-Bar Truss ($A_1 = A_3 \neq A_2$)

	$P_x^{(1)}$	$P_y^{(1)}$
$W_x^{(1)}$	$\frac{1}{A_1}\frac{L\sqrt{2}}{E}$	0
$W_y^{(1)}$	0	$\frac{1}{A_1 + A_2\sqrt{2}}\frac{L\sqrt{2}}{E}$
N_1	$\frac{1}{\sqrt{2}}$	$\frac{A_1}{(A_1 + A_2\sqrt{2})\sqrt{2}}$
N_2	0	$\frac{A_2\sqrt{2}}{A_1 + A_2\sqrt{2}}$
N_3	$-\frac{1}{\sqrt{2}}$	$\frac{A_3}{(A_1 + A_2\sqrt{2})\sqrt{2}}$

Table 12.2. Influence Coefficients for Three-Bar Truss ($A_1 = A_2 = A_3 = A$)

	$P_x^{(1)}$	$P_y^{(1)}$
$W_x^{(1)}$	$\frac{L\sqrt{2}}{EA}$	0
$W_y^{(1)}$	0	$\frac{1}{1+\sqrt{2}}\frac{L\sqrt{2}}{EA}$
N_1	$\frac{1}{\sqrt{2}}$	$\frac{1}{(1+\sqrt{2})\sqrt{2}}$
N_2	0	$\frac{\sqrt{2}}{1+\sqrt{2}}$
N_3	$-\frac{1}{\sqrt{2}}$	$\frac{1}{(1+\sqrt{2})\sqrt{2}}$

It is clear from Table 12.1 that $P_x^{(1)}$ causes no axial force in element 2. Hence, the symmetric truss behaves under a horizontal load as if it were isostatic: the internal forces in elements 1 and 3 do not depend on the cross-sectional areas.

That property holds for a truss of constant cross section $A = A_1 = A_2 = A_3$ as well. The stiffness matrix of such a truss is

$$\mathbf{K} = \frac{EA\sqrt{2}}{2L}\begin{bmatrix} 1 & 0 \\ 0 & 1+\sqrt{2} \end{bmatrix},$$

its inverse is

$$\mathbf{K}^{-1} = \frac{L\sqrt{2}}{EA}\begin{bmatrix} 1 & 0 \\ 0 & 1/1+\sqrt{2} \end{bmatrix},$$

and the influence coefficients for the displacements and internal forces are given in Table 12.2.

Exercise 12.2

Find the stiffness matrix of the truss in Fig. 11.10 by (a) aggregation of submatrices and by (b) direct calculation of the triple product (12.11).

Use the matrix $\mathbf{C}^{\mathrm{T}}$ given in Table S.1 of Chapter 20 and the matrix $\mathbf{E}$ given in the solution of Exercise 11.5 in Chapter 20.

Exercise 12.3

Find elastic displacements and bending moments for the beam in Fig. 11.11 by the direct stiffness method. Use the matrix $\mathbf{C}^{\mathrm{T}}$ given in Table S.2 in Chapter 20, and take $E = 200$ MPa, $I = 50.0\ \mathrm{cm}^4$ when calculating the elasticity matrix.

12.2.4. Flexibility Method

Solving the constraint of problem (12.6) with respect to s and substituting the result given by Eq. (11.7) into the expression for complementary energy, we can replace (12.6) by the equivalent unconstrained problem

$$\min\{\tfrac{1}{2}\mathbf{z}^{\mathrm{T}}\mathbf{F}\mathbf{z} + \mathbf{d}^{\mathrm{T}}\mathbf{z} + b\}. \tag{12.16}$$

Here

$$\mathbf{F} = \mathbf{C}^{0}\mathbf{E}^{-1}\mathbf{C}^{0\mathrm{T}} \tag{12.17}$$

is the flexibility matrix of the structure, and the matrices

$$\mathbf{d} = \mathbf{C}^{0}\mathbf{E}^{-1}\mathbf{C}^{-\mathrm{T}}\mathbf{p}_*, \tag{12.18}$$

$$b = \mathbf{p}_*^{\mathrm{T}}\mathbf{C}^{-}\mathbf{E}^{-1}\mathbf{C}^{-\mathrm{T}}\mathbf{p}_* \tag{12.19}$$

depend on the given load. The condition that the gradient of the function minimized in (12.16) must vanish at the extremum $\mathbf{z}_*$ gives the canonical equation of the flexibility method:

$$\mathbf{F}\mathbf{z} + \mathbf{d} = \mathbf{0}. \tag{12.20}$$

After the redundant stresses $\mathbf{z}_*$ are obtained from the solution of Eq. (12.20), we can compute the total generalized stress

$$\mathbf{s}_* = \mathbf{C}^{-\mathrm{T}}\mathbf{p}_* + \mathbf{C}^{0\mathrm{T}}\mathbf{z}_* \tag{12.21}$$

[compare Eq. (11.7)]. As soon as $\mathbf{s}_*$ is known, we can obtain the kinematic quantities $\mathbf{q}_*$ and $\mathbf{w}_*$ just as we did for isostatic structures, except that we must replace the ordinary inverse $\mathbf{C}^{-1}$ in Eq. (12.9) by the generalized inverse $\mathbf{C}^{-}$.

As an example of an application of the flexibility method, we take the portal frame in Fig. 12.2a. According to the computational scheme of Fig. 12.2b, we have $m = 3 \times 2 = 6$ and $n = 3$ (we neglect axial elongations of the bars). Thus, nodes 2 and 4 have a common horizontal displacement

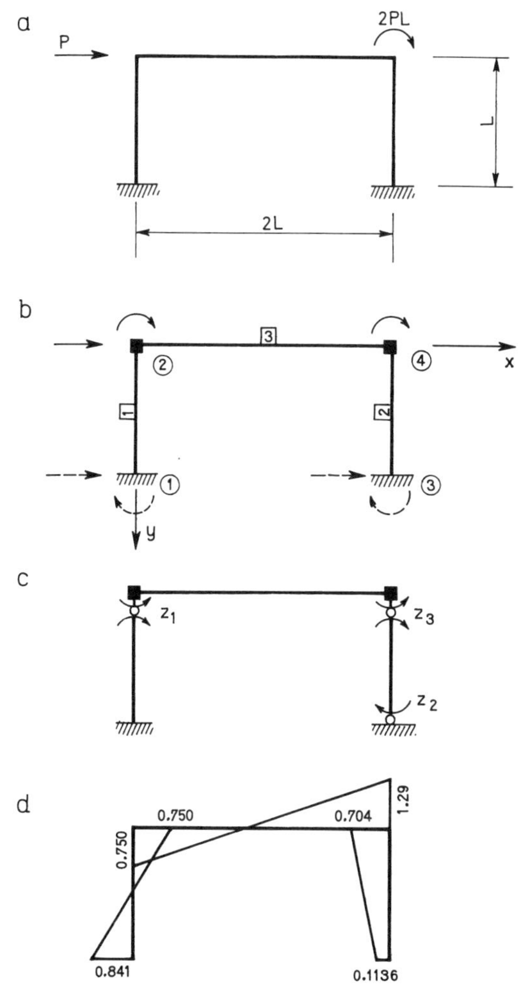

Figure 12.2. Portal Frame: (a) Layout, Dimensions, and Loading. (b) Computational Scheme. (c) Primary System. (d) Bending Moments.

and no vertical displacements. The degree of statical redundancy is $k = 6 - 3 = 3$.

We take P and L as the reference quantities P_0 and L_0, respectively. The nondimensional generalized parameters defined by Eqs. (11.10) enter the matrices of state variables:

$$\mathbf{w} = \{w_x^{(2)}, w_z^{(2)}, w_z^{(4)}\},$$

$$\mathbf{p} = \{p_x^{(2)}, p_z^{(2)}, p_z^{(4)}\},$$

$$\mathbf{q} = \{\theta_a^{(1)}, \theta_b^{(1)}, \theta_a^{(2)}, \theta_b^{(2)}, \theta_a^{(3)}, \theta_b^{(3)}\},$$

$$\mathbf{s} = \{m_a^{(1)}, m_b^{(1)}, m_a^{(2)}, m_b^{(2)}, m_a^{(3)}, m_b^{(3)}\}.$$

The rotations of the ends of beams measured with respect to their chords and caused by the ith unit displacement enter the ith column of the compatibility matrix:

$$\mathbf{C} = \begin{bmatrix} -1 & 0 & 0 \\ -1 & 1 & 0 \\ -1 & 0 & 0 \\ -1 & 0 & 1 \\ 0 & 1 & 0 \\ 0 & 0 & 1 \end{bmatrix}.$$

Its transpose

$$\mathbf{C}^{\mathrm{T}} = \begin{bmatrix} -1 & -1 & -1 & -1 & 0 & 0 \\ 0 & 1 & 0 & 0 & 1 & 0 \\ 0 & 0 & 0 & 1 & 0 & 1 \end{bmatrix}$$

collects the coefficients of the equilibrium equations. The first row reflects the requirement that the sum of the shear forces in elements 1 and 2 must be equal to the horizontal load applied to node 2. The remaining two rows express the balance of moments acting on nodes 2 and 4.

All three elements are made of the same material with Young's modulus E_0. The columns have a common moment of inertia I_0, but the horizontal beam differs in that respect: $I_3 = 1.5I_0$. According to Eq. (11.29), we have the following nondimensional flexural stiffnesses of elements: $e_M^{(1)} = e_M^{(2)} = E_0 I_0 / P_0 L_0^2$, $e_M^{(3)} = 0.75 E_0 I_0 / P_0 L_0^2$. The elasticity matrix of the frame includes three submatrices of the type given in Section 11.5.3:

$$\mathbf{E} = \frac{E_0 I_0}{P_0 L_0^2} \begin{bmatrix} 4.0 & 2.0 & 0 & 0 & 0 & 0 \\ 2.0 & 4.0 & 0 & 0 & 0 & 0 \\ 0 & 0 & 4.0 & 2.0 & 0 & 0 \\ 0 & 0 & 2.0 & 4.0 & 0 & 0 \\ 0 & 0 & 0 & 0 & 3.0 & 1.5 \\ 0 & 0 & 0 & 0 & 1.5 & 3.0 \end{bmatrix}.$$

Its inverse is

$$\mathbf{E}^{-1} = \frac{P_0 L_0^2}{E_0 I_0} \begin{bmatrix} 0.333 & -0.167 & 0 & 0 & 0 & 0 \\ -0.167 & 0.333 & 0 & 0 & 0 & 0 \\ 0 & 0 & 0.333 & -0.167 & 0 & 0 \\ 0 & 0 & -0.167 & 0.333 & 0 & 0 \\ 0 & 0 & 0 & 0 & 0.444 & -0.222 \\ 0 & 0 & 0 & 0 & -0.222 & 0.444 \end{bmatrix}.$$

Applying the automatic procedure described in Borkowski (1987) to the equilibrium matrix $\mathbf{C}^{\mathrm{T}}$, we obtain its generalized inverse and its kernel:

$$\mathbf{C}^{-\mathrm{T}} = \begin{bmatrix} -1 & 0 & 0 \\ 0 & 0 & 0 \\ 0 & 0 & 0 \\ 0 & 0 & 0 \\ 0 & 1 & 0 \\ 0 & 0 & 1 \end{bmatrix}, \qquad \mathbf{C}^{0\mathrm{T}} = \begin{bmatrix} -1 & -1 & -1 \\ 1 & 0 & 0 \\ 0 & 1 & 0 \\ 0 & 0 & 1 \\ -1 & 0 & 0 \\ 0 & 0 & -1 \end{bmatrix}.$$

These matrices correspond to the isostatic primary system shown in Fig. 12.2c. For example, the entries of the first column of $\mathbf{C}^{-\mathrm{T}}$ are the bending moments induced in that system by the load $p_x^{(2)} = 1$. The entries of the first column of $\mathbf{C}^{0\mathrm{T}}$ are the moments caused by $m_b^{(1)} = 1$. The bending moments $m_b^{(1)}$, $m_a^{(2)}$, and $m_b^{(2)}$ were chosen as the redundant parameters: $\mathbf{z} = \{m_b^{(1)}, m_a^{(2)}, m_b^{(2)}\}$.

Calculating the triple product defined by Eq. (12.17), we obtain the flexibility matrix

$$\mathbf{F} = \frac{P_0 L_0^2}{E_0 I_0} \begin{bmatrix} 1.444 & 0.500 & 0.278 \\ 0.500 & 0.667 & 0.167 \\ 0.278 & 0.167 & 1.111 \end{bmatrix}.$$

The free terms of the canonical equations

$$\mathbf{d} = \frac{P_0 L_0^2}{e_0 I_0}\{0.944, 0.333, -0.556\}$$

follow from Eq. (12.18). Solving Eq. (12.20), we obtain the redundant moments

$$\mathbf{z}_* = \{-0.750, -0.1136, 0.704\}.$$

Given $\mathbf{p}_*$ and $\mathbf{z}_*$, the final distribution of bending moments

$$\mathbf{s}_* = \{-0.841, -0.750, -0.1136, 0.704, 0.750, 1.295\}$$

follows from the general solution (12.20) of the equilibrium equation. These nondimensional bending moments are shown in Fig. 12.2d.

Multiplying $\mathbf{E}^{-1}$ by $\mathbf{s}_*$, we obtain the net rotations of the ends of elements:

$$\mathbf{q}_* = \frac{P_0 L_0^2}{E_0 I_0}\{-4.864, -4.682, 0.954, 2.591, 4.193, 5.011\}.$$

Finally, the product of $\mathbf{C}$ and $\mathbf{q}_*$ gives us the nodal displacements:

$$\mathbf{w}_* = \frac{P_0 L_0^2}{e_0 I_0}\{4.864, 4.193, 5.011\}.$$

The first entry of $\mathbf{w}_*$ is the nondimensional horizontal displacement of node 2. To obtain the true shift, we multiply that entry by L_0. The other two entries are the rotations of nodes 2 and 4.

Exercise 12.4

Solve the truss in Fig. 11.10 by the flexibility method. Use the matrices **C** and **E** obtained from solving Exercises 11.1 and 11.5, respectively; see Chapter 20.

Exercise 12.5

Solve the beam in Fig. 11.11 by the flexibility method. Use the matrices $\mathbf{C}^{\mathrm{T}}$ and **E** from Exercise 12.3.

12.2.5. Concluding remarks

The complete set of $2m + n$ equations (12.1) is reducible either to the n equations (12.12) of the direct stiffness method or to the k equations

(12.20) of the flexibility method. The direct stiffness method is usually chosen for computer codes because it is easier to implement than the flexibility method. The main advantage of the former is that much of the work can be done at the element level. Thus, the local stiffnesses $\bar{\mathbf{K}}_i$ of typical elements can be calculated in advance. The global matrix $\mathbf{K}$ is then assembled from such submatrices transformed into the global reference frame and placed according to the connectivity matrix $\mathbf{B}$ (see the procedure in Section 11.3). After the set of equations (12.12) has been solved, the resulting displacement matrix $\mathbf{w}_*$ is decomposed into the submatrices corresponding to individual elements. We then calculate the strains $\bar{\mathbf{q}}_*$ and the stresses $\bar{\mathbf{s}}_*$ from the local relations (11.12) and (11.25), respectively. These subvectors give us the desired solution $\mathbf{q}_*$, $\mathbf{s}_*$.

Thus, in computer implementations of the direct stiffness method, the large global matrices $\mathbf{C}$ and $\mathbf{E}$ need never be generated and stored. The main demand for computer memory comes from the symmetric matrix $\mathbf{K}$: we must store its $n(n+1)/2$ entries. This number can be further reduced, since the bandwidth of $\mathbf{K}$ is usually considerably smaller than n.

On the contrary, the global stiffness matrix $\mathbf{F}$ cannot be assembled element by element as with $\mathbf{K}$. The only way to obtain $\mathbf{F}$ automatically is to perform the matrix multiplications prescribed by Eq. (12.17). That means that matrices $\mathbf{C}^0$ and $\mathbf{E}^{-1}$ must be available. We can assemble the latter rather easily from the submatrices related to elements, but we obtain the former by using Jordan eliminations on $\mathbf{C}$. Although often $k \ll n$, the necessity to store large matrices $\mathbf{C}$, $\mathbf{C}^-$, $\mathbf{C}^0$, and $\mathbf{E}^{-1}$, as well as the time-consuming matrix multiplications needed to obtain $\mathbf{F}$ and $\mathbf{d}$, prevent greater use of the flexibility method.

12.3. Optimization of Trusses—Alternative Formulations

We consider only the deterministic optimal design of elastic trusses subjected to static load. As stated in Volume 1, we understand the structural optimization problem to be the minimization of structural cost under behavioral, geometrical, and/or technological constraints. The cost itself is understood in a generalized sense: for example, it can be the weight or the volume of the structure. It is defined as a given function $\Gamma = \Gamma(\psi_i)$ of the design variables ψ_i, $i = 1, 2, \ldots, t$. Additionally, we have the state variables $\mathbf{s}_*$, $\mathbf{q}_*$, $\mathbf{w}_*$ that describe the response of the optimized structure to the given load $\mathbf{p}_*$.

According to the classification introduced in Chapter 2 of Volume 1, two types of problems are to be distinguished: type A, the optimization of the cross-sectional areas for a fixed layout, and type B, where the layout is subject to optimization as well. We will specify the formulations of both types of optimization problems for trusses.

In the type A optimization problem, the positions of all nodes and the pattern of connections between them are given. We are looking for the optimal values A_{j*} of the cross-sectional areas of the individual elements that minimize the volume V of the truss. Let the lengths L_j and the cross-sectional areas A_j of those elements be arranged into the m-dimensional matrices $\mathbf{l}$ and $\mathbf{a}$, respectively. Then the volume of the truss is equal to the scalar product of those matrices:

$$V = \mathbf{l}^{\mathrm{T}}\mathbf{a}. \tag{12.22}$$

For practical reasons we usually introduce a degree of regularity into the optimal structure. Hence the number t of independent design variables ψ_i is usually less than the number of structural members m. Let the design variables be represented by the t-dimensional matrix $\boldsymbol{\psi}$. Its entries and the cross-sectional areas of the truss are related by the equation

$$\mathbf{a} = \mathbf{G}^{\mathrm{T}}\boldsymbol{\psi}, \tag{12.23}$$

where $\mathbf{G}$ is a given $t \times m$ matrix, called the *linkage matrix*. For example, we might optimize the truss in Fig. 12.5a under the assumption that the horizontal, vertical, and diagonal elements have common cross-sectional areas A_h, A_v, and A_d, respectively. Then $\boldsymbol{\psi} = \{A_h, A_v, A_d\}$, and the linkage matrix is

$$\mathbf{G} = \begin{bmatrix} 1 & 1 & 1 & 1 & 1 & 1 & 1 & 0 & 0 & 0 & 0 & 0 & 0 & 0 & 0 & 0 & 0 & 0 \\ 0 & 0 & 0 & 0 & 0 & 0 & 0 & 1 & 1 & 1 & 1 & 0 & 0 & 0 & 0 & 0 & 0 & 0 \\ 0 & 0 & 0 & 0 & 0 & 0 & 0 & 0 & 0 & 0 & 0 & 1 & 1 & 1 & 1 & 1 & 1 & 1 \end{bmatrix}. \tag{12.24}$$

Assuming $\Gamma = V$ and using (12.22) and (12.23), we obtain the linear cost function

$$\Gamma = \mathbf{c}^{\mathrm{T}}\boldsymbol{\psi}, \tag{12.25}$$

where

$$\mathbf{c} = \mathbf{G}\mathbf{l} \tag{12.26}$$

is the given matrix of cost factors.

To restrict the values of the design variables to a desired range, we introduce technological constraints

$$\boldsymbol{\psi}^- \leqslant \boldsymbol{\psi} \leqslant \boldsymbol{\psi}^+, \tag{12.27}$$

where the t-dimensional matrices $\boldsymbol{\psi}^-$, $\boldsymbol{\psi}^+$ are composed of given lower and upper bounds imposed on each design variable. Such constraints eliminate optimal designs with vanishing or excessively large cross-sectional areas of elements. On the other hand, a range that is too narrow can diminish the effect of optimization or even produce no solution.

The volume of the truss is to be minimized subject to behavioral constraints. The latter ensure that the values of stresses, strains, and displacements induced in the optimal truss by the given load do not exceed certain prescribed limits. The most common constraint of that type comes from the concept of admissible stress: the true stress of the jth member,

$$\sigma_j = N_j/A_j, \tag{12.28}$$

should not exceed σ_j^+ in tension and σ_j^- in compression. Introducing the diagonal matrices

$$\begin{aligned} \mathbf{S}^+ &= \lceil \sigma_1^+, \sigma_2^+, \ldots, \sigma_m^+ \rfloor, \\ \mathbf{S}^- &= \lceil \sigma_1^-, \sigma_2^-, \ldots, \sigma_m^- \rfloor, \end{aligned} \tag{12.29}$$

we can write the stress admissibility constraints for the whole truss in the form of matrix inequalities:

$$\mathbf{S}^-\mathbf{a} \leqslant \mathbf{s} \leqslant \mathbf{S}^+\mathbf{a}. \tag{12.30}$$

The linkage relation (12.25) allows us to replace (12.30) by

$$\mathbf{D}^-\boldsymbol{\psi} \leqslant \mathbf{s} \leqslant \mathbf{D}^+\boldsymbol{\psi}, \tag{12.31}$$

where

$$\mathbf{D}^- = \mathbf{S}^-\mathbf{G}^{\mathrm{T}}, \qquad \mathbf{D}^+ = \mathbf{S}^+\mathbf{G}^{\mathrm{T}} \tag{12.32}$$

are known matrices.

The second group of behavioral constraints ensures sufficient rigidity of the structure. Namely, the displacement w_i should be no less than w_i^- if it is negative and no greater than w_i^+ if it is positive. Collecting the given bounds on displacements into the n-dimensional matrices $\mathbf{w}^-$ and $\mathbf{w}^+$, we can write the rigidity constraints as

$$\mathbf{w}^- \leqslant \mathbf{w} \leqslant \mathbf{w}^+. \tag{12.33}$$

Thus the optimization of the cross-sectional areas amounts to the following problem:

$$\min\{\mathbf{c}^{\mathrm{T}}\boldsymbol{\psi} \mid \mathbf{D}^{-}\boldsymbol{\psi} \leqslant \mathbf{s} \leqslant \mathbf{D}^{+}\boldsymbol{\psi}, \mathbf{w}^{-} \leqslant \mathbf{w} \leqslant \mathbf{w}^{+}, \boldsymbol{\psi}^{-} \leqslant \boldsymbol{\psi} \leqslant \boldsymbol{\psi}^{+}\}. \quad (12.34)$$

Here $\mathbf{s}$ and $\mathbf{w}$ are obtained from the solution of the elastic analysis problem. For the given layout the compatibility matrix $\mathbf{C}$ is constant. The elasticity matrix $\mathbf{E}$ depends on the cross-sectional areas and, hence, on the design variables. Thus problem (12.34) is nonlinear with respect to the unknowns ψ_i.

So far we have considered the optimization of the cross-sectional areas for a given layout of the truss. Now let us see what happens when the layout is to be optimized. We keep the same cost function (volume) and the same behavioral constraints. The difference arises in the selection of the design variables.

In the simplest version of the type B problem, the layout of the truss is given except for the positions of certain nodes. These nodes are called *free*, and their coordinates become elements of the matrix $\boldsymbol{\psi}$. For example, taking the vertical coordinates of the nodes 7, 8, and 9 of the truss shown in Fig. 12.3a as the design variables, we can look for the optimal shape of

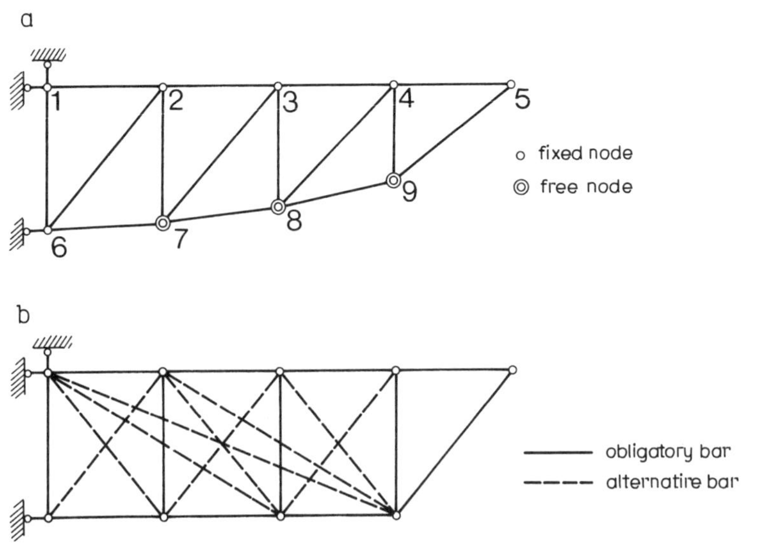

Figure 12.3. Alternative Problems of Layout Optimization: (a) Optimal Positions of Free Nodes Are Found under Given Connections. (b) Optimal Connections Are Chosen under Fixed Positions of Nodes.

that truss. Note that the lengths of diagonal elements depend nonlinearly upon the positions of the free nodes. Hence the cost function of such a problem is nonlinear.

The behavioral constraints remain analogous to those introduced in the type A problem. However, the stress admissibility constraints can be simplified. Since both the cross-sectional areas A_j and the stress limits σ_j^-, σ_j^+ are given, we can compute in advance the bounds on axial forces,

$$s_j^- = \sigma_j^- A_j, \qquad s_j^+ = \sigma_j^+ A_j, \tag{12.35}$$

and collect them into the m-dimensional matrices $\mathbf{s}^-$ and $\mathbf{s}^+$. This allows us to replace (12.31) by

$$\mathbf{s}^- \leqslant \mathbf{s} \leqslant \mathbf{s}^+. \tag{12.36}$$

Hence, the optimization of the positions of free nodes leads to the following problem:

$$\min\{\Gamma(\boldsymbol{\psi})\,|\,\mathbf{s}^- \leqslant \mathbf{s} \leqslant \mathbf{s}^+, \mathbf{w}^- \leqslant \mathbf{w} \leqslant \mathbf{w}^+, \boldsymbol{\psi}^- \leqslant \boldsymbol{\psi} \leqslant \boldsymbol{\psi}^+\}. \tag{12.37}$$

The entries of $\mathbf{s}$ and $\mathbf{w}$ are implicit nonlinear functions of the design variables. Such a dependence comes through the compatibility matrix $\mathbf{C}$: its entries are influenced by the variable coordinates of the free nodes. The elasticity matrix $\mathbf{E}$ depends on the design variables as well, because of the variable lengths L_j that appear in the axial stiffnesses of individual elements.

Instead of looking for the optimal positions of nodes, we can optimize the pattern of connections. In this version all nodes are fixed *a priori*, whereas the best layout is to be selected among the finite set of possible connections between them. Figure 12.3b shows such a set for the cantilever truss (obviously nonadvantageous connections have been eliminated for the sake of simplicity). Formally, this type of variable layout can be associated with the variable connectivity matrix $\mathbf{B}$ or with some kind of Boolean representation of the fact that a certain element is present in the truss. However, the simplest solution is to begin with the structure containing all possible elements and to let them eventually disappear in the process of solution.

Simultaneous optimization of the cross-sectional areas and the layout is also possible. Several examples of such solutions are given in Section 12.5.3.

Most structures are subject, in practice, to multiple loads. In our approach, then, instead of the single-column $\mathbf{p}_*$ we should use the $n \times r$ matrix $\mathbf{P}_*$, where r is the number of loading cases. The admissibility conditions for stresses and displacements must be checked separately for each loading, which increases the number of constraints and the number of computations.

12.4. Direct NLP Approach

We showed in the previous section that for elastic trusses A- and B-type formulations of the optimization problem lead to NLP problems. Moreover, in many cases the behavioral constraints cannot be expressed explicitly in terms of the design variables. Therefore a direct solution by NLP algorithms is possible for only a few simple structures.

As an example, consider the optimization of the three-bar truss in Fig. 11.1a. That truss is to be optimized under the following initial data:

$$L = 2.0\ \text{m}, \qquad E = 2\times 10^5\ \text{MPa}, \qquad \sigma^+ = 200\ \text{MPa}, \qquad \sigma^- = -150\ \text{MPa},$$

$$\text{Loading case 1:} \quad P_x^{(1)} = 200\ \text{kN}, \qquad P_y^{(1)} = 100\ \text{kN},$$

$$\text{Loading case 2:} \quad P_x^{(1)} = -200\ \text{kN}, \qquad P_y^{(1)} = 100\ \text{kN}.$$

Here σ^+ and σ^- are the admissible stresses in tension and in compression, and are assumed to be identical for each bar. The three-bar truss subjected to slightly different loading was first optimized by Schmit (1960). Since then it has been cited as a standard example in almost every book on structural optimization.

We assume that the truss is symmetric (i.e., $A_1 = A_3$). Then $t = 2$ and $\boldsymbol{\psi} = \{A_1, A_2\}$. The cost (i.e., volume) of the truss is

$$\Gamma = V = 4\sqrt{2}A_1 + 2A_2, \tag{12.38}$$

which follows from Eq. (12.22) with $l = \{2\sqrt{2}, 2, 2\sqrt{2}\}$,

$$\mathbf{G} = \begin{bmatrix} 1 & 0 & 1 \\ 0 & 1 & 0 \end{bmatrix}, \qquad \mathbf{c} = \{4\sqrt{2}, 2\}.$$

We take the stress admissibility conditions as the behavioral constraints. According to definition (12.31), these conditions are

$$\sigma^- \begin{bmatrix} 1 & 0 \\ 0 & 1 \\ 1 & 0 \end{bmatrix} \begin{bmatrix} A_1 \\ A_2 \end{bmatrix} \leqslant \begin{bmatrix} s_1 \\ s_2 \\ s_3 \end{bmatrix} \leqslant \sigma^+ \begin{bmatrix} 1 & 0 \\ 0 & 1 \\ 1 & 0 \end{bmatrix} \begin{bmatrix} A_1 \\ A_2 \end{bmatrix}, \tag{12.39}$$

since in this simple case $\mathbf{D}^- = \sigma^- \mathbf{G}^{\mathrm{T}}$ and $\mathbf{D}^+ = \sigma^+ \mathbf{G}^{\mathrm{T}}$.

In general, all six inequalities should be checked separately for each loading case. However, the symmetry of both the loading and structure allows us to consider only one loading case. Moreover, we can anticipate the sign of the stress in each member. It is clear that, under the given loading, elements 1 and 2 are in tension and element 3 is compressed. Excluding passive constraints from the set (12.39), we end up with three inequalities:

$$\frac{s_1}{A_1} \leqslant \sigma^+, \quad \frac{s_2}{A_2} \leqslant \sigma^+, \quad -\frac{s_3}{A_3} \leqslant -\sigma^-. \tag{12.40}$$

The explicit formulas for the axial forces s_1, s_2, s_3 are given in Table 12.1. Substituting those expressions together with the given values of admissible stresses and loads into the inequalities (12.40), we change the dimension of the unknown cross-sectional areas from square meters into square centimeters. That yields the following final constraints:

$$\begin{aligned} \frac{\sqrt{2}}{A_1} + \frac{1}{\sqrt{2}(A_1 + A_2\sqrt{2})} &\leqslant 2.0, \\ \frac{\sqrt{2}}{A_1 + A_2\sqrt{2}} &\leqslant 2.0, \\ \frac{\sqrt{2}}{A_1} - \frac{1}{\sqrt{2}(A_1 + A_2\sqrt{2})} &\leqslant 1.5. \end{aligned} \tag{12.41}$$

Since no technological constraints were introduced, the optimization problem (12.34) simply means finding the minimum of (12.38) under the constraints (12.41). The solution is

$$\boldsymbol{\psi}_* = \{9.958, 1.582\}\ \text{cm}^2, \qquad V_* = 5949\ \text{cm}^3.$$

This result can be compared with the volume of the truss having constant cross section $A = A_1 = A_2 = A_3$. It is easy to see that, to satisfy the constraint $\sigma_1 = \sigma^+$, $A = 8.54\ \text{cm}^2$. Then

$$V_{\text{const}} = (2\sqrt{2} + 1)AL = 6539\ \text{cm}^3,$$

which is about 9% greater than the volume of the optimal truss. A gain of such an order is typical for the optimization of cross-sectional areas. We can reduce the volume even more by simultaneously optimizing the layout.

So far we have considered only stress admissibility. If, for some reason, the horizontal displacement of node 1 does not exceed, say, $L/1000 = 2.0$ mm, then the additional constraint

$$\frac{L\sqrt{2}}{E}\frac{1}{A_1}P_x^{(1)} \leqslant \frac{L}{1000} \tag{12.42}$$

must be added to inequalities (12.41) (see the expression of $W_x^{(1)}$ in Table 12.1). The optimal solution is then $\boldsymbol{\psi}_* = \{14.142, 0\}$ cm^2, corresponding to the volume $V_* = 7998$ cm^3. Thus the requirement of high rigidity can overrule the strength constraints and can lead to an entirely different optimal solution.

Since the horizontal load is twice that of the vertical load, that loading is carried mainly by the skew elements 1 and 3. Let us check, therefore, whether a statically determinate truss consisting solely of those two members is better than the computed redundant optimal truss.

The stresses in the members of the isostatic truss are

$$\begin{aligned} \sigma_1 &= \frac{\sqrt{2}}{2}(P_x^{(1)} + P_y^{(1)})\frac{1}{A_1}, \\ \sigma_3 &= \frac{\sqrt{2}}{2}(-P_x^{(1)} + P_y^{(1)})\frac{1}{A_3}. \end{aligned} \tag{12.43}$$

Optimizing such a truss is trivial. If only stress admissibility constraints are considered and if $A_1 = A_3$, then the solution is $A_* = 10.607$ cm^2, with $V_* = 6000$ cm^3. Hence, the isostatic truss is slightly heavier (more expensive) than the optimal redundant truss. The crucial point is that multiple loading cases were considered. For the single load $P_x^{(1)} = 200$ kN and $P_y^{(1)} = 100$ kN, the absolute optimal solution is isostatic and fully stressed: $A_{1*} = 10.607$ cm^2, $A_{3*} = 4.714$ cm^2, and $V_* = 4333$ cm^3.

12.5. Sequential LP Approach

To overcome the difficulties mentioned in the previous section, we often use an iterative procedure based upon the sequential linear approximation [compare, e.g., Zienkiewicz and Campbell (1973) and Pedersen (1981)]. Instead of directly solving the NLP problem (12.34) or (12.37), we solve a sequence of properly defined LP problems.

12.5.1. Mathematical Models

For now, we consider only the A-type optimization problem (12.34). Assume that an admissible design $\boldsymbol{\psi}_0$ is known together with the corresponding state $\mathbf{w}_0$, $\mathbf{s}_0$ of the structure. Our aim is to find an improved design $\boldsymbol{\psi}_*$ that is in state $\mathbf{w}_*$, $\mathbf{s}_*$ under load $\mathbf{p}_*$. Expanding the unknown displacements and stresses in Taylor series about $\mathbf{w}_0$, $\mathbf{s}_0$ and retaining the first two terms of the series, we obtain linear approximations

$$\begin{aligned} \mathbf{w} &\simeq \mathbf{w}_0 + \boldsymbol{\nabla} w_0(\boldsymbol{\psi} - \boldsymbol{\psi}_0), \\ \mathbf{s} &\simeq \mathbf{s}_0 + \boldsymbol{\nabla} s_0(\boldsymbol{\psi} - \boldsymbol{\psi}_0). \end{aligned} \tag{12.44}$$

Here $\boldsymbol{\nabla} w_0$ and $\boldsymbol{\nabla} s_0$ are the known values assumed at $\mathbf{w}_0$, $\mathbf{s}_0$ by the matrices of gradients

$$\boldsymbol{\nabla} w = \begin{bmatrix} \partial w_1/\partial\psi_1, \partial w_1/\partial\psi_2, \ldots, \partial w_1/\partial\psi_t \\ \partial w_2/\partial\psi_1, \partial w_2/\partial\psi_2, \ldots, \partial w_2/\partial\psi_t \\ \cdots \\ \partial w_n/\partial\psi_1, \partial w_n/\partial\psi_2, \ldots, \partial w_n/\partial\psi_t \end{bmatrix}, \tag{12.45}$$

$$\boldsymbol{\nabla} s = \begin{bmatrix} \partial s_1/\partial\psi_1, \partial s_1/\partial\psi_2, \ldots, \partial s_1/\partial\psi_t \\ \partial s_2/\partial\psi_1, \partial s_2/\partial\psi_2, \ldots, \partial s_2/\partial\psi_t \\ \cdots \\ \partial s_m/\partial\psi_1, \partial s_m/\partial\psi_2, \ldots, \partial s_m/\partial\psi_t \end{bmatrix}. \tag{12.46}$$

Introducing the expressions (12.44) into the constraints of the original problem (12.34), we obtain its local linear approximation:

$$\begin{aligned} \min\{\mathbf{c}^{\mathrm{T}}\boldsymbol{\psi} \,|\, \mathbf{D}^{-}\boldsymbol{\psi} \leqslant \mathbf{s}_0 + \boldsymbol{\nabla} s_0(\boldsymbol{\psi} - \boldsymbol{\psi}_0) \leqslant \mathbf{D}^{+}\boldsymbol{\psi}, \\ \mathbf{w}^{-} \leqslant \mathbf{w}_0 + \boldsymbol{\nabla} w_0(\boldsymbol{\psi} - \boldsymbol{\psi}_0) \leqslant \mathbf{w}^{+}, \boldsymbol{\psi}^{-} \leqslant \boldsymbol{\psi} \leqslant \boldsymbol{\psi}^{+}\}. \end{aligned} \tag{12.47}$$

After this problem has been solved, we take its solution $\boldsymbol{\psi}_*$ as the current design $\boldsymbol{\psi}_0$ and start the next iteration.

Now, linear approximation could take us far away from the exact solution if we imposed no limitations on the change of design

$$\delta\boldsymbol{\psi} = \boldsymbol{\psi} - \boldsymbol{\psi}_0. \tag{12.48}$$

In order to keep that change small, we must introduce move limits

$$-\delta\boldsymbol{\psi}^{-} \leqslant \delta\boldsymbol{\psi} \leqslant \delta\boldsymbol{\psi}^{+}. \tag{12.49}$$

The entries of the matrices $\delta\boldsymbol{\psi}^-$, $\delta\boldsymbol{\psi}^+$ are given positive constants that define the domain of possible variations in the space of design variables (Fig. 12.4). Usually the move limits are made large at the beginning of the iterative process and reduced successively as the current point $\boldsymbol{\psi}_0$ approaches the solution.

Both the technological constraints and the move limits can be taken into account by the variable bounds

$$\boldsymbol{\psi}_0^- \leq \boldsymbol{\psi} \leq \boldsymbol{\psi}_0^+, \tag{12.50}$$

adjusted at the beginning of each iteration. As Fig. 12.4 clearly shows, that adjustment can have the form

$$\begin{aligned} \psi_{0i}^- &= \max[\psi_i^-, (\psi_{0i} - \delta\psi_i^-)], \\ \psi_{0i}^+ &= \min[\psi_i^+, (\psi_{0i} + \delta\psi_i^+)]. \end{aligned} \tag{12.51}$$

The disadvantage of the model (12.47) is that the only information passed to the subsequent iteration is $\boldsymbol{\psi}_0$. This circumstance can lead to numerical instability—e.g., to oscillations of the LP solution between two adjacent vertices of the admissible domain. The move limits are supposed to prevent this, but it is better to take a more consistent mathematical model from the beginning. Such is the concept of optimal redesign: given a current admissible solution $\boldsymbol{\psi}_0$, we are looking for a small change $\delta\boldsymbol{\psi}_*$ of that design that reduces the cost Γ.

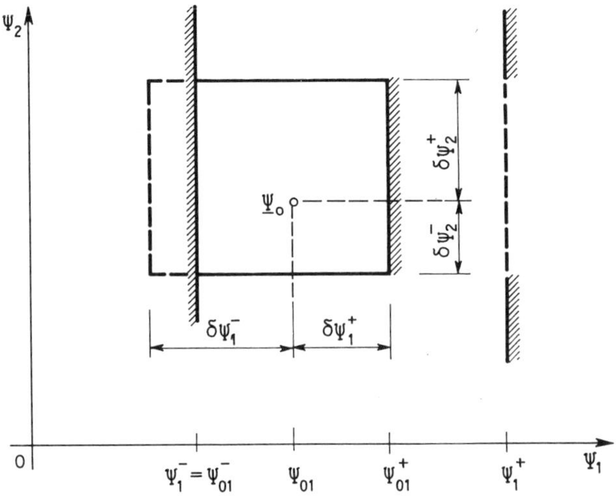

Figure 12.4. Move Limits, Absolute Bounds, and Relative Bounds on Design Variables.

Let us derive the optimal redesign counterpart of the problem

$$\min\{\mathbf{c}^{\mathrm{T}}\boldsymbol{\psi} \mid \mathbf{w}^- \leqslant \mathbf{w} \leqslant \mathbf{w}^+\}. \tag{12.52}$$

In other words, we want to optimize the cross-sectional areas under rigidity constraints. Introducing the matrices

$$\mathbf{A}_w = \begin{bmatrix} \mathbf{I} \\ -\mathbf{I} \end{bmatrix}, \qquad \hat{\mathbf{w}} = \begin{bmatrix} \mathbf{w}^- \\ -\mathbf{w}^+ \end{bmatrix}, \tag{12.53}$$

where $\mathbf{I}$ is the identity matrix of order n, we can convert (12.52) into the more convenient form

$$\min\{\mathbf{c}^{\mathrm{T}}\boldsymbol{\psi} \mid \mathbf{A}_w\mathbf{w} \geqslant \hat{\mathbf{w}}\}. \tag{12.54}$$

The change of cost introduced by the modification (12.48) of the current design is equal to $\delta\Gamma = \mathbf{c}^{\mathrm{T}}\delta\boldsymbol{\psi}$. Replacing $\mathbf{w}$ in the constraint of problem (12.54) by its linear approximation (12.44), we express that constraint as

$$\mathbf{A}_w(\mathbf{w}_0 + \boldsymbol{\nabla} w_0 \delta\boldsymbol{\psi}) \geqslant \hat{\mathbf{w}},$$

or

$$\mathbf{Q}_w \delta\boldsymbol{\psi} \geqslant \delta\hat{\mathbf{w}}, \tag{12.55}$$

where

$$\mathbf{Q}_w = \mathbf{A}_w \boldsymbol{\nabla} w_0 \tag{12.56}$$

and

$$\delta\hat{\mathbf{w}} = \hat{\mathbf{w}} - \mathbf{A}_w\mathbf{w}_0. \tag{12.57}$$

Note that the entries of $\delta\hat{\mathbf{w}}$ indicate how close the displacements of the current structure are to their bounds.

The linear matrix inequality (12.55) is the incremental form of the constraint on displacements. It must be supplemented by the move limits in order to ensure sufficient accuracy. Hence, the optimal redesign problem looked for is

$$\min\{\mathbf{c}^{\mathrm{T}}\delta\boldsymbol{\psi} \mid \mathbf{Q}_w\delta\boldsymbol{\psi} \geqslant \delta\hat{\mathbf{w}}, -\delta\boldsymbol{\psi}_0^- \leqslant \delta\boldsymbol{\psi} \leqslant \delta\boldsymbol{\psi}_0^+\}. \tag{12.58}$$

After we find its solution $\delta\boldsymbol{\psi}_*$, we can update the design, $\boldsymbol{\psi}_* = \boldsymbol{\psi}_0 + \delta\boldsymbol{\psi}_*$, calculate the new state $\mathbf{w}_*$, $\mathbf{s}_*$, and take it as the current point of the next iteration.

The result is slightly different when we try to convert the problem with stress constraints

$$\min\{\mathbf{c}^{\mathrm{T}}\boldsymbol{\psi} \mid \mathbf{D}^{-}\boldsymbol{\psi} \leqslant \mathbf{s} \leqslant \mathbf{D}^{+}\boldsymbol{\psi}\} \tag{12.59}$$

into incremental form. First, we write it as

$$\min\{\mathbf{c}^{\mathrm{T}}\boldsymbol{\psi} \mid \mathbf{A}_s\mathbf{s} - \mathbf{A}_\psi\boldsymbol{\psi} \geqslant \mathbf{0}\}, \tag{12.60}$$

with

$$\mathbf{A}_s = \begin{bmatrix} \mathbf{I} \\ -\mathbf{I} \end{bmatrix}, \qquad \mathbf{A}_\psi = \begin{bmatrix} -\mathbf{D}^{-} \\ \mathbf{D}^{+} \end{bmatrix}, \tag{12.61}$$

$\mathbf{I}$ being the identity matrix of order m. Introducing linear approximations of $\mathbf{s}$ and $\boldsymbol{\psi}$ into the inequality constraint of the problem (12.57) and rearranging the terms, we obtain that constraint in the form

$$\mathbf{Q}'_s\delta\boldsymbol{\psi} \geqslant \delta\hat{\mathbf{s}}, \tag{12.62}$$

where

$$\mathbf{Q}_s = \mathbf{A}_s\boldsymbol{\nabla} s_0, \tag{12.63}$$

$$\mathbf{Q}'_s = \dot{\mathbf{q}}_s - \mathbf{A}_\psi, \tag{12.64}$$

and

$$\delta\hat{\mathbf{s}} = \mathbf{A}_\psi\boldsymbol{\psi}_0 - \mathbf{A}_s\mathbf{s}_0. \tag{12.65}$$

This time the right side of inequality (12.62) indicates the margin between the actual and admissible stresses.

Therefore, the optimal redesign problem under the presence of stress constraints has the following model:

$$\min\{\mathbf{c}^{\mathrm{T}}\delta\boldsymbol{\psi} \mid \mathbf{Q}'_s\delta\boldsymbol{\psi} \geqslant \delta\hat{\mathbf{s}},\ -\delta\boldsymbol{\psi}_0^{-} \leqslant \delta\boldsymbol{\psi} \leqslant \delta\boldsymbol{\psi}_0^{+}\}. \tag{12.66}$$

The optimal redesign problem that simultaneously considers constraints on displacements and stresses can be composed from the special cases (12.57) and (12.66).

Similar models can be derived for optimizing the layout. The only difference is that the cost function must be linearized locally and that the design variables include the coordinates of the free nodes.

12.5.2. Gradients of Displacements and Stresses

The SLP method requires us to calculate the gradient matrices ∇w_0 and/or ∇s_0 at each step of the iteration. We begin by evaluating ∇w_0, and we let $d\mathbf{w}_0^{(i)}$ be the ith column of that matrix. According to definition (12.45), the entries of $d\mathbf{w}_0^{(i)}$ are the derivatives of displacements with respect to the ith design variable.

Differentiating Eq. (12.12) with respect to ψ_i, we obtain

$$\mathbf{K}_0\, d\mathbf{w}_0^{(i)} = d\mathbf{p}_0^{(i)} - d\mathbf{K}_0^{(i)}\, \mathbf{w}_0. \tag{12.67}$$

Here $d\mathbf{K}_0^{(i)}$ and $d\mathbf{p}_0^{(i)}$ are the derivatives of the stiffness matrix and the load matrix with respect to ψ_i. The subscript 0 indicates that the relevant quantity has been taken at the point $\boldsymbol{\psi}_0$. Setting

$$d\hat{\mathbf{p}}_0^{(i)} = d\mathbf{p}_0^{(i)} - d\mathbf{K}_0^{(i)}\, \mathbf{w}_0, \tag{12.68}$$

we can rewrite Eq. (12.67) as

$$\mathbf{K}_0\, d\mathbf{w}_0^{(i)} = d\hat{\mathbf{p}}_0^{(i)}. \tag{12.69}$$

Solving this matrix equation gives us the ith column of the gradient matrix ∇w_0.

Therefore, to calculate the ith derivative of displacements requires us to solve the canonical equation of the direct stiffness method with the right side replaced by $d\hat{\mathbf{p}}_0^{(i)}$. Computationally, this is advantageous because to calculate the derivatives we can use the same solver of linear equations that we use for analyzing the structure. Moreover, if that solver implements the Choleski factorization scheme, then the solution corresponding to the additional right side is found by a simple backward substitution [see e.g., McGuire and Gallagher (1979)].

If the self-weight of the structure is neglected, then $d\mathbf{p}_0^{(i)} = \mathbf{0}$. Otherwise, the nodal forces p_j are linear functions of the design variables, and, hence, $d\mathbf{p}_0^{(i)}$ is constant.

The derivative $d\mathbf{K}^{(i)}$ of the stiffness matrix can be assembled element by element. If the layout is fixed, then, from differentiating Eq. (12.11), we have

$$d\mathbf{K}^{(i)} = \mathbf{C}^{\mathrm{T}}\, d\mathbf{E}^{(i)}\, \mathbf{C}. \tag{12.70}$$

For trusses the entries of $\mathbf{E}$ are linear functions of the cross-sectional areas [compare Eq. (11.27)]. Thus $d\mathbf{E}^{(i)}$ is constant, and, hence, $d\mathbf{K}^{(i)}$ is constant as well. This circumstance obviously accelerates the calculations.

We now turn to the gradient of generalized stresses ∇s_0. We can find the ith column of that gradient by differentiating Eq. (12.14) with respect to ψ_i:

$$d\mathbf{s}_0^{(i)} = d\mathbf{E}_0^{(i)}\,\mathbf{C}\mathbf{w}_0 + \mathbf{E}_0\mathbf{C}\,d\mathbf{w}_0^{(i)}. \tag{12.71}$$

We have assumed $\mathbf{C} = \text{const}$—i.e., the layout is fixed. Equation (12.71) enables us to compute the stress derivative if we know the displacement derivative $d\mathbf{w}_0^{(i)}$.

We now recall the three-bar truss in Fig. 11.1a. For $A_1 = A_3$, we want to find the first columns of the matrices ∇w_0 and ∇s_0, i.e., the matrices

$$d\mathbf{w}^{(1)} = \left\{\frac{\partial w_x^{(1)}}{\partial A_1}, \frac{\partial w_y^{(1)}}{\partial A_1}\right\}, \qquad d\mathbf{s}^{(1)} = \left\{\frac{\partial s_1}{\partial A_1}, \frac{\partial s_2}{\partial A_1}, \frac{\partial s_3}{\partial A_1}\right\}$$

evaluated at the point $\boldsymbol{\psi}_0 = \{10.0, 2.0\}$. We can easily check that these are admissible for the load $\mathbf{p}_* = \{200, 100\}$ [kN].

Differentiating the stiffness matrix

$$\mathbf{K} = \frac{E}{L\sqrt{2}}\begin{bmatrix} A_1 & 0 \\ 0 & A_1 + A_2\sqrt{2} \end{bmatrix}$$

with respect to A_1, we obtain

$$d\mathbf{K}^{(1)} = \frac{E}{L\sqrt{2}}\begin{bmatrix} 1 & 0 \\ 0 & 1 \end{bmatrix}.$$

Substituting given values of the cross-sectional areas into $\mathbf{K}$, we have

$$\mathbf{K}_0 = \frac{E\sqrt{2}}{2L}\begin{bmatrix} 10.0 & 0.00 \\ 0.00 & 12.83 \end{bmatrix}.$$

The solution of Eq. (12.12) gives the displacements $\mathbf{w}_0 = \{0.2828, 0.1102\}$ [cm]. The nodal forces do not depend on the cross-sectional areas. Hence, $d\mathbf{p}^{(1)} = \mathbf{0}$ and

$$d\hat{\mathbf{p}}_0^{(1)} = -d\mathbf{K}_0^{(1)}\,\mathbf{w}_0 = -\frac{E\sqrt{2}}{2L}\begin{bmatrix} 1 & 0 \\ 0 & 1 \end{bmatrix}\begin{bmatrix} 0.2828 \\ 0.1102 \end{bmatrix} = -\frac{E\sqrt{2}}{2L}\begin{bmatrix} 0.2828 \\ 0.1102 \end{bmatrix}.$$

Substituting these results into Eq. (12.69), we obtain two linear equations:

$$\frac{E\sqrt{2}}{2L}\begin{bmatrix} 10.0 & 0.00 \\ 0.00 & 12.83 \end{bmatrix}\begin{bmatrix} \partial w_x^{(1)}/\partial A_1 \\ \partial w_y^{(1)}/\partial A_1 \end{bmatrix} = -\frac{E\sqrt{2}}{2L}\begin{bmatrix} 0.2828 \\ 0.1102 \end{bmatrix}.$$

Solving these equations gives us the required derivative

$$d\mathbf{w}_0^{(1)} = \{-0.0283, -0.0086\}.$$

As initial data for calculating the stress derivative, we have the matrices

$$\mathbf{E} = \frac{E}{L}\begin{bmatrix} 0.707A_1 & 0 & 0 \\ 0 & A_2 & 0 \\ 0 & 0 & 0.707A_1 \end{bmatrix}$$

and

$$\mathbf{C} = \begin{bmatrix} 0.707 & 0.707 \\ 0.000 & 1.000 \\ -0.707 & 0.707 \end{bmatrix}.$$

The derivative of elasticity matrix with respect to A_1 is

$$d\mathbf{E}^{(1)} = \frac{E}{L}\begin{bmatrix} 0.707 & 0.000 & 0.000 \\ 0.000 & 0.000 & 0.000 \\ 0.000 & 0.000 & 0.707 \end{bmatrix}.$$

We can now calculate the stress derivatives according to Eq. (12.71):

$$\begin{aligned} d\mathbf{s}_0^{(1)} &= \frac{E}{L}\begin{bmatrix} 0.707 & 0.000 & 0.000 \\ 0.000 & 0.000 & 0.000 \\ 0.000 & 0.000 & 0.707 \end{bmatrix}\begin{bmatrix} 0.707 & 0.707 \\ 0.000 & 1.000 \\ -0.707 & 0.707 \end{bmatrix}\begin{bmatrix} 0.2828 \\ 0.1102 \end{bmatrix} \\ &\quad -\frac{E}{L}\begin{bmatrix} 7.071 & 0.000 & 0.000 \\ 0.000 & 2.000 & 0.000 \\ 0.000 & 0.000 & 7.071 \end{bmatrix}\begin{bmatrix} 0.707 & 0.707 \\ 0.000 & 1.000 \\ -0.707 & 0.707 \end{bmatrix}\begin{bmatrix} 0.0283 \\ 0.0086 \end{bmatrix} \\ &= \frac{E}{L}\begin{bmatrix} 0.1965 \\ 0.0000 \\ -0.0864 \end{bmatrix} - \frac{E}{L}\begin{bmatrix} 0.1845 \\ 0.0172 \\ 0.0985 \end{bmatrix} \end{aligned}$$

$$= \frac{E}{L}\begin{bmatrix} 0.0121 \\ -0.0172 \\ 0.0121 \end{bmatrix}.$$

The derivatives that enter the second columns of the gradients ∇w_0 and ∇s_0 can be computed in like manner. Besides being necessary for the SLP method, those gradients can be useful for refining the design accomplished "manually" (i.e., by sequential reanalysis on a microcomputer). For example, the entries of $d\mathbf{w}_0^{(1)}$ tell us that by increasing A_1 we can reduce the horizontal displacement much more significantly than the vertical one.

12.5.3. Applications

Despite its conceptual simplicity, the SLP method requires numerical experience and skill. Both were possessed by the group of researchers at the Technical University of Denmark in Lyngby, headed by P. Pedersen. Several results obtained by that group are cited here (see also Pedersen (1981, 1983)).

Pedersen (1969) investigated the optimal layout of the cantilever truss. Though the SLP technique was not formally derived at that time, Pedersen used it in that paper. First the cross-sectional areas were optimized for the initial configuration depicted in Fig. 12.5a. It turned out that when the stability constraints for compressed members (i.e., Euler buckling loads) and the self-weight of the structure are taken into account, the mass of the truss could not be less than 922 kg. Then the optimal configuration for the given set of nodes was looked for. The result is shown in Fig. 12.5b. It reduced the mass to 685 kg (26% gain). Finally, that configuration was further optimized with respect to the positions of nodes 7, 8, 9. After 20 iterations the solution depicted in Fig. 12.5c was obtained. Its mass is 582 kg, which is 37% less than the initial design. Thus the optimization of layout produces much more significant results than mere adjustment of the cross-sectional areas.

The second example concerns the bridge in Fig. 12.6a, taken from Pedersen (1972). The bridge is loaded successively at five nodes fixed along the roadway. Thus, five loading cases are considered.

The configuration of the truss is preassigned, whereas the cross-sectional areas of bars and the positions of nodes situated on the arch are to be optimized. The mass of the bridge is minimized under the following constraints:

Stresses should not exceed admissible values (possible buckling in the elastic or plastic range is taken into account).

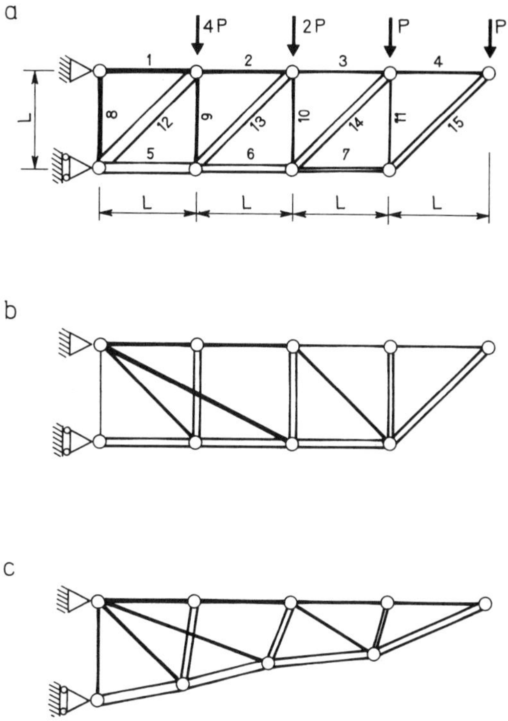

Figure 12.5. Optimal Design of Cantilever Truss: (a) Initial Configuration. (b) Optimal Configuration for Initial Nodes. (c) Final Solution.

Maximum displacement should not be greater than a given value.
The cross-sectional areas and the positions of free nodes should meet technological bounds.

The results are summarized in Figs. 12.6b–e. When the stability and rigidity constraints were neglected, the solution (b) was obtained. The total mass of that truss is 11% less than the mass obtained by optimizing cross sections for the initial layout under the same constraints. The fully stressed truss is slightly heavier than the optimal one. Note the shift of the upper nodes toward the supports, a shift observed in all optimal solutions.

As expected, the optimization under rigidity constraints with neglected stability requirements produces the relatively high truss shown in Fig. 12.6c. If these constraints are reversed, then the optimal truss tends to be low (Fig. 12.6d), since shorter bars are more resistant against buckling. The true optimal solution (Fig. 12.6e) lies between the solutions (c) and (d). The gain of mass against the solution (a) is about 10%.

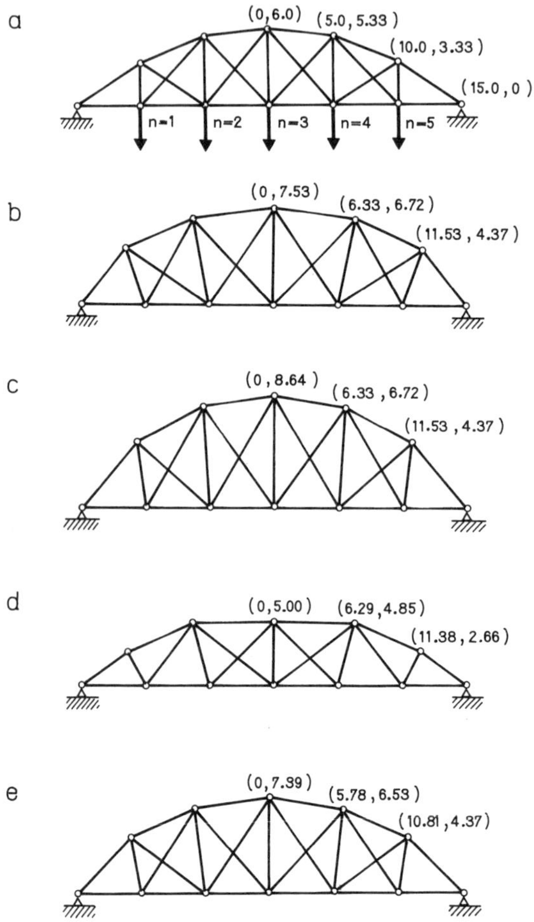

Figure 12.6. Optimal Design of Bridge: (a) Initial Layout and Loading. (b) Solution without Stability and Rigidity Constraints. (c) Solution with Rigidity Constraint but without Stability Requirements. (d) Solution with Stability Constraints but without Rigidity Requirements. (e) Complete Solution.

Pedersen (1972) also gave an example of the optimal design of a truss supporting the antenna of a radiotelescope. Intensive studies on optimization of such structures were conducted at the Technical University of Siegen in West Germany [Eschenauer (1986)]. Since those projects were substantially related to multiobjective optimization, they will be reported in Chapter 18.

13

Optimization of Plastic Structures

13.1. Introduction

In the previous chapter we considered structures that remain linearly elastic under a given loading. Demanding that the true stress at any point of the structure not exceed the admissible value and taking that value below the yield stress of the material, we were able to optimize the distribution of material or the layout of the structure. Such an approach prevailed for a long time in design methodology and is still used today in many areas of engineering.

The concept of admissible stresses, though conceptually simple, has a significant flaw: it is based upon the linear elastic model of the material. Hence, it cannot deal with the resistance of the structure to possible overloading that inevitably leads to nonlinear response. Many structural systems have sufficient redundancy to allow local yielding of the material without significant reduction of their overall rigidity and stability. On the other hand, a structure designed according to the admissible stress philosophy might have very low resistance against overloading. Both circumstances stimulated the development of an entirely different design approach based upon the theory of plasticity. Originated in the late 1930s by Gvozdez (1934, 1949), it was further developed by Prager (1956b), Neal (1956), Hodge (1959), Massonet and Save (1965), and many other researchers. These methods for evaluation of the ultimate (or collapse) load of a given structure were soon complemented by the development of an analytical approach to optimal plastic design. Its foundations were laid by Prager (1956a, 1959). Later a systematic treatment of optimal plastic design using the theory of mathematical programming was given by Čyras and his co-workers [Čyras (1969, 1983), Čyras *et al.* (1974)].

In the following section we use the equivalence scheme introduced in Chapter 10 to derive the two fundamental theorems of ultimate load analysis. These theorems allow us to find the ultimate load for a given structure, so they can be considered as part of the structure analysis. Then in Section 13.3 we formulate the problem of optimal plastic design and show that it can be reduced to a dual pair of LP problems. Section 13.4 gives several modifications of the mathematical model and discusses numerical aspects. Examples of the optimal design of trusses and frames are given in Section 13.5.

13.2. Ultimate Load Analysis

13.2.1. Fundamental Theorems

Let the load increase monotonically according to the relation

$$\mathbf{p} = \mu \mathbf{p}_0, \tag{13.1}$$

where $\mathbf{p}_0$ denotes the reference loading and μ is the load factor. Our aim is to find the ultimate value μ_* of that factor together with the stress state $\mathbf{s}_*$ corresponding to incipient plastic collapse and the mechanism $\dot{\mathbf{q}}_*$, $\dot{\mathbf{w}}_*$ of that collapse.

We now consider the governing relations that constitute the complete set for such a problem. First, we must take into account the constitutive law. Since we are interested solely in the state of ultimate equilibrium, the elastic properties of the structure are immaterial and we can use the rigid, perfectly plastic model (11.54). That model implies that the stress state $\mathbf{s}_*$ must be admissible; i.e., it must satisfy the inequality

$$\mathbf{N}^{\mathrm{T}}\mathbf{s} \leqslant \mathbf{k}, \tag{13.2}$$

with a given right side $\mathbf{k}$ that characterizes the plastic moduli of the structure. At the same time the stress state $\mathbf{s}_*$ must be in equilibrium with the ultimate load $\mathbf{p}_* = \mu_* \mathbf{p}_0$; hence, it must obey the equation

$$\mathbf{C}^{\mathrm{T}}\mathbf{s} = \mu \mathbf{p}_0. \tag{13.3}$$

The strain rates $\dot{\mathbf{q}}_*$ must follow from the associated flow rule and simultaneously be kinematically compatible with the displacement rates $\dot{\mathbf{w}}_*$

that occur in the collapse mechanism. Substituting the right side of Eq. (11.54) into the kinematic equation (11.1), we obtain the condition of kinematic compatibility in terms of $\dot{\boldsymbol{\lambda}}$ and $\dot{\mathbf{w}}$:

$$\mathbf{N}\dot{\boldsymbol{\lambda}} = \mathbf{C}\dot{\mathbf{w}}. \tag{13.4}$$

We can now fill the template (10.32). The candidates to be x-type unknowns are the kinematic quantities $\dot{\mathbf{q}}$ and $\dot{\mathbf{w}}$. Since the plastic strain rates $\dot{\mathbf{q}}$ are completely defined by the product of matrices $\mathbf{N}$ and $\dot{\boldsymbol{\lambda}}$, we take $\mathbf{x} = \{\dot{\boldsymbol{\lambda}}, \dot{\mathbf{w}}\}$. The static quantities should be treated as y-type unknowns, so we assume $\mathbf{y} = \{\mathbf{s}, \mu\}$.

Under such assumptions, relations (13.2)–(13.4) can be put into the following table:

$$\begin{array}{rcc:cccll}
 & \dot{\boldsymbol{\lambda}}^{\mathrm{T}} & \dot{\mathbf{w}}^{\mathrm{T}} & \mathbf{s}^{\mathrm{T}} & \mu & 1 & & \\
\hline
\nabla f_\lambda = & \mathbf{0} & \mathbf{0} & -\mathbf{N}^{\mathrm{T}} & \mathbf{0} & \mathbf{k} & \geqslant \mathbf{0} & (l) \\
\nabla f_w = & \mathbf{0} & \mathbf{0} & \mathbf{C}^{\mathrm{T}} & -\mathbf{p}_0 & \mathbf{0} & = \mathbf{0} & (n) \\
\hdashline
\nabla f_s = & -\mathbf{N} & \mathbf{C} & \mathbf{0} & \mathbf{0} & \mathbf{0} & = \mathbf{0} & (m) \\
\hline
 & (l) & (n) & (m) & (1) & (1) & &
\end{array}$$

However, the table is not complete: since we have four unknowns $\dot{\boldsymbol{\lambda}}$, $\dot{\mathbf{w}}$, $\mathbf{s}$, and μ, there must be a fourth row of the table corresponding to the derivative of the potential with respect to μ. The first three entries of that row are defined by the requirement that the global matrix of coefficients be symmetric: they follow from the transposition of the fourth column. The fourth entry is zero, and the fifth entry is an arbitrary constant (e.g., unity). The derivative ∇f_μ must vanish at the solution μ_*, because μ is a free variable. Hence, the final table looks like

$$\begin{array}{rcc:cccll}
 & \dot{\boldsymbol{\lambda}}^{\mathrm{T}} & \dot{\mathbf{w}}^{\mathrm{T}} & \mathbf{s}^{\mathrm{T}} & \mu & 1 & & \\
\hline
\nabla f_\lambda = & \mathbf{0} & \mathbf{0} & -\mathbf{N}^{\mathrm{T}} & \mathbf{0} & \mathbf{k} & \geqslant \mathbf{0} & (l) \\
\nabla f_w = & \mathbf{0} & \mathbf{0} & \mathbf{C}^{\mathrm{T}} & -\mathbf{p}_0 & \mathbf{0} & = \mathbf{0} & (n) \\
\hdashline
\nabla f_s = & -\mathbf{N} & \mathbf{C}^{\mathrm{T}} & \mathbf{0} & \mathbf{0} & \mathbf{0} & = \mathbf{0} & (m) \\
\nabla f_\mu = & 0 & -\mathbf{p}_0 & \mathbf{0} & 0 & 1 & = 0 & (1) \\
\hline
 & (l) & (n) & (m) & (1) & (1) & &
\end{array} \tag{13.5}$$

The last row of Eq. (13.5) contains the equation

$$\mathbf{p}_0^{\mathrm{T}}\dot{\mathbf{w}} = 1. \tag{13.6}$$

Its physical meaning is simple: we must normalize the power (i.e., rate of work)

$$\dot{W}_0 = \mathbf{p}_0^{\mathrm{T}}\dot{\mathbf{w}} \tag{13.7}$$

produced by the reference load $\mathbf{p}_0$ because, otherwise, the motion of the collapse mechanism would be unrestricted. This shows one advantage of the systematic approach used in this book: even if certain equations have been omitted when constructing the complete set of governing relations of the problem, they come out automatically through the symmetry condition.

Equation (13.5) contains the linear part of the complete set of governing relations. The nonlinear part is the KKT (Karush–Kuhn–Tucker) condition $\dot{\boldsymbol{\lambda}}^{\mathrm{T}}\nabla f_\lambda = 0$ generated by the nonnegative variable $\dot{\boldsymbol{\lambda}} \geqslant \mathbf{0}$. It coincides with the complementarity rule $(11.54)_4$.

Both $\mathbf{A}_{xx}$ and $\mathbf{A}_{yy}$ are now zero matrices, which does not keep us from regarding them formally as being positive and negative semidefinite, respectively. Thus all the prerequisites for applying the equivalence scheme of Section 10.5 are fulfilled. The potential function contains only bilinear and linear terms:

$$f(\dot{\boldsymbol{\lambda}}, \dot{\mathbf{w}}, \mathbf{s}, \mu) = -\dot{\boldsymbol{\lambda}}^{\mathrm{T}}\mathbf{N}^{\mathrm{T}}\mathbf{s} + \dot{\mathbf{w}}^{\mathrm{T}}\mathbf{C}^{\mathrm{T}}\mathbf{s} - \mu\mathbf{p}_0^{\mathrm{T}}\dot{\mathbf{w}} + \mathbf{k}^{\mathrm{T}}\dot{\boldsymbol{\lambda}} + \mu. \tag{13.8}$$

The solution $(\dot{\boldsymbol{\lambda}}_*, \dot{\mathbf{w}}_*, \mathbf{s}_*, \mu_*)$ of the ultimate load problem corresponds to the saddle point

$$\min_{\dot{\mathbf{w}},\dot{\boldsymbol{\lambda}}\geqslant\mathbf{0}} \max_{\mathbf{s},\mu} f(\dot{\boldsymbol{\lambda}}, \dot{\mathbf{w}}, \mathbf{s}, \mu) \tag{13.9}$$

or to the solution of the following dual LP problems:

$$\begin{aligned}&\min\{\mathbf{k}^{\mathrm{T}}\dot{\boldsymbol{\lambda}} \mid \mathbf{C}\dot{\mathbf{w}} - \mathbf{N}\dot{\boldsymbol{\lambda}} = \mathbf{0}, \mathbf{p}_0^{\mathrm{T}}\dot{\mathbf{w}} = 1, \dot{\boldsymbol{\lambda}} \geqslant \mathbf{0}\},\\ &\max\{\mu \mid \mathbf{C}^{\mathrm{T}}\mathbf{s} - \mu\mathbf{p}_0 = \mathbf{0}, \mathbf{N}^{\mathrm{T}}\mathbf{s} \leqslant \mathbf{k}\}.\end{aligned} \tag{13.10}$$

These problems have the following physical interpretation [Prager (1956b)]:

Static Theorem of Limit Analysis. Of all statically admissible load factors, the maximum one corresponds to the ultimate load (problem $(13.10)_2$).

Kinematic Theorem of Limit Analysis. Of all kinematically admissible collapse mechanisms, the true one corresponds to the minimum dissipated energy (problem $(13.10)_1$).

Usually the latter theorem is formulated as follows: the ultimate load factor μ_* is the smallest of factors

$$\mu_k = \dot{D}/\dot{W}_0 \tag{13.11}$$

computed for all kinematically admissible collapse mechanisms. Here $\dot{D}$ stands for the dissipated power, defined as

$$\dot{D} = \dot{\mathbf{q}}^{\mathrm{T}} \cdot \mathbf{s} = \mathbf{k}^{\mathrm{T}} \cdot \dot{\boldsymbol{\lambda}}. \tag{13.12}$$

It follows from (13.11) that $\mu_k = \dot{D}$ when $\dot{W}_0 = 1$. Thus in problem $(13.10)_1$ we are minimizing the kinematic load factor, and both formulations of the kinematic theorem are equivalent.

The static problem $(13.10)_2$ has at least one admissible solution for any given $\mathbf{p}_0$, namely, the trivial solution $\mu_* = 0$, $\mathbf{s}_* = \mathbf{0}$. The maximum of μ is unique and finite. Duality theory tells us that the kinematic problem $(13.10)_1$ also has a solution $\dot{\boldsymbol{\lambda}}_*$, $\dot{\mathbf{w}}_*$ that corresponds to the minimum of dissipated power $\dot{D}_* = \mathbf{k}^{\mathrm{T}} \cdot \dot{\boldsymbol{\lambda}}_*$. However, neither $\mathbf{s}_*$ nor $\dot{\boldsymbol{\lambda}}_*$ and $\dot{\mathbf{w}}_*$ need necessarily be unique, due to the semidefiniteness of submatrices $\mathbf{A}_{xx}$ and $\mathbf{A}_{yy}$ of Eq. (13.5). The linear cost functions f', f'' appearing in the dual problems (13.10) can represent the hyperplanes parallel to certain faces of the admissible domains. Then any point belonging to that face is a solution of the relevant LP problem.

Physically the nonuniqueness of the stress state at incipient collapse comes from the simplified nature of the rigid, perfectly plastic model. Frequently the collapse mechanism of a hyperstatic structure is incomplete: part of the structure collapses, and the rest remains statically indeterminate. Then the distribution of internal forces in the latter part cannot be determined uniquely unless the elastic properties of the material are known.

13.2.2. Pin-Jointed Structures

We must be cautious when applying the ultimate load estimation, based upon the rigid, perfectly plastic model, to trusslike systems, because this approach ignores any possible buckling of compressed members prior to yielding.

According to Euler, the critical load for a pin-ended strut under compression is

$$P_{\mathrm{cr}} = \frac{\pi^2 E I_{\min}}{L^2} \tag{13.13}$$

where $I_{\min}$ is the smallest moment of inertia possessed by the cross section of the bar. Introducing the nondimensional slenderness ratio

$$\nu = L\sqrt{\frac{A}{I_{\min}}}, \tag{13.14}$$

we can express the buckling stress as

$$\sigma_{cr} = \frac{\pi^2 E}{\nu^2}. \tag{13.15}$$

If the inequality $\sigma_0 \leqslant \sigma_{cr}$ holds for each member of the truss, then model (13.10) accurately estimates the ultimate load. Buckling of members that have σ_{cr} slightly less than σ_0 can be taken into account by introducing into the rigid-plastic model a fictitious yield stress equal to σ_{cr}. That is equivalent to approximating the true postcritical behavior of the compressed strut, shown by the solid line in Fig. 13.1a, by a $\sigma = \text{const}$ type of behavior (dashed line in the same figure). Since slender elements exhibit an abrupt reduction of stiffness after buckling (Fig. 13.1b), such an approximation is not acceptable for them. In conclusion, the classical ultimate load analysis is applicable for trusses composed of rather stiff elements that approach yield before they lose stability.

Refer again to the three-bar truss of Fig. 11.1. The matrix model of that truss was derived in Section 11.2.1 and was used in Sections 12.2.3 and 12.4. We noted in the last section that the truss composed of tubes having inner radius $r = 35$ mm and outer radius $r' = 40$ mm is feasible for the load $P_x^{(1)} = 200$ kN, $P_y^{(1)} = 100$ kN for admissible stresses $\sigma^+ = 200$ MPa, $\sigma^- = -100$ MPa. That result was obtained in the frame of elasticity theory. We now want to check how large the safety margin of such a truss is against plastic collapse.

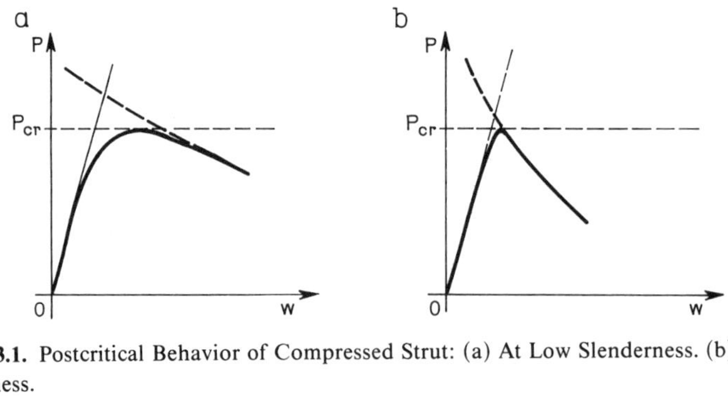

Figure 13.1. Postcritical Behavior of Compressed Strut: (a) At Low Slenderness. (b) At High Slenderness.

First, we compute the geometric characteristics of the cross section:

$$A = \pi(r'^2 - r^2) = 3.14(4.0^2 - 3.5^2) = 11.78\ \mathrm{cm}^2,$$
$$I = 0.25\pi(r'^4 - r^4) = (0.25)(3.14)(4.0^4 - 3.5^4) = 83.2\ \mathrm{cm}^4.$$

Knowing the yield stress $\sigma_0 = 250$ MPa and the cross-sectional area of a generic element, we can compute the yield force in tension:

$$N^0 = (250 \times 10^3)(11.78 \times 10^{-4}) = 294\ \mathrm{kN}.$$

Excluding the possibility of buckling, we take $\tilde{N}^0 = N^0$. Let $P_0 = 100$ kN be the reference force. Then nondimensional loads are $p_x^{(1)} = 2.0$, $p_y^{(1)} = 1.0$, and the nondimensional yield forces are $n^0 = \tilde{n}^0 = 2.94$.

The stress admissibility conditions for a generic element are given by relations (11.35) and (11.37)–(11.39). Assembling them according to the scheme (11.51), we obtain the global matrices

$$\mathbf{f} = \{f_1^{\mathrm{I}}, f_1^{\mathrm{II}}, f_2^{\mathrm{I}}, f_2^{\mathrm{II}}, f_3^{\mathrm{I}}, f_3^{\mathrm{II}}\},$$
$$\dot{\boldsymbol{\lambda}} = \{\dot{\lambda}_1^{\mathrm{I}}, \dot{\lambda}_1^{\mathrm{II}}, \dot{\lambda}_2^{\mathrm{I}}, \dot{\lambda}_2^{\mathrm{II}}, \dot{\lambda}_3^{\mathrm{I}}, \dot{\lambda}_3^{\mathrm{II}}\},$$
$$\mathbf{k} = \{2.94, 2.94, 2.94, 2.94, 2.94, 2.94\},$$
$$\mathbf{N} = \left[\begin{array}{cc|cc|cc} 1 & -1 & 0 & 0 & 0 & 0 \\ \hline 0 & 0 & 1 & -1 & 0 & 0 \\ \hline 0 & 0 & 0 & 0 & 1 & -1 \end{array}\right].$$

Other matrices appearing in model (13.10) were derived previously:

$$\mathbf{s} = \{n_1, n_2, n_3\},$$
$$\dot{\mathbf{w}} = \{\dot{w}_x^{(1)}, \dot{w}_y^{(1)}\},$$
$$\mathbf{C}^{\mathrm{T}} = \begin{bmatrix} 0.707 & 0.000 & -0.707 \\ 0.707 & 1.000 & 0.707 \end{bmatrix}.$$

Depending on the available simplex solver, the data might refer either to problem $(13.10)_1$ or to $(13.10)_2$. But does it not matter, since the solutions of both dual problems are obtained simultaneously. The code used for solving the examples for this book regarded the maximization problem as the primal one. Hence, the parameters fed into the computer for evaluating the ultimate load of the three-bar truss were four free unknowns (three entries of $\mathbf{s}$ and the load factor μ), two equality constraints (equilibrium equations), and six inequality constraints (admissibility conditions). The initial data for the problem (13.10) are given in Table 13.1 (compare Table 10.1).

The strain rates

$$\dot{\mathbf{q}} = \{0.354, 0.000, -0.354\}$$

in the elements of the truss follow from Eq. (11.54). They correspond to the collapse mechanism shown in Fig. 13.2a. Multiplying the entries of $\dot{\mathbf{w}}$

Table 13.1. Ultimate Load of Three-Bar Truss—Buckling Neglected

a. Static Approach

	μ	n_1	n_2	n_3	
	1	2	3	4	
1	2.0	−0.707	0.0	0.707	= 0.0
2	1.0	−0.707	−1.0	−0.707	= 0.0
3	0.0	1.0	0.0	0.0	⩽ 2.94
4	0.0	−1.0	0.0	0.0	⩽ 2.94
5	0.0	0.0	1.0	0.0	⩽ 2.94
6	0.0	0.0	−1.0	0.0	⩽ 2.94
7	0.0	0.0	0.0	1.0	⩽ 2.94
8	0.0	0.0	0.0	−1.0	⩽ 2.94
9	1.0	0.0	0.0	0.0	Costs
10	2.08	2.94	2.08	−2.94	Solution

b. Kinematic Approach

	$\dot{w}_x^{(1)}$	$\dot{w}_y^{(1)}$	$\dot{\lambda}_1^{\mathrm{I}}$	$\dot{\lambda}_1^{\mathrm{II}}$	$\dot{\lambda}_2^{\mathrm{I}}$	$\dot{\lambda}_2^{\mathrm{II}}$	$\dot{\lambda}_3^{\mathrm{I}}$	$\dot{\lambda}_3^{\mathrm{II}}$	
	1	2	3	4	5	6	7	8	
1	2.0	1.0	0.0	0.0	0.0	0.0	0.0	0.0	= 1.0
2	−0.707	−0.707	1.0	−1.0	0.0	0.0	0.0	0.0	= 0.0
3	0.0	−1.0	0.0	0.0	1.0	−1.0	0.0	0.0	= 0.0
4	0.707	−0.707	0.0	0.0	0.0	0.0	1.0	−1.0	= 0.0
5	0.0	0.0	2.94	2.94	2.94	2.94	2.94	2.94	Costs
6	0.500	0.0	0.354	0.0	0.0	0.0	0.0	0.354	Solution

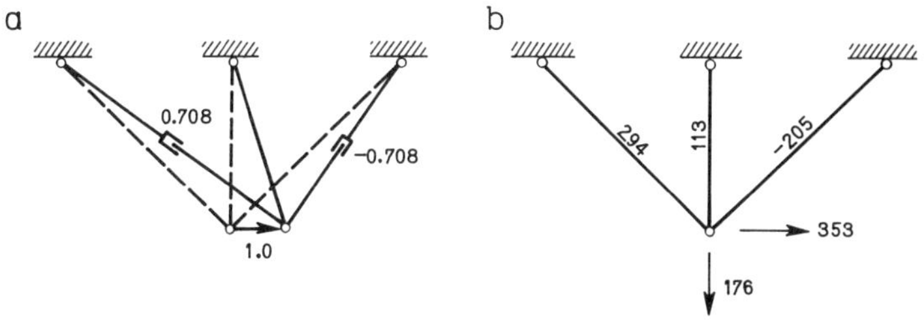

Figure 13.2. Plastic Collapse of Truss: (a) Collapse Mechanism. (b) Axial Forces.

and $\dot{\mathbf{q}}$ by the reference length $L_0 = 2.0$ m, we could obtain dimensioned quantities. However, it is not reasonable to do that, because only the ratios of kinematic parameters, not their values, are important: the motion of a collapse mechanism is given only up to a common scaling factor. In the sequel, all kinematic solutions are kept nondimensional. They are normalized in the sense that a certain characteristic component of $\dot{\mathbf{w}}$ is taken equal to unity.

The external forces acting on node 1 and the axial forces in the elements of truss are shown in Fig. 13.2b. The former were obtained by multiplying the entries of $\mathbf{p}_0$ by $\mu_* P_0$, the latter by multiplying the entries of $\mathbf{s}_*$ by P_0. Thus, we found that the considered truss can sustain overloading up to 100% ($\mu_* = 2.08$). We now check whether this result is correct from a stability point of view.

According to Eq. (13.14), the slenderness ratio of the compressed element 3 is

$$\nu_3 = 283\sqrt{\frac{11.78}{83.2}} = 106.$$

Knowing the slenderness ratio and Young's modulus, we can calculate from Eq. (13.15) the buckling stress of that element:

$$\sigma_{cr} = \frac{(3.14)^2(2.0)(10)}{106} = 174 \text{ MPa}.$$

Taking this as the fictitious yield stress, we compute

$$\tilde{N}^0 = (174 \times 10^3)(11.78 \times 10^{-4}) = 205 \text{ kN}.$$

Substituting $\tilde{n}^0 = 2.05$ for the entries of $\mathbf{k}$ that correspond to compressive axial forces, we obtain the modified model (13.10) given in Table 13.2. Its

Table 13.2. Ultimate Load of Three-Bar Truss—Buckling Taken into Account

a. Static Approach

	μ	n_1	n_2	n_3	
	1	2	3	4	
1	2.0	−0.707	0.0	0.707	= 0.0
2	1.0	−0.707	−1.0	−0.707	= 0.0
3	0.0	1.0	0.0	0.0	≤ 2.94
4	0.0	−1.0	0.0	0.0	≤ 2.05
5	0.0	0.0	1.0	0.0	≤ 2.94
6	0.0	0.0	−1.0	0.0	≤ 2.05
7	0.0	0.0	0.0	1.0	≤ 2.94
8	0.0	0.0	0.0	−1.0	≤ 2.05
9	1.0	0.0	0.0	0.0	Costs
10	1.766	2.94	1.133	−2.05	Solution

b. Kinematic Approach

	$\dot{w}_x^{(1)}$	$\dot{w}_y^{(1)}$	$\dot{\lambda}_1^{\mathrm{I}}$	$\dot{\lambda}_1^{\mathrm{II}}$	$\dot{\lambda}_2^{\mathrm{I}}$	$\dot{\lambda}_2^{\mathrm{II}}$	$\dot{\lambda}_3^{\mathrm{I}}$	$\dot{\lambda}_3^{\mathrm{II}}$	
	1	2	3	4	5	6	7	8	
1	2.0	1.0	0.0	0.0	0.0	0.0	0.0	0.0	= 1.0
2	−0.707	−0.707	1.0	−1.0	0.0	0.0	0.0	0.0	= 0.0
3	0.0	−1.0	0.0	0.0	1.0	−1.0	0.0	0.0	= 0.0
4	0.707	−0.707	0.0	0.0	0.0	0.0	1.0	−1.0	= 0.0
5	0.0	0.0	2.94	2.05	2.94	2.05	2.94	2.05	Costs
6	0.50	0.0	0.354	0.0	0.0	0.0	0.0	0.354	Solution

solution shows that the overloading factor cannot be greater than $\mu_* = 1.76$. Thus, neglecting the buckling phenomenon, we overestimated the load-carrying capacity of the truss by about 15%.

13.2.3. Framed Structures

Flexural systems such a continuous beams, grillages, and frames are much more suitable than trusses for the application of model (13.10). Usually bending prevails in those systems, which allows us to neglect the influence of axial and shear forces upon yielding. Then the simplest stress admissibility condition (11.41) holds, and the evaluation of the ultimate load is similar to that for trusses, except that there are two internal forces for each element instead of one.

The situation is different for multistory skeleton frames of tall buildings and for single-story frames that have to carry traveling cranes in industrial plants. In such cases the columns of the frame are subjected to considerable axial forces that diminish their capacity to withstand bending moment. Hence, the continuous interaction between M and N must be taken into account for such structures.

As an example, we take the frame in Fig. 11.3a. Let $L_1 = 2.0$ m, $L_2 = 3.0$ m, and let the single loading consist of the force $P_x = 10.0$ kN applied horizontally to node 3. Solving that frame elastically for $EI = \text{const}$, we find that the maximum bending moment occurs at the built-in end of the column: $M_{\max} = M_a^{(1)} = 13.33$ kN m. If $\sigma^{\pm} = 200.0$ MPa is taken as the admissible stress, then the I section 140PE meets the condition

$$\sigma_{\max} = \frac{M_{\max}}{W_x} = \frac{13.33 \times 10^{-3}}{77.3 \times 10^{-6}} = 172 \text{ MPa} < 200 \text{ MPa}.$$

We now evaluate the safety margin that the frame manufactured from such an I section possesses against plastic collapse. We know the yield stress of the material $\sigma_0 = 300$ MPa, the cross-sectional area $A = 16.4 \text{ cm}^2$, and the static moment $S = 44.2 \text{ cm}^3$. Therefore, we compute

$$M^0 = 2(300 \times 10^3)(44.2 \times 10^{-6}) = 26.5 \text{ kN m},$$

$$\text{N}^0 = (300 \times 10^3)(16.4 \times 10^{-4}) = 492 \text{ kN}.$$

According to Eq. (11.50), the reference length should be

$$L_0 = \frac{26.5}{492} = 0.054 \text{ m}.$$

We take the reference force $P_0 = 10.0$ kN. Then the nondimensional plastic moduli are $m^0 = n^0 = 49.2$.

It is reasonable to expect that under given load the axial forces are of secondary importance from the point of view of plastic collapse. Therefore, we adopt the bending-only model of the frame introduced in Section 11.2.2. Transforming it into nondimensional form, we obtain the matrices

$$\mathbf{s} = \{m_a^{(1)}, m_b^{(1)}, m_a^{(2)}\},$$

$$\mathbf{q} = \{\theta_a^{(1)}, \theta_b^{(1)}, \theta_a^{(2)}\},$$

$$\mathbf{p}_0 = \{p_x^{(2)}, p_z^{(2)}\} = \{1.0, 0.0\},$$

$$\dot{\mathbf{w}} = \{\dot{w}_x^{(2)}, \dot{w}_z^{(2)}\},$$

$$\mathbf{C}^{\mathrm{T}} = \begin{bmatrix} -0.027 & -0.027 & 0.0 \\ 0.0 & 1.0 & 1.0 \end{bmatrix}.$$

The entries of the first row of $\mathbf{C}^{\mathrm{T}}$ came from the fact that $1/L_1 = L_0/L_1 = 0.054/2.0 = 0.027$.

The bending moment at the right end of element 2 has been eliminated from the foregoing matrix model, since we know that $m_b^{(2)} = 0$. To keep the matrix $\mathbf{N}$ consistent with that assumption, we have to include in that matrix a full 2×4 submatrix $\bar{\mathbf{N}}_1$ of the type (11.43) and a part $[1, -1]$ of the submatrix $\bar{\mathbf{N}}_2$. The result is

$$\mathbf{N} = \left[\begin{array}{cccc|cc} 1 & -1 & 0 & 0 & 0 & 0 \\ 0 & 0 & 1 & -1 & 0 & 0 \\ \hline 0 & 0 & 0 & 0 & 1 & -1 \end{array}\right].$$

The corresponding matrices of state parameters are

$$\mathbf{f} = \{f_{a1}^{\mathrm{I}}, f_{a_1}^{\mathrm{II}}, f_{b1}^{\mathrm{I}}, f_{b1}^{\mathrm{II}}, f_{a2}^{\mathrm{I}}, f_{a2}^{\mathrm{II}}\},$$

$$\dot{\boldsymbol{\lambda}} = \{\dot{\lambda}_{a1}^{\mathrm{I}}, \dot{\lambda}_{a1}^{\mathrm{II}}, \dot{\lambda}_{b1}^{\mathrm{I}}, \dot{\lambda}_{b1}^{\mathrm{II}}, \dot{\lambda}_{a2}^{\mathrm{I}}, \dot{\lambda}_{a2}^{\mathrm{II}}\},$$

$$\mathbf{k} = \{49.2, 49.2, 49.2, 49.2, 49.2, 49.2\}.$$

Thus we have all matrices that appear in the evaluation of the ultimate load according to model (13.10). The initial simplex tables for problems $(13.10)_2$ and $(13.10)_1$ are Table 13.3a and Table 13.3b, respectively. The last rows of those tables contain the solutions $\mathbf{s}_*$, μ_* and $\dot{\mathbf{w}}_*$, $\dot{\boldsymbol{\lambda}}_*$.

It follows from $\mu_* = 2.65$ that the frame under consideration can survive large overloading before it collapses. There are two reasons for such a considerable safety margin: first, the yield moment of the cross section is

Table 13.3. Ultimate Load of Frame—Axial Forces Neglected

a. Static Approach

	$\rightarrow$	$m_a^{(1)}$	$m_b^{(1)}$	$m_a^{(2)}$	
	1	2	3	4	
1	1.0	0.027	0.027	0.0	= 0.0
2	0.0	0.0	−1.0	−1.0	= 0.0
3	0.0	1.0	0.0	0.0	≤ 49.2
4	0.0	−1.0	0.0	0.0	≤ 49.2
5	0.0	0.0	1.0	0.0	≤ 49.2
6	0.0	0.0	−1.0	0.0	≤ 49.2
7	0.0	0.0	0.0	1.0	≤ 49.2
8	0.0	0.0	0.0	−1.0	≤ 49.2
9	1.0	0.0	0.0	0.0	Costs
10	2.65	−49.2	−49.2	49.2	Solution

b. Kinematic Approach

	$\dot{w}_x^{(2)}$	$\dot{w}_z^{(2)}$	$\dot{\lambda}_{a1}^{\mathrm{I}}$	$\dot{\lambda}_{a1}^{\mathrm{II}}$	$\dot{\lambda}_{b1}^{\mathrm{I}}$	$\dot{\lambda}_{b2}^{\mathrm{II}}$	$\dot{\lambda}_{a2}^{\mathrm{I}}$	$\dot{\lambda}_{a2}^{\mathrm{II}}$	
	1	2	3	4	5	6	7	8	
1	1.0	0.0	0.0	0.0	0.0	0.0	0.0	0.0	= 1.0
2	0.027	0.0	1.0	−1.0	0.0	0.0	0.0	0.0	= 0.0
3	0.027	−1.0	0.0	0.0	1.0	−1.0	0.0	0.0	= 0.0
4	0.0	−1.0	0.0	0.0	0.0	0.0	1.0	−1.0	= 0.0
5	0.0	0.0	49.2	49.2	49.2	49.2	49.2	49.2	Costs
6	1.0	0.0	0.0	0.027	0.0	0.027	0.0	0.0	Solution

1.7 times greater than the bending moment that produces the stress $\sigma^{\pm}$ in the outermost fibers; second, two cross sections must be simultaneously at yield in order to bring the frame to plastic collapse.

Multiplying **N** by $\dot{\boldsymbol{\lambda}}_*$, we obtain $\dot{\mathbf{q}}_* = \{-0.027, -0.027, 0.000\}$, which are the plastic strain rates in the collapse mechanism shown in Fig. 13.3a. The bending moments and the axial forces at incipient collapse are given in Figs. 13.3b, c. They are in equilibrium with the ultimate load $P_* = 2.65 \times 10.0 = 26.5$ kN.

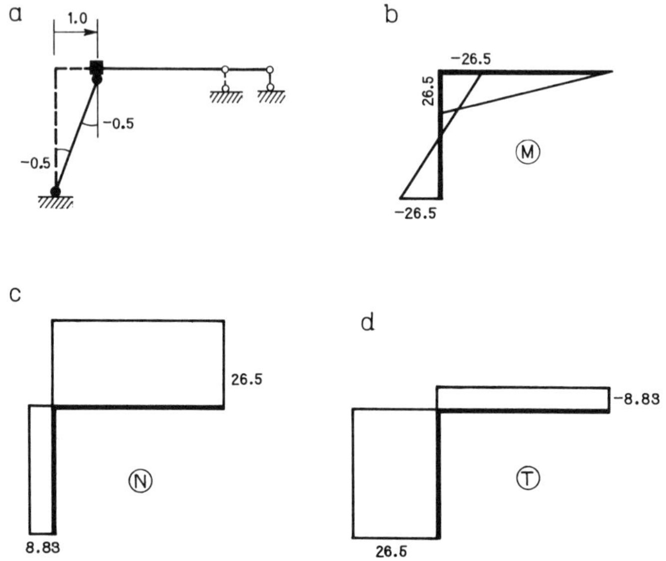

Figure 13.3. Plastic Collapse of Frame (Influence of Axial Forces upon Yielding Neglected): (a) Collapse Mechanism. (b) Bending Moments. (c) Axial Forces. (d) Shear Forces.

Imagine now that node 2 is additionally loaded by the vertical force $P_y = 100.0\,\text{kN}$. We want to know how much the ultimate load factor decreases under such a load; that is, we want to know the safety margin of the frame. We shall use the $m + n$ model derived in Section 11.3.

Keeping the same reference values L_0, P_0 as before, we have

$$\mathbf{s} = \{m_a^{(1)}, m_b^{(1)}, n_1, m_a^{(2)}, m_b^{(2)}, n_2\},$$

$$\dot{\mathbf{q}} = \{\theta_a^{(1)}, \theta_b^{(1)}, \dot{\varepsilon}_1, \dot{\theta}_a^{(2)}, \dot{\theta}_b^{(2)}, \dot{\varepsilon}_2\},$$

$$\mathbf{p}_0 = \{p_x^{(2)}, p_y^{(2)}, p_z^{(2)}, p_x^{(3)}, p_z^{(3)}\} = \{0.0, 10.0, 0.0, 1.0, 0.0\},$$

$$\dot{\mathbf{w}} = \{\dot{w}_x^{(2)}, \dot{w}_y^{(2)}, \dot{w}_z^{(2)}, \dot{w}_x^{(3)}, \dot{w}_z^{(3)}\},$$

$$\mathbf{C}^{\mathrm{T}} = \begin{bmatrix} -0.027 & -0.027 & 0.0 & 0.0 & 0.0 & -1.0 \\ 0.0 & 0.0 & -1.0 & 0.018 & 0.018 & 0.0 \\ 0.0 & 1.0 & 0.0 & 0.0 & 0.0 & 0.0 \\ 0.0 & 0.0 & 0.0 & 0.0 & 0.0 & 1.0 \\ 0.0 & 0.0 & 0.0 & 0.0 & 1.0 & 0.0 \end{bmatrix}.$$

The latter matrix comes from the transposition of the matrix **C** defined by Eq. (11.24), with $1/l_1 = 0.027$ as before and $1/l_2 = L_0/L_2 = 0.054/3.0 = 0.018$.

We now must choose the interaction curve. To demonstrate the influence of this selection on the results, we solve the problem twice, taking the rhombic and hexagonal curves (Figs. 11.18b and 11.19b, respectively).

When we take the rhombic curve, **N** is composed of two identical 3×8 submatrices, $\bar{\mathbf{N}}_1$ and $\bar{\mathbf{N}}_2$, of type (11.47). Hence, the total number of constraints in the static problem (13.10) is 21 (5 equalities + 16 inequalities). However, we know in advance the signs of the internal forces: the bending moments $m_a^{(1)}$, $m_b^{(1)}$ and the axial force n_1 will be negative for the given load, whereas $m_a^{(2)}$, $m_b^{(2)}$, and n_z are expected to remain positive. We also know that $m_b^{(2)}$ must vanish, but we prefer to keep that information hidden in the fourth row of $\mathbf{C}^{\mathrm{T}}$.

Under these circumstances, it is clear which edges of the interaction curve can become active at plastic collapse. For the internal forces in element 1, it is edge III, for generalized stresses in element 2, it is edge I. Therefore, we take

$$\mathbf{f} = \{f_{a1}^{\mathrm{III}}, f_{b1}^{\mathrm{III}}, f_{a2}^{\mathrm{I}}, f_{b2}^{\mathrm{I}}\},$$

$$\dot{\boldsymbol{\lambda}} = \{\dot{\lambda}_{a1}^{\mathrm{III}}, \dot{\lambda}_{b1}^{\mathrm{III}}, \dot{\lambda}_{a2}^{\mathrm{I}}, \dot{\lambda}_{b2}^{\mathrm{I}}\},$$

$$\mathbf{k} = \{49.2, 49.2, 49.2, 49.2\},$$

$$\mathbf{N} = \begin{bmatrix} -1 & 0 & 0 & 0 \\ 0 & -1 & 0 & 0 \\ -1 & -1 & 0 & 0 \\ \hline 0 & 0 & 1 & 0 \\ 0 & 0 & 0 & 1 \\ 0 & 0 & 1 & 1 \end{bmatrix}.$$

Table 13.4 contains data for such a model. Note that the ultimate load factor $\mu_* = 1.744$ is 34% less than the factor computed for bending only. The product $\mathbf{N}\dot{\boldsymbol{\lambda}}_*$ gives

$$\dot{\mathbf{q}}_* = \{-0.017, -0.018, -0.035, 0.000, 0.000, 0.000\}.$$

The corresponding collapse mechanism is depicted in Fig. 13.4a. This time the plastic hinges that develop in sections a and b of element 1 are generalized: they exhibit both rotational and axial components of the strain rate. That is confirmed by the fact that node 2 has a nonzero vertical component of the displacement rate.

The diagrams of internal forces at incipient collapse are shown in Figs. 13.4b, c. The values of the bending moments at the hinges are now considerably less than $M^0 = 26.5$ kN m, due to the influence of the axial force on the yielding of the cross sections.

Table 13.4. Ultimate Load of Frame—Axial Forces Taken into Account According to Rhombic Interaction Curve

a. Static Approach

	μ	$m_a^{(1)}$	$m_b^{(1)}$	n_1	$m_a^{(2)}$	$m_b^{(2)}$	n_2	
	1	2	3	4	5	6	7	
1	0.0	0.027	0.027	0.0	0.0	0.0	1.0	= 0.0
2	10.0	0.0	0.0	1.0	−0.018	−0.018	0.0	= 0.0
3	0.0	0.0	−1.0	0.0	−1.0	0.0	0.0	= 0.0
4	1.0	0.0	0.0	0.0	0.0	0.0	−1.0	= 0.0
5	0.0	0.0	0.0	0.0	0.0	−1.0	0.0	= 0.0
6	0.0	−1.0	0.0	−1.0	0.0	0.0	0.0	⩽ 49.2
7	0.0	0.0	−1.0	−1.0	0.0	0.0	0.0	⩽ 49.2
8	0.0	0.0	0.0	0.0	1.0	0.0	1.0	⩽ 49.2
9	0.0	0.0	0.0	0.0	0.0	1.0	1.0	⩽ 49.2
10	1.0	0.0	0.0	0.0	0.0	0.0	0.0	Costs
11	1.744	−32.4	−32.4	−16.85	32.4	0.0	1.744	Solution

b. Kinematic Approach

	$\dot{w}_x^{(2)}$	$\dot{w}_y^{(2)}$	$\dot{w}_z^{(2)}$	$\dot{w}_x^{(3)}$	$\dot{w}_z^{(3)}$	$\dot{\lambda}_{a1}^{\text{III}}$	$\dot{\lambda}_{b1}^{\text{III}}$	$\dot{\lambda}_{a2}^{\text{I}}$	$\dot{\lambda}_{b2}^{\text{I}}$	
	1	2	3	4	5	6	7	8	9	
1	0.0	10.0	0.0	1.0	0.0	0.0	0.0	0.0	0.0	= 1.0
2	0.027	0.0	0.0	0.0	0.0	−1.0	0.0	0.0	0.0	= 0.0
3	0.027	0.0	−1.0	0.0	0.0	0.0	−1.0	0.0	0.0	= 0.0
4	0.0	1.0	0.0	0.0	0.0	−1.0	−1.0	0.0	0.0	= 0.0
5	0.0	−0.018	−1.0	0.0	0.0	0.0	0.0	1.0	0.0	= 0.0
6	0.0	−0.018	0.0	0.0	−1.0	0.0	0.0	0.0	1.0	= 0.0
7	1.0	0.0	0.0	−1.0	0.0	0.0	0.0	1.0	1.0	= 0.0
8	0.0	0.0	0.0	0.0	0.0	49.2	49.2	49.2	49.2	Costs
9	0.646	0.035	−0.001	0.646	0.0	0.017	0.018	0.0	0.0	Solution

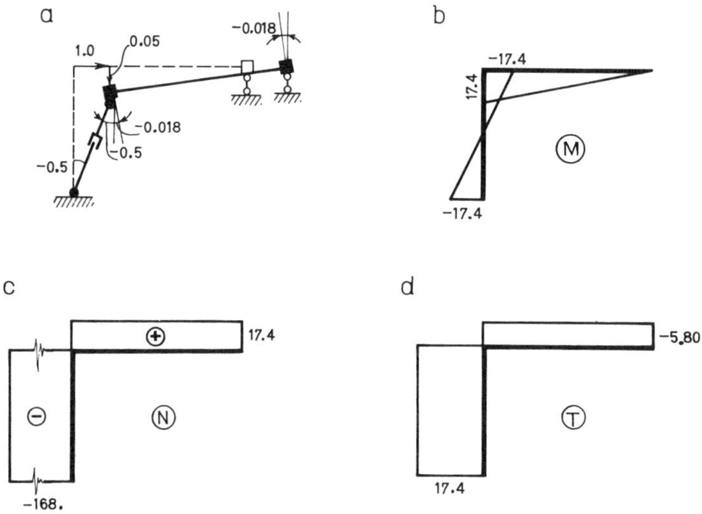

Figure 13.4. Plastic Collapse of Frame (Interaction between Bending Moments and Axial Forces Taken into Account According to Rhombic Yield Curve): (a) Collapse Mechanism. (b) Bending Moments. (c) Axial Forces. (d) Shear Forces.

Finally, let us solve the same example by using the hexagonal interaction curve. The matrices **s**, $\mathbf{p}_0$, $\dot{\mathbf{q}}$, $\dot{\mathbf{w}}$, and **C** remain the same. The full-size matrix **N** will have 24 rows [generated by two submatrices of type (11.49)]. Knowledge of the signs of the internal forces allows us to reduce that number to 8. It is clear that only edges IV and V can be active for element 1, whereas edges I and II must be used for element 2. Taking $\alpha = 0.15$, we compute the components of the outward normal for edge II [see Eqs. (11.48)]:

$$\xi = \frac{1}{1.0 + (1.0 - 0.15)} = 0.762,$$

$$\eta = (1.0 - 0.15)0.762 = 0.648.$$

The distance of that edge from the origin is $k^{\mathrm{II}} = 0.762 \times 49.2 = 37.5$. Thus, we have the following description of the frame:

$$\mathbf{f} = \{f_{a1}^{\mathrm{III}}, f_{a1}^{\mathrm{IV}}, f_{b1}^{\mathrm{III}}, f_{b1}^{\mathrm{IV}}, f_{a2}^{\mathrm{I}}, f_{a2}^{\mathrm{II}}, f_{b2}^{\mathrm{I}}, f_{b2}^{\mathrm{II}}\},$$

$$\dot{\boldsymbol{\lambda}} = \{\dot{\lambda}_{a1}^{\mathrm{III}}, \dot{\lambda}_{a1}^{\mathrm{IV}}, \dot{\lambda}_{b1}^{\mathrm{III}}, \dot{\lambda}_{b1}^{\mathrm{IV}}, \dot{\lambda}_{a2}^{\mathrm{I}}, \dot{\lambda}_{a2}^{\mathrm{II}}, \dot{\lambda}_{b2}^{\mathrm{I}}, \dot{\lambda}_{b2}^{\mathrm{II}}\},$$

$$\mathbf{k} = \{37.5, 49.2, 37.5, 49.2, 49.2, 37.5, 49.2, 37.5\},$$

Table 13.5. Ultimate Load of Frame—Axial Forces Taken Into Account According to Hexagonal Interaction Curve

a. Static Approach

	μ	$m_a^{(1)}$	$m_b^{(1)}$	n_1	$m_a^{(2)}$	$m_b^{(2)}$	n_2	
	1	2	3	4	5	6	7	
1	0.0	0.027	0.027	0.0	0.0	0.0	1.0	= 0.0
2	10.0	0.0	0.0	1.0	−0.018	−0.018	0.0	= 0.0
3	0.0	0.0	−1.0	0.0	−1.0	0.0	0.0	= 0.0
4	1.0	0.0	0.0	0.0	0.0	0.0	−1.0	= 0.0
5	0.0	0.0	0.0	0.0	0.0	−1.0	0.0	= 0.0
6	0.0	−0.648	0.0	−0.762	0.0	0.0	0.0	$\leqslant$ 37.5
7	0.0	−1.0	0.0	0.0	0.0	0.0	0.0	$\leqslant$ 49.2
8	0.0	0.0	−0.648	−0.762	0.0	0.0	0.0	$\leqslant$ 37.5
9	0.0	0.0	−1.0	0.0	0.0	0.0	0.0	$\leqslant$ 49.2
10	0.0	0.0	0.0	0.0	1.0	0.0	0.0	$\leqslant$ 49.2
11	0.0	0.0	0.0	0.0	0.648	0.0	0.762	$\leqslant$ 37.5
12	0.0	0.0	0.0	0.0	0.0	1.0	0.0	$\leqslant$ 49.2
13	0.0	0.0	0.0	0.0	0.0	0.648	0.762	$\leqslant$ 37.5
14	1.0	0.0	0.0	0.0	0.0	0.0	0.0	Costs
15	1.934	−35.9	−35.9	−18.7	35.9	0.0	1.934	Solution

b. Kinematic Approach

	$\dot{w}_x^{(2)}$	$\dot{w}_y^{(2)}$	$\dot{w}_z^{(2)}$	$\dot{w}_x^{(3)}$	$\dot{w}_z^{(3)}$	$\dot{\lambda}_{a1}^{III}$	$\dot{\lambda}_{a1}^{IV}$	$\dot{\lambda}_{b1}^{III}$	$\dot{\lambda}_{b1}^{IV}$	$\dot{\lambda}_{a2}^{I}$	$\dot{\lambda}_{a2}^{II}$	$\dot{\lambda}_{b2}^{I}$	$\dot{\lambda}_{b2}^{II}$	
	1	2	3	4	5	6	7	8	9	10	11	12	13	
1	0.0	10.0	0.0	1.0	0.0	0.0	0.0	0.0	0.0	0.0	0.0	0.0	0.0	= 0.0
2	0.027	0.0	0.0	0.0	0.0	−0.648	−1.0	0.0	0.0	0.0	0.0	0.0	0.0	= 0.0
3	0.027	0.0	−1.0	0.0	0.0	0.0	0.0	−0.648	−1.0	0.0	0.0	0.0	0.0	= 0.0
4	0.0	1.0	0.0	0.0	0.0	−0.762	0.0	−0.762	0.0	0.0	0.0	0.0	0.0	= 0.0
5	0.0	−0.018	−1.0	0.0	0.0	0.0	0.0	0.0	0.0	1.0	0.648	0.0	0.0	= 0.0
6	0.0	−0.018	0.0	0.0	−1.0	0.0	0.0	0.0	0.0	0.0	0.0	1.0	0.648	= 0.0
7	1.0	0.0	0.0	−1.0	0.0	0.0	0.0	0.0	0.0	0.0	0.762	0.0	0.762	= 0.0
9	0.0	0.0	0.0	0.0	0.0	37.5	49.2	37.5	49.2	49.2	37.5	49.2	37.5	Costs
9	0.607	0.039	−0.001	0.607	−0.001	0.025	0.0	0.026	0.0	0.0	0.0	0.0	0.0	Solution

$$\mathbf{N} = \left[\begin{array}{cccccccc} -0.648 & -1.0 & 0.0 & 0.0 & 0.0 & 0.0 & 0.0 & 0.0 \\ 0.0 & 0.0 & -0.648 & -1.0 & 0.0 & 0.0 & 0.0 & 0.0 \\ -0.762 & 0.0 & -0.762 & 0.0 & 0.0 & 0.0 & 0.0 & 0.0 \\ \hdashline 0.0 & 0.0 & 0.0 & 0.0 & 1.0 & 0.648 & 0.0 & 0.0 \\ 0.0 & 0.0 & 0.0 & 0.0 & 0.0 & 0.762 & 1.0 & 0.648 \\ 0.0 & 0.0 & 0.0 & 0.0 & 0.0 & 0.0 & 0.0 & 0.762 \end{array}\right].$$

The data for this model are given in Table 13.5. The ultimate load factor, as expected, is slightly greater than that obtained via the rhombic interaction curve: $\mu_* = 1.934$. The collapse mechanism and the distribution of internal forces are practically the same as shown in Figs. 13.4a, b, c.

Exercise 13.1

Find the ultimate load factor μ_* for the truss in Fig. 11.10. Take the yield axial forces to be $N^0 = 90.0\,\text{kN}$, $\tilde{N}^0 = 30.0\,\text{kN}$ common for all bars, and use the matrix $\mathbf{C}^{\mathrm{T}}$ given in Table S.1 in Chapter 20.

Exercise 13.2

Find the ultimate load factor μ_* for the beam in Fig. 11.11. Assume a uniform rectangular cross section with height $H = 10.0$ cm and width $B = 4.0$ cm. Take the yield stress of the material to be $\sigma_0 = 100$ MPa, and use the matrix $\mathbf{C}^{\mathrm{T}}$ given in Table S.2 in Chapter 20.

13.3. Optimal Plastic Design

13.3.1. General Formulation

In the previous section we explained how the safety margin or the ultimate load factor can be calculated for a given structure. Now we are going to deal with the opposite problem: given the safety margin or the ultimate load, find the optimal structure based on some criterion. According to the general concept of the theory of load-carrying capacity, we neglect the elastic deformability of the structure and adopt the rigid, perfectly plastic model for the material. This model lets us keep the model in the frame of LP theory and puts certain limitations upon practical applicability of the optimal solutions obtained. Usually these solutions serve as useful guides at the preliminary stage of structural design. Before coming to the final design, we have to consider many additional constraints, such as stiffness under service loads or local and global stability, that are not taken into account in optimal plastic design. Nevertheless, the possibility of obtaining optimal solutions via plasticity at low computational cost deserves consideration.

Before we come to real optimization, we formulate a scaling problem closely related to that of the ultimate load factor. Imagine that we have a structure with its plastic properties described by the given matrix $\mathbf{k}_0$. That structure is able to carry the given load $\mathbf{p}_*$. If we try to diminish proportionally the entries of $\mathbf{k}_0$, then sooner or later the structure will collapse. It might be instructive to know how far such a proportional deterioration of the plastic moduli can proceed. Let

$$\mathbf{k} = \rho \mathbf{k}_0, \tag{13.16}$$

where ρ is the scaling factor. Given the load $\mathbf{p}_*$ and the reference design $\mathbf{k}_0$, we want to find the least value of the scaling factor ρ_* that corresponds to plastic collapse of the structure.

Exploiting the analogy with the problem of the ultimate load factor, we put the governing relations into a table:

$$\begin{array}{rcc:cccll}
 & \dot{\boldsymbol{\lambda}}^{\mathrm{T}} & \dot{\mathbf{w}}^{\mathrm{T}} & \mathbf{s}^{\mathrm{T}} & \rho & 1 & & \\
\hline
\nabla f_\lambda = & \mathbf{0} & \mathbf{0} & -\mathbf{N}^{\mathrm{T}} & \mathbf{k}_0 & \mathbf{0} & \geqslant \mathbf{0} & (l) \\
\nabla f_w = & \mathbf{0} & \mathbf{0} & \mathbf{C}^{\mathrm{T}} & \mathbf{0} & -\mathbf{p}_* & = \mathbf{0} & (n) \\
\hdashline
\nabla f_s = & -\mathbf{N} & \mathbf{C} & \mathbf{0} & \mathbf{0} & \mathbf{0} & = \mathbf{0} & (m) \\
\hline
 & (l) & (n) & (m) & (1) & (1) & &
\end{array}$$

This table must be supplemented by a fourth row, generated by the derivative ∇f_ρ. Following the same arguments as in the ultimate load problem, we obtain the final table:

$$\begin{array}{rcc:cccll}
 & \dot{\boldsymbol{\lambda}}^{\mathrm{T}} & \dot{\mathbf{w}}^{\mathrm{T}} & \mathbf{s}^{\mathrm{T}} & \rho & 1 & & \\
\hline
\nabla f_\lambda = & \mathbf{0} & \mathbf{0} & -\mathbf{N}^{\mathrm{T}} & \mathbf{k}_0 & \mathbf{0} & \geqslant \mathbf{0} & (l) \\
\nabla f_w = & \mathbf{0} & \mathbf{0} & \mathbf{C}^{\mathrm{T}} & \mathbf{0} & -\mathbf{p}_* & = \mathbf{0} & (n) \\
\hdashline
\nabla f_s = & -\mathbf{N} & \mathbf{C} & \mathbf{0} & \mathbf{0} & \mathbf{0} & = \mathbf{0} & (m) \\
\nabla f_\rho = & \mathbf{k}_0^{\mathrm{T}} & \mathbf{0} & \mathbf{0} & 0 & -1 & = 0 & (1) \\
\hline
 & (l) & (n) & (m) & (1) & (1) & &
\end{array} \tag{13.17}$$

The last row contains the equation

$$\mathbf{k}_0^{\mathrm{T}} \dot{\boldsymbol{\lambda}} = 1, \tag{13.18}$$

which normalizes the dissipated power $\dot{D}$ and, hence, allows us to obtain the finite motion of the collapse mechanism.

The linear part of the governing relations must be supplemented by the complementarity rule Eq. $(11.54)_4$. That rule can be obtained as the KKT condition $\dot{\boldsymbol{\lambda}}^{\mathrm{T}}\boldsymbol{\nabla} f_\lambda = 0$ corresponding to $\dot{\boldsymbol{\lambda}} \geqslant \mathbf{0}$. Applying further the equivalence scheme of Section 10.5, we obtain

the potential function

$$f(\dot{\boldsymbol{\lambda}}, \dot{\mathbf{w}}, \mathbf{s}, \rho) = -\dot{\boldsymbol{\lambda}}^{\mathrm{T}}\mathbf{N}^{\mathrm{T}}\mathbf{s} + \dot{\mathbf{w}}^{\mathrm{T}}\mathbf{C}^{\mathrm{T}}\mathbf{s} + \rho\mathbf{k}_0^{\mathrm{T}}\dot{\boldsymbol{\lambda}} - \mathbf{p}_*^{\mathrm{T}}\dot{\mathbf{w}} - \rho, \tag{13.19}$$

the saddle point problem

$$\min_{\dot{\mathbf{w}},\dot{\boldsymbol{\lambda}}\geqslant\mathbf{0}} \max_{\mathbf{s},\rho} f(\dot{\boldsymbol{\lambda}}, \dot{\mathbf{w}}, \mathbf{s}, \rho), \tag{13.20}$$

and the dual pair of LP problems

$$\min\{-\mathbf{p}_*^{\mathrm{T}}\dot{\mathbf{w}} \,|\, \mathbf{C}\dot{\mathbf{w}} - \mathbf{N}\dot{\boldsymbol{\lambda}} = \mathbf{0}, \mathbf{k}_0^{\mathrm{T}}\dot{\boldsymbol{\lambda}} = 1, \dot{\boldsymbol{\lambda}} \geqslant \mathbf{0}\},$$
$$\max\{-\rho \,|\, \rho\mathbf{k}_0 - \mathbf{N}^{\mathrm{T}}\mathbf{s} \geqslant \mathbf{0}, \mathbf{C}^{\mathrm{T}}\mathbf{s} = \mathbf{p}_*\},$$

or, equivalently,

$$\begin{aligned} &\min\{\rho \,|\, \rho\mathbf{k}_0 - \mathbf{N}^{\mathrm{T}}\mathbf{s} \geqslant \mathbf{0}, \mathbf{C}^{\mathrm{T}}\mathbf{s} = \mathbf{p}_*\}, \\ &\max\{\mathbf{p}_*^{\mathrm{T}}\dot{\mathbf{w}} \,|\, \mathbf{C}\dot{\mathbf{w}} - \mathbf{N}\dot{\boldsymbol{\lambda}} = \mathbf{0}, \mathbf{k}_0^{\mathrm{T}}\dot{\boldsymbol{\lambda}} = 1, \dot{\boldsymbol{\lambda}} \geqslant \mathbf{0}\}. \end{aligned} \tag{13.21}$$

The structure can be brought to plastic collapse in two ways: increasing the load or decreasing the plastic resistance. If this is done via scaling factors μ and ρ for the same reference matrices $\mathbf{p}_0$ and k_0, then the extreme values of those factors are mutually reciprocal: $\rho_* = 1/\mu_*$.

The scaling problem (13.21), though computationally equivalent to the ultimate load problem (13.10), allowed us to make the transfer from analysis to design of the structure. We are ready now to drop the proportionality rule (13.16) and to regard each component of $\mathbf{k}$ as an independent variable. Then we must replace the minimization of ρ by the minimization of a cost function $\Gamma = \Gamma(\mathbf{k})$. To keep things linear, we take

$$\Gamma = \mathbf{b}^{\mathrm{T}}\mathbf{k}, \tag{13.22}$$

where $\mathbf{b} \in E^l$ contains given cost factors.

Let us try to modify the table of Eq. (13.17) in such a way that it corresponds to the optimization problem. First, instead of the scalar ρ, we now have the matrix of unknown plastic resistances $\mathbf{k}$. That matrix should appear explicitly in the stress admissibility conditions: hence, $\mathbf{k}_0$ must be replaced in the first row of the table by the unit matrix $\mathbf{I}$ of order l. The matrix of cost factors $\mathbf{b}$ should replace unity in the last row, because we expect $\mathbf{b}$ to appear in the expression for the minimized function.

After such changes, Eq. (13.17) is replaced by the following one:

$$
\begin{array}{r|ccccc|ll}
 & \dot{\boldsymbol{\lambda}}^{\mathrm{T}} & \dot{\mathbf{w}}^{\mathrm{T}} & \mathbf{s}^{\mathrm{T}} & \mathbf{k}^{\mathrm{T}} & 1 & & \\
\hline
\boldsymbol{\nabla} f_\lambda = & \mathbf{0} & \mathbf{0} & -\mathbf{N}^{\mathrm{T}} & \mathbf{I} & \mathbf{0} & \geqslant \mathbf{0} & (l) \\
\boldsymbol{\nabla} f_w = & \mathbf{0} & \mathbf{0} & \mathbf{C}^{\mathrm{T}} & \mathbf{0} & -\mathbf{p}_* & = \mathbf{0} & (n) \\
\hdashline
\boldsymbol{\nabla} f_s = & -\mathbf{N} & \mathbf{C} & \mathbf{0} & \mathbf{0} & \mathbf{0} & = \mathbf{0} & (m) \\
\boldsymbol{\nabla} f_k = & \mathbf{I} & \mathbf{0} & \mathbf{0} & \mathbf{0} & -\mathbf{b} & = \mathbf{0} & (l) \\
\hline
 & (l) & (n) & (m) & (l) & (1) & &
\end{array}
\qquad (13.23)
$$

Applying further the equivalence scheme of Section 10.5, we obtain the potential function

$$f(\dot{\boldsymbol{\lambda}}, \dot{\mathbf{w}}, \mathbf{s}, \mathbf{k}) = -\dot{\boldsymbol{\lambda}}^{\mathrm{T}}\mathbf{N}^{\mathrm{T}}\mathbf{s} + \dot{\boldsymbol{\lambda}}^{\mathrm{T}}\mathbf{I}\mathbf{k} + \dot{\mathbf{w}}^{\mathrm{T}}\mathbf{C}^{\mathrm{T}}\mathbf{s} - \mathbf{p}_*^{\mathrm{T}}\dot{\mathbf{w}} - \mathbf{b}^{\mathrm{T}}\mathbf{k}, \qquad (13.24)$$

the saddle point problem

$$\min_{\dot{\mathbf{w}}, \dot{\boldsymbol{\lambda}} \geqslant \mathbf{0}} \max_{\mathbf{s}, \mathbf{k} \geqslant \mathbf{0}} f(\dot{\boldsymbol{\lambda}}, \dot{\mathbf{w}}, \mathbf{s}, \mathbf{k}), \qquad (13.25)$$

and the dual pair of LP problems

$$
\begin{aligned}
&\min\{\mathbf{b}^{\mathrm{T}}\mathbf{k} \,|\, \mathbf{k} - \mathbf{N}^{\mathrm{T}}\mathbf{s} \geqslant \mathbf{0}, \mathbf{C}^{\mathrm{T}}\mathbf{s} = \mathbf{p}_*, \mathbf{k} \geqslant \mathbf{0}\}, \\
&\max\{\mathbf{p}_*^{\mathrm{T}}\dot{\mathbf{w}} \,|\, \mathbf{C}\dot{\mathbf{w}} - \mathbf{N}\dot{\boldsymbol{\lambda}} = \mathbf{0}, \mathbf{0} \leqslant \dot{\boldsymbol{\lambda}} \leqslant \mathbf{b}\}.
\end{aligned}
\qquad (13.26)
$$

Consider the physical interpretation of the preceding LP problems. In $(13.26)_1$ we are looking for the design $\mathbf{k}_*$ that minimizes the cost Γ and is statically admissible: there is a stress state $\mathbf{s}_*$ equilibrated with the load $\mathbf{p}_*$ and satisfying the yield inequalities (13.2). The solution $\mathbf{k}_*$, $\mathbf{s}_*$ of problem $(13.26)_1$ exists for any $\mathbf{p}_*$ because $\mathbf{s}_*$ is supplied by the general solution of the equilibrium equation [Eq. (11.7)], and positive entries of $\mathbf{k}_*$ can be taken sufficiently large to ensure stress admissibility. It is reasonable to assume that the cost factors $\mathbf{b}$ are nonnegative. Then the minimum $\Gamma_* = \mathbf{b}^T\mathbf{k}_*$ is bounded from below by zero.

The theory of duality tells us that problem $(13.26)_2$ also has the solution $\dot{\mathbf{w}}$, $\dot{\boldsymbol{\lambda}}_*$ and, moreover, that $\mathbf{p}_*^T\dot{\mathbf{w}} = \mathbf{b}^T\mathbf{k}_*$. Obviously, $\dot{\mathbf{w}}_*$ describes the collapse mechanism associated with the design $\mathbf{k}_*$. The motion of that mechanism, though, is restricted differently than in the model $(13.21)_2$. Now each plastic multiplier $\dot{\lambda}_i$ has an upper bound equal to the cost factor b_i.

Remembering that the scalar product $\mathbf{k}_*^T \cdot \dot{\boldsymbol{\lambda}}_*$ gives the power dissipated during collapse, and that the product $\mathbf{p}_*^T \cdot \dot{\mathbf{w}}_*$ stands for the power produced by the load, we can associate the following interpretation with the LP problems (13.26) [Čyras (1969)]:

Static Theorem of Optimal Plastic Design. Of all statically admissible designs, the optimal one corresponds to the minimum dissipated energy [problem $(13.26)_1$].

Kinematic Theorem of Optimal Plastic Design. Of all kinematically admissible collapse mechanisms, the true one for an optimal structure corresponds to the maximum external power [problem $(13.26)_2$].

For models (13.10) or (13.21) we were looking for the safety margin of a given structure. Model (13.26) enables us to find an optimal distribution of plastic resistances $\mathbf{k}_*$, if the ultimate load $\mathbf{p}_*$ is given. In practice, we have a service load $\mathbf{p}_0$ and we choose a value of the overloading factor μ_*. We thus obtain $\mathbf{p}_*$ according to Eq. (13.1).

13.3.2. Design Variables, Optimality Criterion, and Minimum Volume

In plasticity as in elasticity, we expect the number of true design variables to be much smaller than the number of yield conditions l. Therefore, we let $\mathbf{k}$ be uniquely defined by the t-dimensional matrix of design variables $\boldsymbol{\psi}$:

$$\mathbf{k} = \mathbf{G}^T\boldsymbol{\psi}, \tag{13.27}$$

where $\mathbf{G}$ is a given $t \times l$ matrix.

Assuming the cost to be a linear function of the design variables [compare Eq. (12.25)] and taking into account Eq. (13.22), we obtain the relation between the cost factors b_i related to the plastic resistances k_i and the cost factors c_j related to the design variables ψ_j:

$$\mathbf{c} = \mathbf{G}\mathbf{b}. \tag{13.28}$$

We are ready now to replace **k** by $\boldsymbol{\psi}$ in the table of Eq. (13.23). The result is

$$\begin{array}{r|cc|ccc|ll} & \dot{\boldsymbol{\lambda}}^{\mathrm{T}} & \dot{\mathbf{w}}^{\mathrm{T}} & \mathbf{s}^{\mathrm{T}} & \boldsymbol{\psi}^{\mathrm{T}} & 1 & & \\ \hline \nabla f_{\lambda} = & \mathbf{0} & \mathbf{0} & -\mathbf{N}^{\mathrm{T}} & \mathbf{G}^{\mathrm{T}} & \mathbf{0} & \geqslant \mathbf{0} & (l) \\ \nabla f_{w} = & \mathbf{0} & \mathbf{0} & \mathbf{C}^{\mathrm{T}} & \mathbf{0} & -\mathbf{p}_{*} & = \mathbf{0} & (n) \\ \hline \nabla f_{s} = & -\mathbf{N} & \mathbf{C} & \mathbf{0} & \mathbf{0} & \mathbf{0} & = \mathbf{0} & (m) \\ \nabla f_{\psi} = & \mathbf{G} & \mathbf{0} & \mathbf{0} & \mathbf{0} & -\mathbf{c} & = \mathbf{0} & (t) \\ \hline & (l) & (n) & (m) & (t) & (1) & & \end{array} \qquad (13.29)$$

According to the equivalence scheme of Section 10.5, this table corresponds to

the potential function

$$f(\dot{\boldsymbol{\lambda}}, \dot{\mathbf{w}}, \mathbf{s}, \boldsymbol{\psi}) = -\dot{\boldsymbol{\lambda}}^{\mathrm{T}}\mathbf{N}^{\mathrm{T}}\mathbf{s} + \dot{\boldsymbol{\lambda}}^{\mathrm{T}}\mathbf{G}^{\mathrm{T}}\boldsymbol{\psi} + \dot{\mathbf{w}}^{\mathrm{T}}\mathbf{C}^{\mathrm{T}}\mathbf{s} - \mathbf{p}_{*}^{\mathrm{T}}\dot{\mathbf{w}} - \mathbf{c}^{\mathrm{T}}\boldsymbol{\psi}, \qquad (13.30)$$

the saddle point problem

$$\min_{\dot{\mathbf{w}}, \dot{\boldsymbol{\lambda}} \geqslant \mathbf{0}} \max_{\mathbf{s}, \boldsymbol{\psi} \geqslant \mathbf{0}} f(\dot{\boldsymbol{\lambda}}, \dot{\mathbf{w}}, \mathbf{s}, \boldsymbol{\psi}), \qquad (13.31)$$

and the dual pair of LP problems

$$\begin{aligned} &\min\{\mathbf{c}^{\mathrm{T}}\boldsymbol{\psi} \mid \mathbf{G}^{\mathrm{T}}\boldsymbol{\psi} - \mathbf{N}^{\mathrm{T}}\mathbf{s} \geqslant \mathbf{0}, \mathbf{C}^{\mathrm{T}}\mathbf{s} = \mathbf{p}_{*}, \boldsymbol{\psi} \geqslant \mathbf{0}\}, \\ &\max\{\mathbf{p}_{*}^{\mathrm{T}}\dot{\mathbf{w}} \mid \mathbf{C}\dot{\mathbf{w}} - \mathbf{N}\dot{\boldsymbol{\lambda}} = \mathbf{0}, \mathbf{G}\dot{\boldsymbol{\lambda}} \leqslant \mathbf{c}, \dot{\boldsymbol{\lambda}} \geqslant \mathbf{0}\}. \end{aligned} \qquad (13.32)$$

Note that the upper bounds c_j in the kinematic problem $(13.32)_2$ are now imposed not on individual entries of $\dot{\boldsymbol{\lambda}}$ but on their linear combinations:

$$\mathbf{G}\dot{\boldsymbol{\lambda}} \leqslant \mathbf{c}. \qquad (13.33)$$

That constraint leads to the optimality criterion for optimal plastic design.

As shown in Volume 1, this criterion says that the optimal structure is characterized by uniform density of the dissipated energy [compare Eq. (3.25) in Volume 1]. The density of dissipated energy $\dot{d}_i$, evaluated in that part of the structure governed by the ith design variable, is defined as the ratio of the dissipated energy $\dot{D}_i$ to the cost Γ_i. Hence, the optimality criterion is

$$\dot{d}_i = \dot{D}_i / \Gamma_i = \text{const.} \qquad (13.34)$$

Using Eqs. (13.12), (13.22), (13.27), and (13.28), we obtain

$$\dot{d}_i = \frac{\psi_i \sum_{j=1}^{l} G_{ij}\dot{\lambda}_j}{\psi_i c_i}.$$

The KKT condition says that if $\psi_{i*} > 0$ then the ith condition (13.33) must be satisfied as an equality:

$$\sum_{j=1}^{l} G_{ij}\dot{\lambda}_j = c_i.$$

Hence, $\dot{d}_{i*} = 1$ in all nonvanishing elements of the optimal structure.

In the previous section we introduced somewhat arbitrarily the linear cost function Γ. It turned out that such a function is a measure of the dissipated energy. Our aim now is to find the relations between the minimization of Γ and such common goals of optimal structural design as the minimization of the volume or weight of the structure.

In dealing with elastic structures in Section 12.3, we first expressed the volume as a linear function of the cross-sectional areas [Eq. (13.16)]. Then we decreased the number of unknowns by the linkage relation (13.17). The same type of reasoning applies for optimal plastic design, though it is convenient to take the plastic resistances instead of the cross-sectional areas as the design variables.

We begin with the minimum volume design of trusses. To simplify derivations, we assume that the structure is made of a homogeneous material with yield stress σ_0. Given σ_0 and the length L_i of the ith element, we can express the cross-sectional area A_i and, hence, the volume V_i of that element linearly in terms of the yield force N_i^0:

$$V_i = A_i L_i = (L_1/\sigma_0) N_i^0. \tag{13.35}$$

Taking

$$V_0 = A_0 L_0 = P_0 L_0/\sigma_0 \tag{13.36}$$

as the reference volume, we convert Eq. (13.35) into the nondimensional form

$$v_i = V_i/V_0 = l_i n_i^0, \tag{13.37}$$

where $l_i = L_i/L_0$ and $n_i^0 = N_i^0/P_0$.

Let $\mathbf{l} = \{l_i\}$ and $\mathbf{n}^0 = \{n_i^0\}$, $i = 1, 2, \ldots, m$. Then the total volume of the truss can be written as

$$v = \mathbf{l}^{\mathrm{T}}\mathbf{n}^0. \tag{13.38}$$

It is easy to establish a linear relation between $\mathbf{n}^0$ and $\mathbf{k}$:

$$\mathbf{k} = \mathbf{R}^{\mathrm{T}}\mathbf{n}^0. \tag{13.39}$$

Here $\mathbf{R}^{\mathrm{T}}$ is a quasidiagonal matrix composed of the submatrices $\bar{\mathbf{k}}_i$ related to individual elements:

$$\mathbf{R}^{\mathrm{T}} = \lceil \bar{k}_1, \bar{k}_2, \ldots, \bar{k}_m \rfloor. \tag{13.40}$$

Moreover, not every yield force is to be regarded as an independent design variable. Therefore, we introduce the linkage relation

$$\mathbf{n}^0 = \mathbf{S}^{\mathrm{T}}\boldsymbol{\psi}, \tag{13.41}$$

similar to Eq. (12.23). The linkage matrix $\mathbf{S}$ has the dimensions $t \times m$ and is composed of zeros or ones [compare Eq. (12.24)].

Substituting the right side of Eq. (13.41) into Eq. (13.38) and taking $\Gamma = v$, we obtain the cost function defined by Eq. (12.25), with

$$\mathbf{c} = \mathbf{S}\mathbf{l}. \tag{13.42}$$

Further, it follows from Eqs. (13.39) and (13.41) that the matrix $\mathbf{G}$ in Eq. (13.27) is the product of the matrices $\mathbf{R}$ and $\mathbf{S}$:

$$\mathbf{G} = \mathbf{S}\mathbf{R}. \tag{13.43}$$

We shall see that for simple structures such a multiplication can be avoided, because it is easy to construct $\mathbf{G}$ by inspection.

The foregoing derivation remains valid for flexural systems with a linear relation between the area and the yield moment of the cross section. For example, such a relation holds for the ideal sandwich cross section shown in the last row of Table 11.2 if the distance H between outer layers is fixed. Let the thickness H' of such a layer also be given; the width B_i of the cross section is subject to optimization. Taking

$$M_i^0 = \sigma_0 H B_i H'$$

from Table 11.2, we calculate the area

$$A_i = 2B_iH' = \frac{2}{\sigma_0 H} M_i^0$$

and the volume

$$V_i = \frac{2L_i}{\sigma_0 H} M_i^0$$

of the outer layers of the ith element. Transforming the last relation into nondimensional form, we obtain

$$v_i = \frac{2L_0}{H} l_i m_i^0 .$$

The constant $2L_0/H$ is immaterial for optimization. Therefore, we can drop it and apply formulas (13.38)–(13.42), with $\mathbf{n}^0$ replaced by $\mathbf{m}^0$.

The same formulas apply for the sandwich cross section with constant H and B and variable H', as well as for the rectangular cross section with constant H and variable B. If the structure is made of I sections or box sections, then model (13.32) can be used for the minimum volume design with the same degree of accuracy obtained when we replace the exact yield curves by the piecewise linear approximations of rhombic or hexagonal shape.

Multiplying V_i by the density of the material, we obtain the weight of the ith element of the structure. If the material is homogeneous, then that density is identical for all elements, and the total weight differs from the total volume by a constant factor. Therefore, the minimization of volume is equivalent to the minimization of weight for such structures. However, for structures made of nonhomogeneous material, the ratios between densities should be taken into account when the costs c_j are calculated.

13.3.3. Example of a Truss

Let us consider the three-bar truss of Fig. 11.1. In Section 13.2.2 we calculated the ultimate load of such a truss manufactured from tubes. Keeping the same data regarding the load and the plastic resistances $p_x^{(1)} = 2.0$, $p_y^{(1)} = 1.0$, $n^0 = 2.94$, and $\tilde{n}^0 = 2.05$, we first formulate the scaling problem (13.21). Since all matrices remain the same, it is a straightforward task to fill the initial simplex Tables 13.6a, b. The solutions are given in the last rows of the tables. Note that the collapse mechanism is identical to

Table 13.6. Scaling of Three-Bar Truss

a. Static Approach

	ρ	n_1	n_2	n_3	
	1	2	3	4	
1	0.0	0.707	0.0	−0.707	= 2.0
2	0.0	0.707	1.0	0.707	= 1.0
3	2.94	−1.0	0.0	0.0	⩾ 0.0
4	2.05	1.0	0.0	0.0	⩾ 0.0
5	2.94	0.0	−1.0	0.0	⩾ 0.0
6	2.05	0.0	1.0	0.0	⩾ 0.0
7	2.94	0.0	0.0	−1.0	⩾ 0.0
8	2.05	0.0	0.0	1.0	⩾ 0.0
9	1.0	0.0	0.0	0.0	Min
10	0.567	1.667	0.643	−1.162	Solution

b. Kinematic Approach

	$\dot{w}_x^{(1)}$	$\dot{w}_y^{(1)}$	$\dot{\lambda}_1^{\rm I}$	$\dot{\lambda}_1^{\rm II}$	$\dot{\lambda}_2^{\rm I}$	$\dot{\lambda}_2^{\rm II}$	$\dot{\lambda}_3^{\rm I}$	$\dot{\lambda}_3^{\rm II}$	
	1	2	3	4	5	6	7	8	
1	0.0	0.0	2.94	2.05	2.94	2.05	2.94	2.05	= 1.0
2	0.707	0.707	−1.0	1.0	0.0	0.0	0.0	0.0	= 0.0
3	0.0	1.0	0.0	0.0	−1.0	1.0	0.0	0.0	= 0.0
4	−0.707	0.707	0.0	0.0	0.0	0.0	−1.0	1.0	= 0.0
5	2.0	1.0	0.0	0.0	0.0	0.0	0.0	0.0	Max
6	0.284	0.0	0.20	0.0	0.0	0.0	0.0	0.20	Solution

that obtained via the model (13.10) and depicted in Fig. 13.2a. On the other hand, the stress state equals the stress state of Fig. 13.2b multiplied by $\rho_* = 0.566$, which is the exact reciprocal of the ultimate load factor $\mu_* = 1.766$. That means that there are two possible ways to bring the truss into a state of plastic collapse: either the load has to be increased 1.766 times or the plastic resistances have to be reduced 0.566 times.

Let us now find the optimal design of the symmetric truss. We take n_1^0 and n_2^0 as the design variables, since $n_3^0 = n_1^0$. Hence, $\boldsymbol{\psi} = \{n_1^0, n_2^0\}$ in the

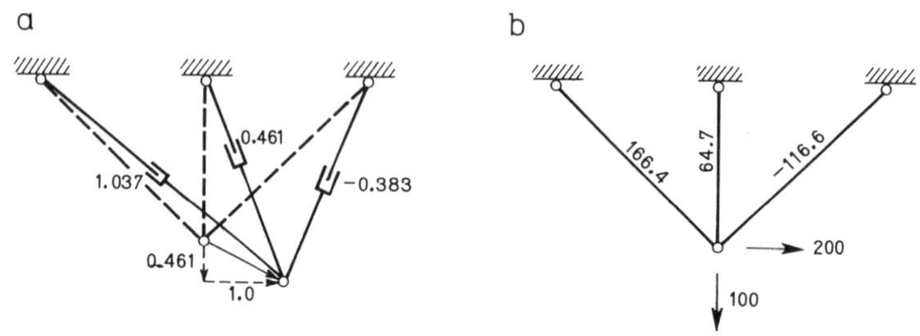

Figure 13.5. Truss Scaled to Plastic Collapse: (a) Collapse Mechanism. (b) Axial Forces.

model (13.32). The matrices **s**, **p**, $\dot{\mathbf{w}}$, $\dot{\boldsymbol{\lambda}}$, **C**, and **N** remain the same as in the scaling problem. We need, however, a new matrix **G** that connects the design variables with the plastic resistances. According to Eq. (13.42), that matrix is the product of the matrices **R** and **S**. The former comes from the relation (13.39) between the entries of

$$\mathbf{k} = \{k_1^{\mathrm{I}}, k_1^{\mathrm{II}}, k_2^{\mathrm{I}}, k_2^{\mathrm{II}}, k_1^{\mathrm{I}}, k_2^{\mathrm{II}}\}$$

and the entries of

$$\mathbf{n}_0^0\{n_1^0, n_2^0, n_3^0\}.$$

The submatrices $\bar{\mathbf{k}}_1, \bar{\mathbf{k}}_2, \bar{\mathbf{k}}_3$ that enter **R** according to Eq. (13.40) were given for the pin-ended bars in Section 11.5.2. Taking into account possible buckling, we have $\xi = 0.7$. Therefore,

$$\mathbf{R} = \left[\begin{array}{cc|cc|cc} 1 & 0.7 & 0 & 0 & 0 & 0 \\ \hline 0 & 0 & 1 & 0.7 & 0 & 0 \\ \hline 0 & 0 & 0 & 0 & 1 & 0.7 \end{array}\right].$$

The matrix **S** relates the design variables to the axial forces [see Eq. (13.41)]. In our case that matrix simply reflects the fact that the yield forces of the elements 1 and 3 are governed by the same ψ_1:

$$\mathbf{S} = \begin{bmatrix} 1 & 0 \\ 0 & 1 \\ 1 & 0 \end{bmatrix}.$$

Performing the required multiplication, we obtain the final linkage matrix:

$$\mathbf{G} = \begin{bmatrix} 1 & 0.7 & 0 & 0 & 1 & 0.7 \\ 0 & 0 & 1 & 0.7 & 0 & 0 \end{bmatrix}.$$

The full derivation of **G** was carried out, though, looking at the computational scheme of the truss, we could assemble **G** directly. The formal way via the intermediate matrices **R**, **S** is convenient for automatic computation.

The matrix of nondimensional lengths of elements is $\mathbf{l} = \{1.414, 1.0, 1.414\}$. Multiplying $\mathbf{l}$ by **S**, we obtain the matrix of cost factors $\mathbf{c} = \{2.828, 1.0\}$. Its first entry is the total length of elements governed by ψ_1, i.e., the double length of element 1. The second entry is equal to l, because there is only one element governed by ψ_2.

Table 13.7 contains the initial data for problem (13.32). Having solved that problem, we obtain the optimal design $\boldsymbol{\psi}_* = \{1.664, 0.647\}$. Incidentally, the stress state at incipient plastic collapse (Fig. 13.5b) is practically the same as for the truss of uniform cross section (compare the last rows of Tables 13.6a and 13.7a). However, the collapse mechanisms are entirely different (compare Figs. 13.2a and 13.5a): node 1 of the optimal truss has a horizontal component of displacement rate as well as a vertical component. The vertical bar of the uniform cross section truss remained rigid at plastic collapse, which enforced horizontal motion of node 1. In the optimal truss all three bars are at yield. That is required by the optimality criterion (13.33) (the first two rows of Table 13.7b). Since both design variables turned out to be positive, these conditions must be fulfilled as equalities. Hence, the multipliers $\dot{\lambda}_2^{\mathrm{I}}$, $\dot{\lambda}_2^{\mathrm{II}}$ that generate the plastic strain rate $\dot{q}_2$ cannot vanish simultaneously.

Now we cancel the symmetry requirement and look for the optimal design without technological constraints. This means that each yield force of the bar is now treated as an independent design variable: $\boldsymbol{\psi} = \{n_1^0, n_2^0, n_3^0\}$. Under such circumstances **S** becomes the unity matrix, the linkage matrix is

$$\mathbf{G} = \left[\begin{array}{cc|cc|cc} 1 & 0.7 & 0 & 0 & 0 & 0 \\ \hline 0 & 0 & 1 & 0.7 & 0 & 0 \\ \hline 0 & 0 & 0 & 0 & 1 & 0.7 \end{array}\right],$$

and the cost factors coincide with the lengths of elements.

The initial data for problem (13.32) are given in Table 13.8. The optimal design turns out to be $\boldsymbol{\psi}_* = \{2.12, 0.0, 1.01\}$. It corresponds to the collapse mechanism in Fig. 13.6a and to the stress state in Fig. 13.6b. These results deserve a few comments. First, the optimal plastic truss is statically determined and fully stressed. The same result was obtained for a single loading in the frame of elasticity theory. Second, the collapse mechanism looks strange: node 1 tends to move up even though the force $p_y^{(1)}$ is pushing it down. This kind of motion is enforced by the optimality criterion related

Table 13.7. Optimal Design of Three-Bar Truss—Symmetry Enforced

a. Static Approach

	n_1^0	n_2^0	n_1	n_2	n_3	
	1	2	3	4	5	
1	0.0	0.0	0.707	0.0	−0.707	= 2.0
2	0.0	0.0	0.707	1.0	0.707	= 1.0
3	1.0	0.0	−1.0	0.0	0.0	⩾ 0.0
4	0.7	0.0	1.0	0.0	0.0	⩾ 0.0
5	0.0	1.0	0.0	−1.0	0.0	⩾ 0.0
6	0.0	0.7	0.0	1.0	0.0	⩾ 0.0
7	1.0	0.0	0.0	0.0	−1.0	⩾ 0.0
8	0.7	0.0	0.0	0.0	1.0	⩾ 0.0
9	2.828	1.0	0.0	0.0	0.0	Min
10	1.664	0.647	1.664	0.647	−1.166	Solution

b. Kinematic Approach

	$\dot{w}_x^{(1)}$	$\dot{w}_y^{(1)}$	$\dot{\lambda}_1^{\text{I}}$	$\dot{\lambda}_1^{\text{II}}$	$\dot{\lambda}_2^{\text{I}}$	$\dot{\lambda}_2^{\text{II}}$	$\dot{\lambda}_3^{\text{I}}$	$\dot{\lambda}_3^{\text{II}}$	
	1	2	3	4	5	6	7	8	
1	0.0	0.0	1.0	0.7	0.0	0.0	1.0	0.7	⩽ 2.828
2	0.0	0.0	0.0	0.0	1.0	0.7	0.0	0.0	⩽ 1.0
3	0.707	0.707	−1.0	1.0	0.0	0.0	0.0	0.0	= 0.0
4	0.0	1.0	0.0	0.0	−1.0	1.0	0.0	0.0	= 0.0
5	−0.707	0.707	0.0	0.0	0.0	0.0	−1.0	1.0	= 0.0
6	2.0	1.0	0.0	0.0	0.0	0.0	0.0	0.0	Max
7	2.17	1.0	2.25	0.0	1.0	0.0	0.0	0.832	Solution

to element 3 (the third row of Table 13.8b). Since $\psi_{3*} > 0$, it follows from the KKT conditions that

$$\dot{\lambda}_3^{\text{I}} + 0.7\dot{\lambda}_3^{\text{II}} = 1.414.$$

However, $\dot{\lambda}_3^{\text{I}}$ must vanish because the relevant yield facet is passive (element 3 is compressed). Therefore,

Table 13.8. Optimal Design of Three-Bar Truss—Each Element Optimized Independently

a. Static Approach

	n_1^0	n_2^0	n_3^0	n_1	n_2	n_3	
	1	2	3	4	5	6	
1	0.0	0.0	0.0	0.707	0.0	−0.707	= 2.0
2	0.0	0.0	0.0	0.707	1.0	0.707	= 1.0
3	1.0	0.0	0.0	−1.0	0.0	0.0	⩾ 0.0
4	0.7	0.0	0.0	1.0	0.0	0.0	⩾ 0.0
5	0.0	1.0	0.0	0.0	−1.0	0.0	⩾ 0.0
6	0.0	0.7	0.0	0.0	1.0	0.0	⩾ 0.0
7	0.0	0.0	1.0	0.0	0.0	−1.0	⩾ 0.0
8	0.0	0.0	0.7	0.0	0.0	1.0	⩾ 0.0
9	1.414	1.0	1.414	0.0	0.0	0.0	Min
10	2.122	0.0	1.010	2.122	0.0	−0.707	Solution

b. Kinematic Approach

	$\dot{w}_x^{(1)}$	$\dot{w}_y^{(1)}$	$\dot{\lambda}_1^{\mathrm{I}}$	$\dot{\lambda}_1^{\mathrm{II}}$	$\dot{\lambda}_2^{\mathrm{I}}$	$\dot{\lambda}_2^{\mathrm{II}}$	$\dot{\lambda}_3^{\mathrm{I}}$	$\dot{\lambda}_3^{\mathrm{II}}$	
1	0.0	0.0	1.0	0.7	0.0	0.0	0.0	0.0	⩽ 1.414
2	0.0	0.0	0.0	0.0	1.0	0.7	0.0	0.0	⩽ 1.0
3	0.0	0.0	0.0	0.0	0.0	0.0	1.0	0.7	⩽ 1.414
4	0.707	0.707	−1.0	1.0	0.0	0.0	0.0	0.0	= 0.0
5	0.0	1.0	0.0	0.0	−1.0	1.0	0.0	0.0	= 0.0
6	−0.707	0.707	0.0	0.0	0.0	0.0	−1.0	1.0	= 0.0
7	2.0	1.0	0.0	0.0	0.0	0.0	0.0	0.0	Max
8	2.43	−0.429	1.414	0.0	0.411	0.840	0.0	2.02	Solution

$$\dot{q}_3 = -\dot{\lambda}_3^{\mathrm{II}} = -\frac{1.414}{0.7} = -2.02.$$

Similar arguments related to the optimality criterion for element 1 (the first row of Table 13.8b) lead to the conclusion that $\dot{q}_1 = 1.414$. Thus, the optimality criterion demands strain rates in elements 1 and 3 that correspond to the negative value of $\dot{w}_y^{(1)}$.

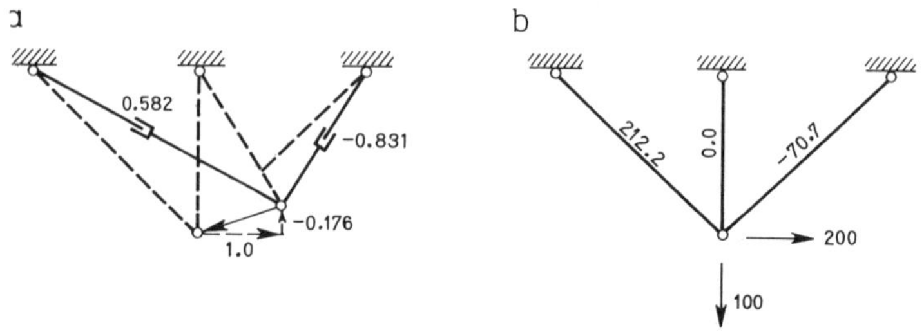

Figure 13.6. Plastic Collapse of Symmetric Optimal Truss: (a) Collapse Mechanism. (b) Axial Forces.

Closer investigation reveals that the solution of the static problem $(13.32)_1$ is degenerate: it contains zero entries n_2^0 and n_2. The theory of linear programming tells us that in such a case the dual problem $(13.32)_2$ has multiple solutions corresponding to the same value of the cost function. Indeed, we see that the plastic multipliers $\dot{\lambda}_2^{\text{I}}$, $\dot{\lambda}_2^{\text{II}}$ can assume any nonnegative values provided that their sum remains equal to $\dot{q}_2 = -0.429$, as required by the optimality criterion.

Multiplying each entry of $\mathbf{n}^0$ by $P_0 = 100$ kN and dividing the result by the yield stress $\sigma_0 = 250$ MPa, if the element is in tension, or by $\sigma_0' = 174$ MPa, if it is in compression, we can calculate the cross-sectional areas of the bars. Then the volume of each design can be found and compared to that of the truss with the uniform cross section. Such a comparison is given in Table 13.9.

The reduction of the volume (weight) of the structure achieved through plastic optimization is significantly more than that obtained via optimal elastic design. However, it is not reasonable to compare directly the volumes of elastic and plastic solutions, because they depend on the assumed ratio between the admissible and yield stresses.

Table 13.9. Summary of Plastic Optimal Design of Three-Bar Truss

Type of solution	A_i (cm^2)			V (cm^3)	ΔV (%)
	$i = 1$	$i = 2$	$i = 3$		
1	2	3	4	5	6
Uniform cross section	11.78	11.78	11.78	9023	0
Symmetric design	11.75	4.57	11.75	7564	16
Free optimum	14.98	0.00	7.13	6253	31

13.3.4. Example of a Frame

As long as the axial force-bending moment interaction is neglected, the optimal plastic design of flexural systems is analogous to that of trusses. Consider, for example, the portal frame depicted in Fig. 12.2a. It was analyzed elastically in Section 12.2.4. We shall use the discretization scheme introduced there in order to formulate and to solve the problems (13.21) and (13.32).

Let us begin with the scaling problem (13.21). The matrices **s**, **p**, **q**, **w**, and **C** were already derived in Chapter 12. Assuming a bending-only model, we take the submatrices $\bar{\mathbf{f}}_i$, $\bar{\mathbf{k}}_i$, and $\bar{\mathbf{N}}_i$ that describe the plastic properties of a generic element from Section 11.5.3. After assembly they generate the global matrices:

$$\mathbf{f} = \{f^{\text{I}}_{a1}, f^{\text{II}}_{a1}, f^{\text{I}}_{b1}, f^{\text{II}}_{b1}, f^{\text{I}}_{a2}, f^{\text{II}}_{a2}, f^{\text{I}}_{b2}, f^{\text{II}}_{b2}, f^{\text{I}}_{a3}, f^{\text{II}}_{a3}, f^{\text{I}}_{b3}, f^{\text{II}}_{b3}\},$$

$$\mathbf{k} = \{1, 1, 1, 1, 1, 1, 1, 1, 1, 1, 1, 1\},$$

$$\dot{\boldsymbol{\lambda}} = \{\dot{\lambda}^{\text{I}}_{a1}, \dot{\lambda}^{\text{II}}_{a1}, \dot{\lambda}^{\text{I}}_{b1}, \dot{\lambda}^{\text{II}}_{b1}, \dot{\lambda}^{\text{I}}_{a2}, \dot{\lambda}^{\text{II}}_{a2}, \dot{\lambda}^{\text{I}}_{b2}, \dot{\lambda}^{\text{II}}_{b2}, \dot{\lambda}^{\text{I}}_{a3}, \dot{\lambda}^{\text{II}}_{a3}, \dot{\lambda}^{\text{I}}_{b3}, \dot{\lambda}^{\text{II}}_{b3}\},$$

$$\mathbf{N} = \left[\begin{array}{cccc|cccc|cccc}
1 & -1 & 0 & 0 & 0 & 0 & 0 & 0 & 0 & 0 & 0 & 0 \\
0 & 0 & 1 & -1 & 0 & 0 & 0 & 0 & 0 & 0 & 0 & 0 \\
\hline
0 & 0 & 0 & 0 & 1 & -1 & 0 & 0 & 0 & 0 & 0 & 0 \\
0 & 0 & 0 & 0 & 0 & 0 & 1 & -1 & 0 & 0 & 0 & 0 \\
\hline
0 & 0 & 0 & 0 & 0 & 0 & 0 & 0 & 1 & -1 & 0 & 0 \\
0 & 0 & 0 & 0 & 0 & 0 & 0 & 0 & 0 & 0 & 1 & -1
\end{array}\right].$$

Table 13.10 contains initial data for problem (13.21) together with its solution. It turns out that the frame having a constant cross section characterized by the yield moment $M^0 = PL$ is exactly at the state of incipient plastic collapse under the external force P and the external moment $2PL$, applied as shown in Fig. 12.2a. Hence, we get the scaling factor equal to unity.

The solution of the kinematic problem $(13.21)_2$ is shown in Fig. 13.7a. It is the extreme case of an incomplete or local collapse mechanism: the entire frame remains intact except in the vicinity of node 4, where two plastic hinges develop. They allow unrestricted rotation of that node, which is regarded as the collapse of the structure.

Figures 13.7b, c, d present the diagrams of internal forces obtained from the solution of problem $(13.21)_1$. The yield moments of elements are indicated in Fig. 13.7b by dashed lines. The stress state obtained is of secondary importance for the designer, because it is only one of the multiple

Table 13.10. Scaling of Portal Frame

a. Static Approach

	ρ	$m_a^{(1)}$	$m_b^{(1)}$	$m_a^{(2)}$	$m_b^{(2)}$	$m_a^{(3)}$	$m_b^{(3)}$	
	1	2	3	4	5	6	7	
1	0.0	−1.0	−1.0	−1.0	−1.0	0.0	0.0	= 1.0
2	0.0	0.0	1.0	0.0	0.0	1.0	0.0	= 0.0
3	0.0	0.0	0.0	0.0	1.0	0.0	1.0	= 2.0
4	1.0	−1.0	0.0	0.0	0.0	0.0	0.0	⩾ 0.0
5	1.0	1.0	0.0	0.0	0.0	0.0	0.0	⩾ 0.0
6	1.0	0.0	−1.0	0.0	0.0	0.0	0.0	⩾ 0.0
7	1.0	0.0	1.0	0.0	0.0	0.0	0.0	⩾ 0.0
8	1.0	0.0	0.0	−1.0	0.0	0.0	0.0	⩾ 0.0
9	1.0	0.0	0.0	1.0	0.0	0.0	0.0	⩾ 0.0
10	1.0	0.0	0.0	0.0	−1.0	0.0	0.0	⩾ 0.0
11	1.0	0.0	0.0	0.0	1.0	0.0	0.0	⩾ 0.0
12	1.0	0.0	0.0	0.0	0.0	−1.0	0.0	⩾ 0.0
13	1.0	0.0	0.0	0.0	0.0	1.0	0.0	⩾ 0.0
14	1.0	0.0	0.0	0.0	0.0	0.0	−1.0	⩾ 0.0
15	1.0	0.0	0.0	0.0	0.0	0.0	1.0	⩾ 0.0
16	1.0	0.0	0.0	0.0	0.0	0.0	0.0	Min
17	1.0	−1.0	−1.0	0.0	1.0	1.0	1.0	Solution

b. Kinematic Approach

	$\dot{w}_x^{(2)}$	$\dot{w}_z^{(2)}$	$\dot{w}_z^{(4)}$	$\dot{\lambda}^{I}_{a1}$	$\dot{\lambda}^{II}_{a1}$	$\dot{\lambda}^{I}_{b1}$	$\dot{\lambda}^{II}_{b1}$	$\dot{\lambda}^{I}_{a2}$	$\dot{\lambda}^{II}_{a2}$	$\dot{\lambda}^{I}_{b2}$	$\dot{\lambda}^{II}_{b2}$	$\dot{\lambda}^{I}_{a3}$	$\dot{\lambda}^{II}_{a3}$	$\dot{\lambda}^{I}_{b3}$	$\dot{\lambda}^{II}_{b3}$	
	1	2	3	4	5	6	7	8	9	10	11	12	13	14	15	
1	0.0	0.0	0.0	1.0	1.0	1.0	1.0	1.0	1.0	1.0	1.0	1.0	1.0	1.0	1.0	= 1.0
2	−1.0	0.0	0.0	−1.0	1.0	0.0	0.0	0.0	0.0	0.0	0.0	0.0	0.0	0.0	0.0	= 0.0
3	−1.0	1.0	0.0	0.0	0.0	−1.0	1.0	0.0	0.0	0.0	0.0	0.0	0.0	0.0	0.0	= 0.0
4	−1.0	0.0	0.0	0.0	0.0	0.0	0.0	−1.0	1.0	0.0	0.0	0.0	0.0	0.0	0.0	= 0.0
5	−1.0	0.0	1.0	0.0	0.0	0.0	0.0	0.0	0.0	−1.0	1.0	0.0	0.0	0.0	0.0	= 0.0
6	0.0	1.0	0.0	0.0	0.0	0.0	0.0	0.0	0.0	0.0	0.0	−1.0	1.0	0.0	0.0	= 0.0
7	0.0	0.0	1.0	0.0	0.0	0.0	0.0	0.0	0.0	0.0	0.0	0.0	0.0	−1.0	1.0	= 0.0
9	1.0	0.0	2.0	0.0	0.0	0.0	0.0	0.0	0.0	0.0	0.0	0.0	0.0	0.0	0.0	Max
9	0.0	0.0	2.0	0.0	0.0	0.0	0.0	0.0	0.0	2.0	0.0	0.0	0.0	2.0	0.0	Solution

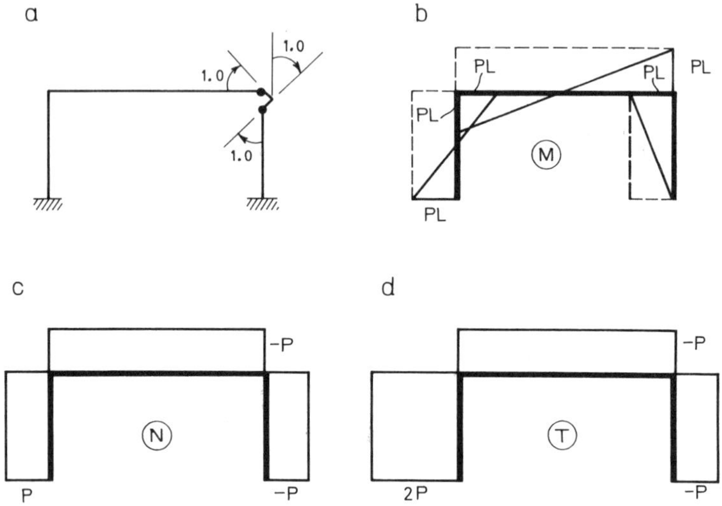

Figure 13.7. Portal Frame Scaled to Plastic Collapse: (a) Collapse Mechanism. (b) Bending and Yield Moments. (c) Axial Forces. (d) Shear Forces.

solutions of the static problem $(13.21)_1$. The physical reason for that is that the bending moments in the hyperstatic part of the frame (cross sections a1, b1, and a3; see Figure 12.2b) cannot be determined uniquely from the equilibrium equations. The mathematical reason is that the solution of problem $(13.21)_2$ is degenerate; that is, it contains zero entries (e.g., $\dot{w}_x^{(2)}$ and $\dot{w}_z^{(2)}$). As we said, the dual of such a solution is not unique.

The nondimensional cost (volume) of the frame of uniform cross section is

$$v_{\text{const}} = (l_1 + l_2 + l_3)m^0 = (1.0 + 1.0 + 2.0)(1.0) = 4.0.$$

Let us try to diminish it through optimization. To achieve the most significant economy, we regard each yield moment as an independent design variable: $\boldsymbol{\psi} = \{m_1^0, m_2^0, m_3^0\}$. Then $\mathbf{S}$ in Eq. (13.41), with $\mathbf{n}^0$ replaced by $\mathbf{m}^0$, is the unit matrix and $\mathbf{R}$ in Eq. (13.39) coincides with

$$\mathbf{G} = \begin{bmatrix} 1 & 1 & 1 & 1 & 0 & 0 & 0 & 0 & 0 & 0 & 0 & 0 \\ 0 & 0 & 0 & 0 & 1 & 1 & 1 & 1 & 0 & 0 & 0 & 0 \\ 0 & 0 & 0 & 0 & 0 & 0 & 0 & 0 & 1 & 1 & 1 & 1 \end{bmatrix}.$$

We take the cost factors equal to the lengths of the elements:

$$\mathbf{c} = \mathbf{l} = \{1.0, 2.0, 1.0\}.$$

That is, we are going to minimize the volume generated by the ideal sandwich cross sections.

The initial data and the solutions of problem (13.32) are given in Table 13.11. Figure 13.8 presents the results obtained graphically. The optimal nondimensional design turns out to be $\boldsymbol{\psi}_* = \{1.0, 2.0, 0.0\}$. Thus the given load is optimally carried by two separate columns: the left one having the yield moment $M_1^0 = PL$ and the right one with the yield moment $M_2^0 = 2PL$. Such a solution, though unusual at first glance, is logical: since beam 3 (see Fig. 12.2b) carries no vertical load, any amount of material assigned to it would be wasted. Moreover, that beam is twice as long as each of the columns, and we demand $M_i^0 = \text{const}$ along each element. Therefore, a positive value of M_3^0 would considerably increase the volume.

This time the solution of the static problem is degenerate. Hence, the collapse mechanism of Fig. 13.8a is not unique. Nonuniqueness comes from the pseudohinges on vanishing element 3: the plastic multipliers $\dot{\lambda}_{a3}^{\mathrm{I}}, \dot{\lambda}_{a3}^{\mathrm{II}}$ and $\dot{\lambda}_{b3}^{\mathrm{I}}, \dot{\lambda}_{b3}^{\mathrm{II}}$, associated with those hinges, can assume arbitrary nonnegative values as long as their sum is equal to unity (compare the similar situation with the vanishing element of the three-bar truss). Despite its nonuniqueness, the collapse mechanism fulfills the optimality criterion: the sum of the plastic multipliers corresponding to each element is equal to the length of that element. In the bending-only model the same rule applies also for the strain rates, i.e., the rates of rotation in plastic hinges.

The nondimensional volume of the optimal frame is

$$v_{\mathrm{opt}} = \mathbf{l}^{\mathrm{T}}\mathbf{m}_*^0 = (1.0)(1.0) + (1.0)(2.0) + (1.0)(0.0) = 3.0,$$

which is 25% less than v_{const}.

Exercise 13.3

Find the optimal design of the truss in Fig. 11.10, given $\mu_* = 2.50$ and $\tilde{N}^0/N^0 = 0.70$. Compare the solutions for (a) free design and (b) design for standardized horizontal bars, vertical bars, and braces. Use the matrix $\mathbf{C}^{\mathrm{T}}$ given in Table S.1 in Chapter 20.

Exercise 13.4

Find the optimal design of the frame in Fig. 11.13, given $\mu_* = 2.50$. Assume the bending-only model and use the matrix $\mathbf{C}^{\mathrm{T}}$ given in Table S.8 in Chapter 20.

Table 13.11. Optimal Design of Portal Frame

a. Static Approach

	m_1^0	m_2^0	m_3^0	$m_a^{(1)}$	$m_b^{(1)}$	$m_a^{(2)}$	$m_b^{(2)}$	$m_a^{(3)}$	$m_b^{(3)}$	
	1	2	3	4	5	6	7	8	9	
1	0.0	0.0	0.0	−1.0	−1.0	−1.0	−1.0	0.0	0.0	= 1.0
2	0.0	0.0	0.0	0.0	1.0	0.0	0.0	1.0	0.0	= 0.0
3	0.0	0.0	0.0	0.0	0.0	0.0	1.0	0.0	1.0	= 2.0
4	1.0	0.0	0.0	−1.0	0.0	0.0	0.0	0.0	0.0	⩾ 0.0
5	1.0	0.0	0.0	1.0	0.0	0.0	0.0	0.0	0.0	⩾ 0.0
6	1.0	0.0	0.0	0.0	−1.0	0.0	0.0	0.0	0.0	⩾ 0.0
7	1.0	0.0	0.0	0.0	1.0	0.0	0.0	0.0	0.0	⩾ 0.0
8	0.0	1.0	0.0	0.0	0.0	−1.0	0.0	0.0	0.0	⩾ 0.0
9	0.0	1.0	0.0	0.0	0.0	1.0	0.0	0.0	0.0	⩾ 0.0
10	0.0	1.0	0.0	0.0	0.0	0.0	−1.0	0.0	0.0	⩾ 0.0
11	0.0	1.0	0.0	0.0	0.0	0.0	1.0	0.0	0.0	⩾ 0.0
12	0.0	0.0	1.0	0.0	0.0	0.0	0.0	−1.0	0.0	⩾ 0.0
13	0.0	0.0	1.0	0.0	0.0	0.0	0.0	1.0	0.0	⩾ 0.0
14	0.0	0.0	1.0	0.0	0.0	0.0	0.0	0.0	−1.0	⩾ 0.0
15	0.0	0.0	1.0	0.0	0.0	0.0	0.0	0.0	1.0	⩾ 0.0
16	1.0	2.0	1.0	0.0	0.0	0.0	0.0	0.0	0.0	Min
17	1.0	2.0	0.0	−1.0	0.0	−2.0	2.0	0.0	0.0	Solution

b. Kinematic Approach

	$\dot{w}_x^{(2)}$	$\dot{w}_z^{(2)}$	$\dot{w}_z^{(4)}$	$\dot{\lambda}^{I}_{a1}$	$\dot{\lambda}^{II}_{a1}$	$\dot{\lambda}^{I}_{b1}$	$\dot{\lambda}^{II}_{b1}$	$\dot{\lambda}^{I}_{a2}$	$\dot{\lambda}^{II}_{a2}$	$\dot{\lambda}^{I}_{b2}$	$\dot{\lambda}^{II}_{b2}$	$\dot{\lambda}^{I}_{a3}$	$\dot{\lambda}^{II}_{a3}$	$\dot{\lambda}^{I}_{b3}$	$\dot{\lambda}^{II}_{b3}$	
	1	2	3	4	5	6	7	8	9	10	11	12	13	14	15	
1	0.0	0.0	0.0	1.0	1.0	1.0	1.0	0.0	0.0	0.0	0.0	0.0	0.0	0.0	0.0	1.0
2	0.0	0.0	0.0	0.0	0.0	0.0	0.0	1.0	1.0	1.0	1.0	0.0	0.0	0.0	0.0	2.0
3	0.0	0.0	0.0	0.0	0.0	0.0	0.0	0.0	0.0	0.0	0.0	1.0	1.0	1.0	1.0	1.0
4	−1.0	0.0	0.0	−1.0	1.0	0.0	0.0	0.0	0.0	0.0	0.0	0.0	0.0	0.0	0.0	= 0.0
5	−1.0	1.0	0.0	0.0	0.0	−1.0	1.0	0.0	0.0	0.0	0.0	0.0	0.0	0.0	0.0	= 0.0
6	−1.0	0.0	0.0	0.0	0.0	0.0	0.0	−1.0	1.0	0.0	0.0	0.0	0.0	0.0	0.0	= 0.0
7	−1.0	0.0	1.0	0.0	0.0	0.0	0.0	0.0	0.0	−1.0	1.0	0.0	0.0	0.0	0.0	= 0.0
8	0.0	1.0	0.0	0.0	0.0	0.0	0.0	0.0	0.0	0.0	0.0	−1.0	1.0	0.0	0.0	= 0.0
9	0.0	0.0	1.0	0.0	0.0	0.0	0.0	0.0	0.0	0.0	0.0	0.0	0.0	−1.0	1.0	= 0.0
10	1.0	0.0	2.0	0.0	0.0	0.0	0.0	0.0	0.0	0.0	0.0	0.0	0.0	0.0	0.0	Max
11	1.0	1.0	1.0	0.0	1.0	0.0	0.0	0.0	1.0	0.0	0.0	1.0	0.0	1.0	0.0	Solution

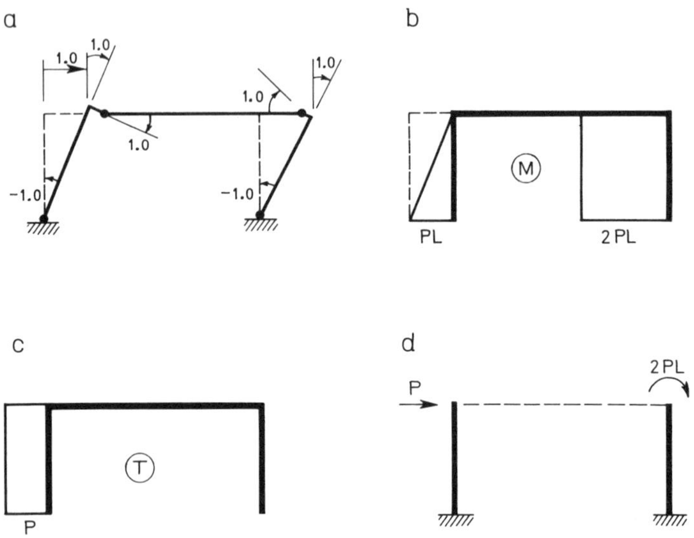

Figure 13.8. Plastic Collapse of Optimal Frame: (a) Collapse Mechanism. (b) Bending and Yield Moments. (c) Shear Forces. (d) Optimal Design.

13.4. Modifications of the Optimal Plastic Design Problem

13.4.1. Technological Constraints

The examples in the previous section showed that optimal plastic solutions tend to be degenerate. Sometimes the information that a certain design variable has zero value in the optimal solution is useful: one is motivated to reconsider the assumed layout of the structure. However, in many cases all members of the structure should be present in the final design because of certain requirements that were not taken into account when problem (13.32) was formulated. On the other hand, it is possible for certain design values to assume optimal values that are too large. Both reasons stimulate us to introduce into the model (13.32) of optimal plastic design the technological constraints defined by Eq. (12.27).

Each inequality $\psi_i \geqslant \psi_i^-$ introduced into the static problem $(13.32)_1$ generates a nonnegative variable $\dot{\gamma}_i^-$ in the kinematic problem $(13.32)_2$. Similarly, each inquality $-\psi_i \geqslant -\psi_i^+$ gives rise to the kinematic variable

$\dot{\gamma}_i^+ \geqslant 0$. Taking Table 10.1 as the template, we can write the modified static problem as follows:

$\mathbf{s}^T$	$\boldsymbol{\psi}^T$		
$\mathbf{C}^T$	$\mathbf{0}$	$=\mathbf{p}$	(n)
$-\mathbf{N}^T$	$\mathbf{G}^T$	$\geqslant \mathbf{0}$	(l)
$\mathbf{0}$	$\mathbf{I}$	$\geqslant \boldsymbol{\psi}^-$	(t)
$\mathbf{0}$	$-\mathbf{I}$	$\geqslant -\boldsymbol{\psi}^+$	(t)
$\mathbf{0}$	$\mathbf{c}^T$	min	

Then the dual problem is found from the transposed table:

$\dot{\mathbf{w}}^T$	$\dot{\boldsymbol{\lambda}}^T$	$\dot{\boldsymbol{\gamma}}^{-T}$	$\dot{\boldsymbol{\gamma}}^{+T}$		
$\mathbf{C}$	$-\mathbf{N}$	$\mathbf{0}$	$\mathbf{0}$	$=\mathbf{0}$	(m)
$\mathbf{0}$	$\mathbf{G}$	$\mathbf{I}$	$-\mathbf{I}$	$=\mathbf{c}$	(t)
$\mathbf{p}^T$	$\mathbf{0}$	$\boldsymbol{\psi}^{-T}$	$\boldsymbol{\psi}^{+T}$	max	

Using our concise notation, we say that the optimal plastic design under technological constraints corresponds to the following dual pair of LP problems:

$$\min\{\mathbf{c}^T\boldsymbol{\psi} \mid \mathbf{C}^T\mathbf{s} = \mathbf{p},\ \mathbf{G}^T\boldsymbol{\psi} - \mathbf{N}^T\mathbf{s} \geqslant \mathbf{0},\ \boldsymbol{\psi}^- \leqslant \boldsymbol{\psi} \leqslant \boldsymbol{\psi}^+\},$$
$$\max\{\mathbf{p}^T\dot{\mathbf{w}} + \boldsymbol{\psi}^{-T}\dot{\boldsymbol{\gamma}}^- - \boldsymbol{\psi}^{+T}\dot{\boldsymbol{\gamma}}^+ \mid \mathbf{C}\dot{\mathbf{w}} - \mathbf{N}\dot{\boldsymbol{\lambda}} = \mathbf{0},$$
$$\mathbf{G}\dot{\boldsymbol{\lambda}} + \dot{\boldsymbol{\gamma}}^- - \dot{\boldsymbol{\gamma}}^+ = \mathbf{c},\ \dot{\boldsymbol{\lambda}} \geqslant \mathbf{0},\ \dot{\boldsymbol{\gamma}}^- \geqslant \mathbf{0},\ \dot{\boldsymbol{\gamma}}^+ \geqslant \mathbf{0}\}. \tag{13.44}$$

Note that the entries of $\boldsymbol{\psi}$ are now formally treated as free variables. Taking the lower bounds ψ_i^- as nonnegative, we ensure the nonnegativeness of each ψ_i. The upper bounds must satisfy the inequalities $\psi_i^+ \geqslant \psi_i^-$; otherwise the constraints of $(13.44)_1$ would be contradictory. The foregoing conditions that make the bounds consistent are necessary but not sufficient for the existence of a solution of the static problem. Taking, for example, upper bounds that are too low, we can exclude the possibility of finding any statically admissible solution. That means simply that the extreme "heavy" design $\boldsymbol{\psi}_* = \boldsymbol{\psi}^+$ will be still too weak to carry the load $\mathbf{p}$.

Now let us consider the kinematic problem $(13.44)_2$. We see that the new variables $\dot{\gamma}_i^-$, $\dot{\gamma}_i^+$ appear in the constraint on plastic multipliers,

$$\mathbf{G}\dot{\boldsymbol{\lambda}} + \dot{\boldsymbol{\gamma}}^- - \dot{\boldsymbol{\gamma}}^+ = \mathbf{c}, \tag{13.45}$$

that replaces the previously used inequality (13.33). In contrast to the latter, the new constraint has the form of a set of equations. We might suspect that Eq. (13.45) should be somehow related to the optimality criterion discussed in the previous section. To clarify that, let us consider three possible cases for the optimal value of the ith design variable:

(a) $\psi_{i*} = \psi_i^-$; i.e., the ith technological constraint is active at the solution. Then the KKT condition $\dot{\gamma}_i^-(\psi_i - \psi_i^-) = 0$ allows $\dot{\gamma}_{i*}^- > 0$. If the rather unusual case $\psi_i^- = \psi_i^+$ is excluded, then the upper bound constraint cannot be active simultaneously with the lower bound. Hence, there follows from the KKT condition $\dot{\gamma}_i^+(-\psi_i + e\psi_i^+) = 0$ that $\dot{\gamma}_{i*}^+$ must vanish.

(b) $\psi_i^- < \psi_{i*} < \psi_i^+$; i.e., the optimal value falls between the bounds. Then both technological constraints are passive, which implies $\dot{\gamma}_{i*}^- = \dot{\gamma}_{i*}^+ = 0$.

(c) $\psi_{i*} = \psi_i^+$; i.e., the upper bound constraint is active. Following arguments similar to those for case (a), we conclude that $\dot{\gamma}_{i*}^+$ can be positive and that $\dot{\gamma}_{i*}^-$ must vanish.

These considerations can be summarized as follows. If the technological constraints are passive for a certain ψ_{i*}, then the additional variables $\dot{\gamma}_{i*}^-$, $\dot{\gamma}_{i*}^+$ in the ith equation (13.45) disappear, and that equation therefore reduces to the optimality criterion (13.33). Otherwise that criterion is not valid. Since we do not know *a priori* whether some bounds are going to be active at the optimum, we cannot apply the methods based upon optimality criteria in the presence of technological constraints.

As an example, let us optimize the three-bar truss under the constraint $n_2^0 \geqslant 0.50$. The data for model (13.44) are given in Table 13.12. Comparing the static version of that table with Table 13.8a, which corresponds to the free optimization of the same truss, we see that the only difference is an additional constraint (the ninth row). In the kinematic approach that constraint corresponds to an additional variable $\dot{\gamma}_2^-$ (the ninth column).

The constrained optimal design is $\boldsymbol{\psi}_* = \{2.37, 0.50, 0.657\}$, and the free optimization gave us $\boldsymbol{\psi}'_* = \{2.12, 0.0, 1.01\}$. The cost (volume) of the constrained solution is about 8% greater than that of the free one. Such a

Table 13.12. Optimal Design of Three-Bar Truss under Technological Constraint $n_2^0 \geqslant 0.50$

a. Static Approach

	n_1^0	n_2^0	n_3^0	n_1	n_2	n_3	
	1	2	3	4	5	6	
1	0.0	0.0	0.0	0.707	0.0	−0.707	= 2.0
2	0.0	0.0	0.0	0.707	1.0	0.707	= 1.0
3	1.0	0.0	0.0	−1.0	0.0	0.0	⩾ 0.0
4	0.7	0.0	0.0	1.0	0.0	0.0	⩾ 0.0
5	0.0	1.0	0.0	0.0	−1.0	0.0	⩾ 0.0
6	0.0	0.7	0.0	0.0	1.0	0.0	⩾ 0.0
7	0.0	0.0	1.0	0.0	0.0	−1.0	⩾ 0.0
8	0.0	0.0	0.7	0.0	0.0	1.0	⩾ 0.0
9	0.0	1.0	0.0	0.0	0.0	0.0	⩾ 0.50
9	1.414	1.0	1.414	0.0	0.0	0.0	Min
10	2.37	0.50	0.657	2.37	−0.35	−0.46	Solution

b. Kinematic Approach

	$\dot{w}_x^{(1)}$	$\dot{w}_y^{(1)}$	$\dot{\lambda}_1^{I}$	$\dot{\lambda}_1^{II}$	$\dot{\lambda}_2^{I}$	$\dot{\lambda}_2^{II}$	$\dot{\lambda}_3^{I}$	$\dot{\lambda}_3^{II}$	$\dot{\gamma}_2^{-}$	
	1	2	3	4	5	6	7	8	9	
1	0.0	0.0	1.0	0.7	0.0	0.0	0.0	0.0	0.0	⩽ 1.414
2	0.0	0.0	0.0	0.0	1.0	0.7	0.0	0.0	1.0	⩽ 1.0
3	0.0	0.0	0.0	0.0	0.0	0.0	1.0	0.7	0.0	⩽ 1.414
4	0.707	0.707	−1.0	1.0	0.0	0.0	0.0	0.0	0.0	= 0.0
5	0.0	1.0	0.0	0.0	−1.0	1.0	0.0	0.0	0.0	= 0.0
6	−0.707	0.707	0.0	0.0	0.0	0.0	−1.0	1.0	0.0	= 0.0
7	2.0	1.0	0.0	0.0	0.0	0.0	0.0	0.0	0.5	Max
8	2.43	−0.429	1.414	0.0	0.0	0.429	0.0	2.02	0.70	Solution

difference would increase with increasing lower bounds on the design variables. On the other hand, decreasing upper bounds would lead finally to the lack of any solution.

The solution of the static problem is no longer degenerate. Therefore, the solution of the kinematic problem given in the last row of Table 13.12b is unique. It corresponds to the same collapse mechanism that was found in Section 13.3 for the free optimal design (see Fig. 13.6a). Note, however, that there is now a single positive plastic multiplier for each element of the truss. The multipliers related to elements 1 and 3 obey the optimality criterion (13.33); e.g., for the first element the relation

$$\dot{\lambda}^{\mathrm{I}}_{1*} + 0.70\dot{\lambda}^{\mathrm{II}}_{1*} = 1.414 + (0.70)(0.0) = 1.414$$

holds.

Since ψ_{2*} is at its lower bound, the dual variable $\dot{\gamma}^{-}_{2*}$ is positive at the solution and

$$\dot{\lambda}^{\mathrm{I}}_{2*} + 0.70\dot{\lambda}^{\mathrm{II}}_{2*} = 0.0 + (0.70)(0.429) = 0.3 < 1.0.$$

Thus, the optimality criterion cannot be applied for element 2.

Exercise 13.5

Find the optimal design of the beam in Fig. 11.11, given $\mu_* = 2.20$. Compare the solutions for (a) free design and (b) design under the constraint $M^0 \geqslant 6.0$ kN m. Use the matrix $\mathbf{C}^{\mathrm{T}}$ given in Table S.2 in Chapter 20.

Exercise 13.6

Find the optimal design of the frame in Fig. 11.12, given $\mu_* = 2.0$. Compare the solutions for (a) free design and (b) design under the constraint $M^0 \geqslant 16.0$ kN m. Use the matrix $\mathbf{C}^{\mathrm{T}}$ given in Table S.5 in Chapter 20.

13.4.2. Reduced Model

The examples presented so far show that the simplex tables for problem (13.32) tend to be large even for relatively simple structures. Their dimensions grow further when the technological constraints are taken into account. Despite various memory-saving techniques in modern simplex solvers (some of those techniques are mentioned in Section 13.5), we might be forced to reduce somehow the dimensions of the problem, especially for solution on a microcomputer.

A natural way to achieve such a reduction is to use the redundant unknowns in the static problem (13.32). Given the compatibility matrix $\mathbf{C}$, we can find its generalized inverse $\mathbf{C}^{-}$ together with the null space matrix $\mathbf{C}^{0}$. Then the stress matrix $\mathbf{s}$ can be expressed in terms of the load $\mathbf{p}$ and the matrix of redundant stresses $\mathbf{z}$ [see Eq. (11.7)].

Introducing that expression into the constraints of $(13.32)_1$, we obtain its reduced version:

$$\min\{\mathbf{c}^T\boldsymbol{\psi} \mid \mathbf{G}^T\boldsymbol{\psi} - \mathbf{Q}^T\mathbf{z} \geqslant \mathbf{u}, \boldsymbol{\psi} \geqslant \mathbf{0}\}, \tag{13.46}$$

where

$$\mathbf{Q} = \mathbf{C}^0\mathbf{N}, \tag{13.47}$$

$$\mathbf{u} = \mathbf{N}^T\mathbf{C}^{-T}\mathbf{p}. \tag{13.48}$$

The dual of problem (13.46) is

$$\max\{\mathbf{u}^T\dot{\boldsymbol{\lambda}} \mid \mathbf{Q}\dot{\boldsymbol{\lambda}} = \mathbf{0}, \mathbf{G}\dot{\boldsymbol{\lambda}} \leqslant \mathbf{c}, \dot{\boldsymbol{\lambda}} \geqslant \mathbf{0}\}. \tag{13.49}$$

The dual can also be obtained by eliminating the displacement rates from the kinematic problem $(13.32)_2$. The equality constraint of the latter can be solved with respect to $\dot{\mathbf{w}}$. Hence, we obtain

$$\dot{\mathbf{w}} = \mathbf{C}^-\dot{\mathbf{q}} = \mathbf{C}^-\mathbf{N}\dot{\boldsymbol{\lambda}}, \tag{13.50}$$

which is valid if the additional strain compatibility condition

$$\mathbf{C}^0\dot{\mathbf{q}} = \mathbf{C}^0\mathbf{N}\dot{\boldsymbol{\lambda}} = \mathbf{0} \tag{13.51}$$

is satisfied [compare Eqs. (11.3) and (11.4)]. Substituting the preceding relations into problem $(13.32)_2$ and using definitions (13.47) and (13.48) of the new matrices, we obtain the same reduced problem (13.47). Its cost function represents the external power expressed in terms of the plastic multipliers, and the single equality constraint is the strain compatibility condition transformed to the same variables. Note that the optimality criterion and the sign constraint remain unaltered by the reduction.

Such a modification produces the following gain: instead of $m + t$ unknowns and $n + l$ constraints of the original problem (13.32), we now have $k + t$ unknowns and l constraints in the reduced one (13.46). When the degree of redundancy k is small, such a reduction can be considerable. Of course, we pay for it with the additional effort to find the matrices $\mathbf{Q}$ and $\mathbf{u}$. This is, however, the usual situation in numerical calculations: saving computer memory usually means waiting longer for the result.

Table 13.13. Optimal Design of Portal Frame—Reduced Problem

a. Static Approach

	m_1^0	m_2^0	m_3^0	$m_b^{(1)}$	$m_a^{(2)}$	$m_b^{(2)}$	
	1	2	3	4	5	6	
1	1.0	0.0	0.0	−1.0	−1.0	−1.0	$\geqslant -1.0$
2	1.0	0.0	0.0	1.0	1.0	1.0	$\geqslant 1.0$
3	1.0	0.0	0.0	1.0	0.0	0.0	$\geqslant 0.0$
4	1.0	0.0	0.0	−1.0	0.0	0.0	$\geqslant 0.0$
5	0.0	1.0	0.0	0.0	1.0	0.0	$\geqslant 0.0$
6	0.0	1.0	0.0	0.0	−1.0	0.0	$\geqslant 0.0$
7	0.0	1.0	0.0	0.0	0.0	1.0	$\geqslant 0.0$
8	0.0	1.0	0.0	0.0	0.0	−1.0	$\geqslant 0.0$
9	0.0	0.0	1.0	−1.0	0.0	0.0	$\geqslant 0.0$
10	0.0	0.0	1.0	1.0	0.0	0.0	$\geqslant 0.0$
11	0.0	0.0	1.0	0.0	0.0	−1.0	$\geqslant 2.0$
12	0.0	0.0	1.0	0.0	0.0	1.0	$\geqslant -2.0$
13	1.0	2.0	1.0	0.0	0.0	0.0	Min
14	1.0	2.0	0.0	0.0	−2.0	2.0	Solution

b. Kinematic Approach

	$\dot{\lambda}^{I}_{a1}$	$\dot{\lambda}^{II}_{a1}$	$\dot{\lambda}^{I}_{b1}$	$\dot{\lambda}^{II}_{b1}$	$\dot{\lambda}^{I}_{a2}$	$\dot{\lambda}^{II}_{a2}$	$\dot{\lambda}^{I}_{b2}$	$\dot{\lambda}^{II}_{b2}$	$\dot{\lambda}^{I}_{a3}$	$\dot{\lambda}^{II}_{a3}$	$\dot{\lambda}^{I}_{b3}$	$\dot{\lambda}^{II}_{b3}$	
	1	2	3	4	5	6	7	8	9	10	11	12	
1	1.0	1.0	1.0	1.0	0.0	0.0	0.0	0.0	0.0	0.0	0.0	0.0	$\leqslant$ 1.0
2	0.0	0.0	0.0	0.0	1.0	1.0	1.0	1.0	0.0	0.0	0.0	0.0	$\leqslant$ 2.0
3	0.0	0.0	0.0	0.0	0.0	0.0	0.0	0.0	1.0	1.0	1.0	1.0	$\leqslant$ 1.0
4	−1.0	1.0	1.0	−1.0	0.0	0.0	0.0	0.0	−1.0	1.0	0.0	0.0	= 0.0
5	−1.0	1.0	0.0	0.0	1.0	−1.0	0.0	0.0	0.0	0.0	0.0	0.0	= 0.0
6	−1.0	1.0	0.0	0.0	0.0	0.0	1.0	−1.0	0.0	0.0	−1.0	1.0	= 0.0
7	−1.0	1.0	2.0	0.0	0.0	0.0	0.0	0.0	0.0	0.0	2.0	−2.0	Max
9	0.0	1.0	0.0	0.0	0.0	1.0	0.0	0.0	1.0	0.0	1.0	0.0	Solution

Let us demonstrate the reduction upon the portal frame optimized in Section 13.3.4. Its matrix of equilibrium equations is

$$\mathbf{C}^{\mathrm{T}} = \begin{bmatrix} -1 & -1 & -1 & -1 & 0 & 0 \\ 0 & 1 & 0 & 0 & 1 & 0 \\ 0 & 0 & 0 & 1 & 0 & 1 \end{bmatrix}.$$

Applying to it the Robinson–Haggenmacher inversion algorithm, we obtain

$$\mathbf{C}^{-} = \begin{bmatrix} -1 & 0 & 0 & 0 & 0 & 0 \\ 0 & 0 & 0 & 0 & 1 & 0 \\ 0 & 0 & 0 & 0 & 0 & 1 \end{bmatrix},$$

$$\mathbf{C}^{0} = \begin{bmatrix} -1 & 1 & 0 & 0 & -1 & 0 \\ -1 & 0 & 1 & 0 & 0 & 0 \\ -1 & 0 & 0 & 1 & 0 & -1 \end{bmatrix}.$$

Before we proceed further, we recall the physical meaning of these matrices. Note that $\mathbf{C}^{-}$ contains a zero 3×3 submatrix and $\mathbf{C}^{0}$ has an identity submatrix of the same size. If a zero column of the generalized inverse matrix has a counterpart column of $\mathbf{C}^{0}$ that incorporates a single nonzero entry equal to unity, then the relevant member of $\mathbf{s}$ has been automatically chosen as the redundant parameter in the description of the stress state according to Eq. (11.7).

In our case $\mathbf{s} = \{m_a^{(1)}, m_b^{(1)}, m_a^{(2)}, m_b^{(2)}, m_a^{(3)}, m_b^{(3)}\}$. The structure of matrices $\mathbf{C}^{-}$ and $\mathbf{C}^{0}$ indicates that the second, third, and fourth components of $\mathbf{s}$ enter the matrix of redundant unknowns: $\mathbf{z} = \{m_b^{(1)}, m_a^{(2)}, m_b^{(2)}\}$. Introducing hinges into the cross sections $b1$, $a2$, $b2$, we obtain the isostatic primary system. The rows of $\mathbf{C}^{-}$ and $\mathbf{C}^{0}$ correspond to the diagrams of bending moments induced in such a system by the unit loads p_i and z_i, respectively.

Now all matrices appearing on the right side of formulas (13.47) and (13.48) are known ($\mathbf{N}$ and $\mathbf{p}$ were defined in Section 13.3.4). Thus, we can compute

$$\mathbf{Q} = \begin{bmatrix} -1 & 1 & 1 & -1 & 0 & 0 & 0 & 0 & -1 & 1 & 0 & 0 \\ -1 & 1 & 0 & 0 & 1 & -1 & 0 & 0 & 0 & 0 & 0 & 0 \\ -1 & 1 & 0 & 0 & 0 & 0 & 1 & -1 & 0 & 0 & -1 & 1 \end{bmatrix},$$

$$\mathbf{u} = \{-1 \quad 1 \quad 0 \quad 0 \quad 0 \quad 0 \quad 0 \quad 0 \quad 0 \quad 0 \quad 2 \quad -2\}.$$

These matrices allow us to fill Table 13.13, which contains the initial data for the reduced problem. The solutions of the reduced problem coincide with the solutions of the original problem (Table 13.11), except that $\mathbf{s}_*$ and $\dot{\mathbf{w}}_*$ must be recovered from Eqs. (11.3), (11.7), and $(11.54)_2$.

Exercise 13.7

Solve the problem treated in Exercise 13.6 by using the reduced model (13.46) and (13.49).

13.5. Further Examples and Numerical Aspects

Here we give some optimal designs of structures typical for civil engineering. Since the dimensions of relevant LP problems prevent their presentation *in extenso*, we restrict ourselves to the final results. After those examples, we make some brief remarks on numerical aspects of the optimal plastic design.

13.5.1. Optimal Design of a Grillage

Grillages occur quite often in various structural systems. They are suitable objects of plastic optimization, because usually no other modes interfere with plastic collapse. The axial forces and, in most cases, the shear forces can be neglected. We should, however, carefully judge whether an interaction between the bending and twisting moments is important.

In general, the twisting moments are small or absent in grillages that do not have rigid nodes at the intersections of beams. An example of such a grillage is shown in Fig. 13.9. We see that one end of each beam is supported by a planar hinge, i.e., a hinge that permits rotation in the plane situated along the axis of the beam. The opposite end is simply supported. The only requirement introduced at the nodes of the grillage is that the intersecting beams must have common vertical displacement. Under such circumstances the bending-only approach is adequate.

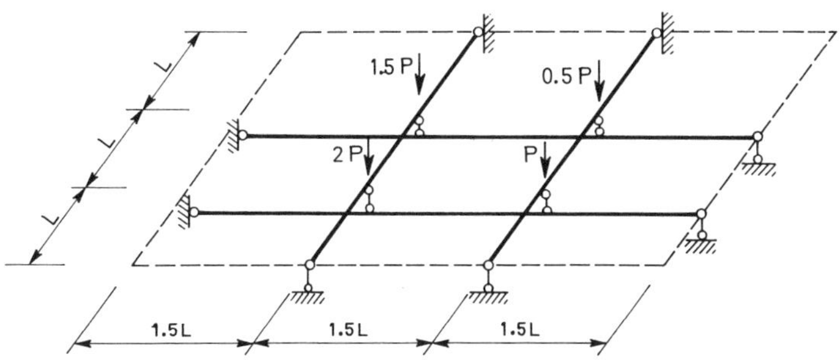

Figure 13.9. Layout and Loading of Grillage.

We now consider optimization of the grillage in Fig. 13.9, which consists of two longitudinal beams of length $4.5L$ and two transverse beams of length $3.0L$. Each beam is prismatic, and its yield moment is subject to optimization.

In the absence of technological constraints, we obtain the results given in Fig. 13.10. The entire load is carried by the transversal beams, and the minimum nondimensional volume of the grillage is 8.0, which is only 40% of the volume $v_{\text{const}} = 13.5$ that would be required if yield moment were constant throughout the structure.

Perhaps we might want only two distinct cross sections: one for longitudinal and one for transversal beams. Moreover, we might impose the technological constraint $m^0 \geqslant 0.5$. Then the minimum volume increases to 11.5, which is still 15% less than v_{const}. The results for the standardized grillage are shown in Fig. 13.11. Note that the sum of the rotations in the plastic hinges that occur in a particular group of beams is equal to the total

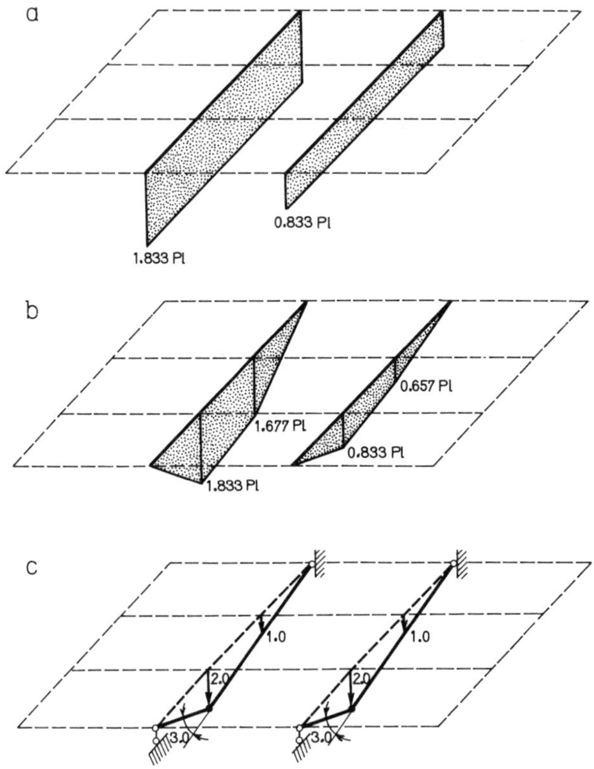

Figure 13.10. Plastic Collapse of Optimal Grillage, Each Beam Optimized Independently: (a) Optimal Yield Moments; (b) Bending Moments at Incipient Collapse; (c) Collapse mechanism.

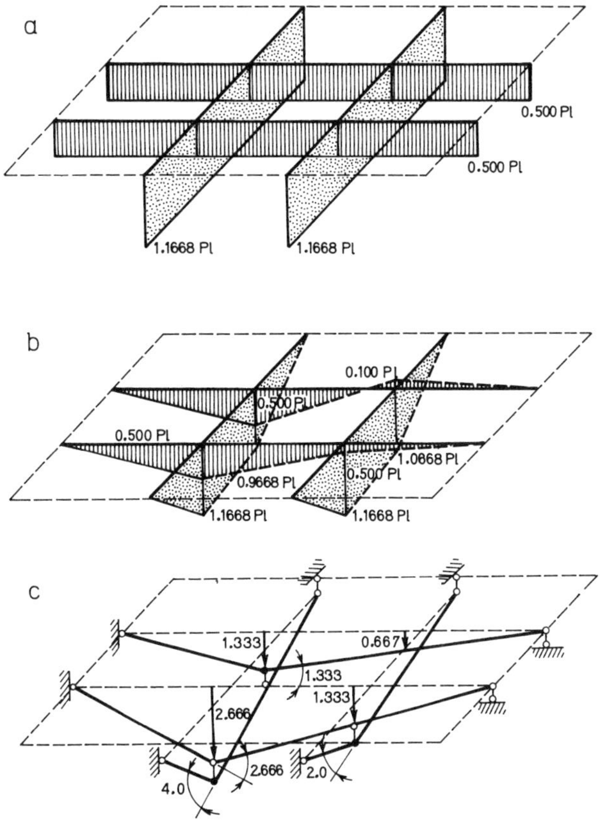

Figure 13.11. Plastic Collapse of Optimal Grillage, Parallel Beams have Uniform Cross Section: (a) Optimal Yield Moments; (b) Bending Moments at Incipient Collapse; (c) Collapse Mechanism.

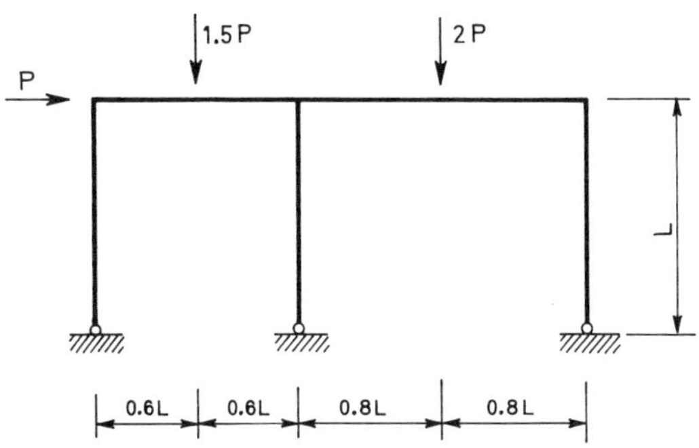

Figure 13.12. Layout and Loading of Two-Bay Frame.

length of those beams. This is enforced by the optimality criterion (13.33). The dashed lines in Fig. 13.11b indicate the values of bending moments that are not uniquely defined by the rigid, perfectly plastic model.

13.5.2. Optimal Design of a Two-bay Frame

The frame in Fig. 13.12 was optimized according to the bending-only version of model (13.32). The results of free optimization are given in Fig. 13.13. Note the small value of the optimal yield moment of the central column and the almost uniform distribution of m^0 for the remaining elements. This solution suggests that we could modify the layout by replacing the central column with a pin-ended strut.

If we demand that all three columns have the same cross section, then the solution of Fig. 13.14 is obtained. The volume of such a frame exceeds that of the free optimum by 10%. This illustrates a typical feature of optimal plastic design: the cost function tends to be flat in the vicinity of the

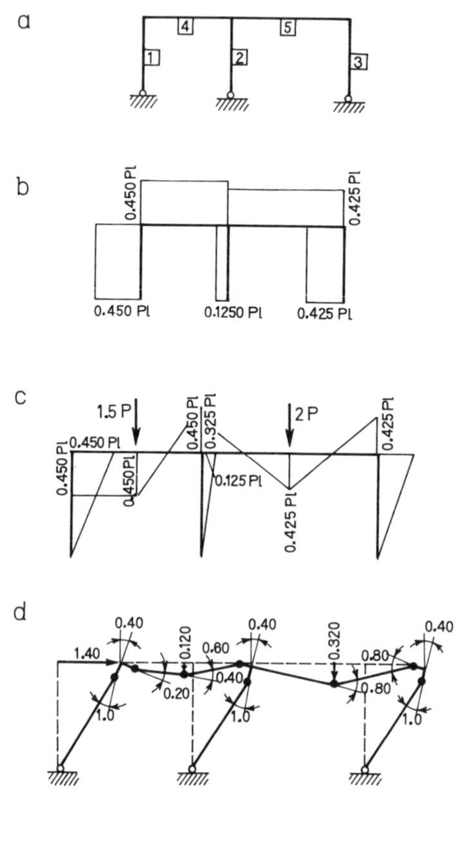

Figure 13.13. Plastic Collapse of Optimal Two-Bay Frame, each Element Optimized Independently: (a) Design Variables; (b) Optimal Yield Moments; (c) Bending Moments at Incipient Collapse; (d) Collapse Mechanism.

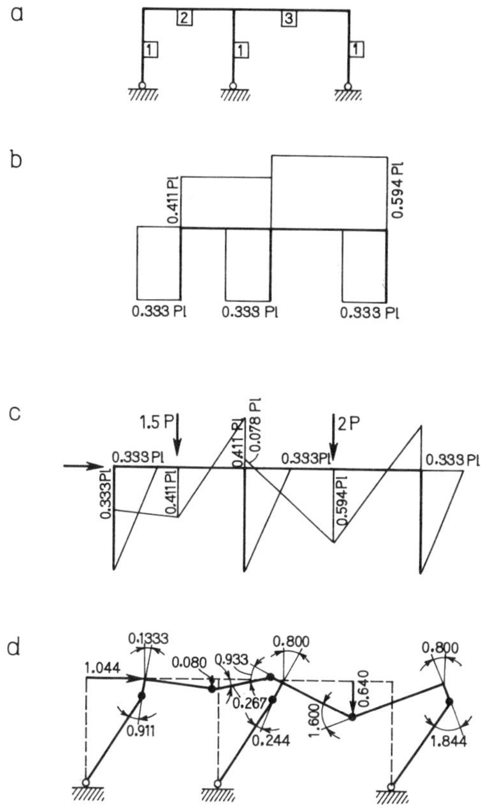

Figure 13.14. Plastic Collapse of Optimal Two-Bay Frame, Columns have Uniform Cross Section: (a) Design Variables; (b) Optimal Yield Moments; (c) Bending Moments at Incipient Collapse; (d) Collapse Mechanism.

minimum. Therefore, substantial changes can be made in the distribution of plastic resistance throughout the structure without a significant effect upon its cost (volume).

13.5.3. Optimal Design of a Three-story Frame

The example of the frame in Fig. 13.15 was taken from Polizzotto and Mazzarella (1979). Its purpose is to illustrate the influence of interaction upon the optimal design. The vertical and horizontal distributed loads acting on the frame were replaced by the system of statically equivalent concentrated forces applied at the nodes (each beam was divided into six segments of equal length; each column, into three segments). Three cases of the optimal design problem were considered:

I. No technological constraints, so the yield moment of each of the 45 characteristic cross sections is optimized independently.

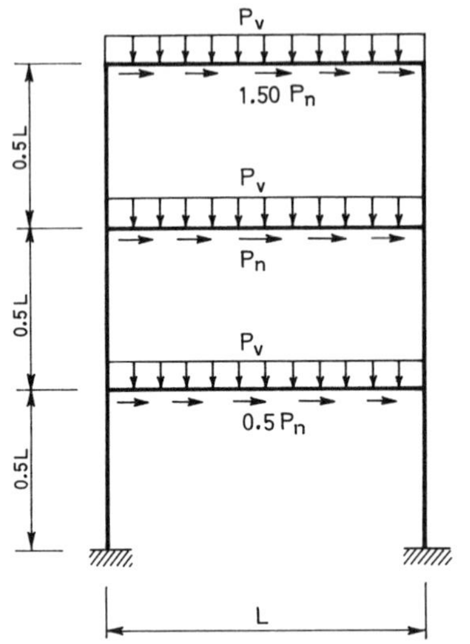

Figure 13.15. Layout and Loading of Three-Story Frame.

II. Prismatic columns and beams.

III. Prismatic columns and beams, with all beams having a common yield moment and columns having nondecreasing yield moment when considered from the top down.

The results for cases I–III are shown in Figs. 13.16–13.18. The minimum weight was found as follows:

	With interaction	Bending only
Case I	22.45	17.53
Case II	30.60	26.34
Case III	31.21	27.22

Note that, despite the general tendency of interaction to increase the required value of the yield moment, there are parts of the structure where the opposite is true.

13.5.4. Numerical Aspects

Since both ultimate load analysis and optimal plastic design supply valuable information for the designer, it is reasonable to include computer programs that solve those problems into the software supporting the design process. Such a program must have a preprocessor that facilitates a convenient input, verification and modification of initial data, a solver of the LP problems, and a post processor that presents the results of computations

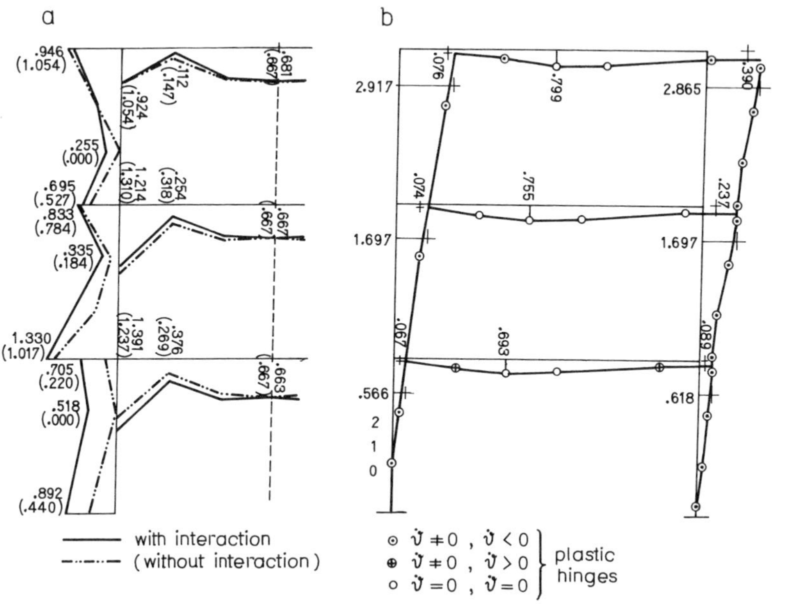

Figure 13.16. Plastic Collapse of Optimal Three-Story Frame in Case I: (a) Optimal Yield Moments; (b) Collapse Mechanism.

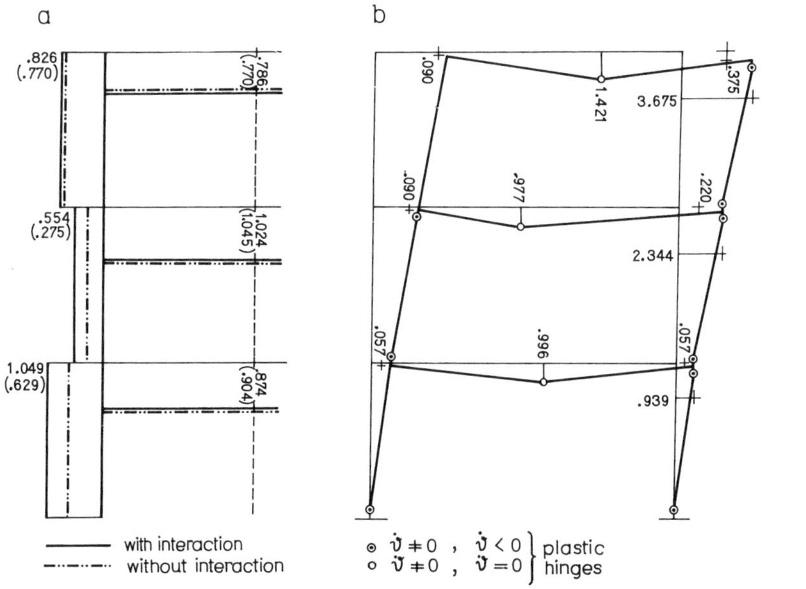

Figure 13.17. Plastic Collapse of Optimal Three-Story Frame in Case II: (a) Optimal Yield Moments; (b) Collapse Mechanism.

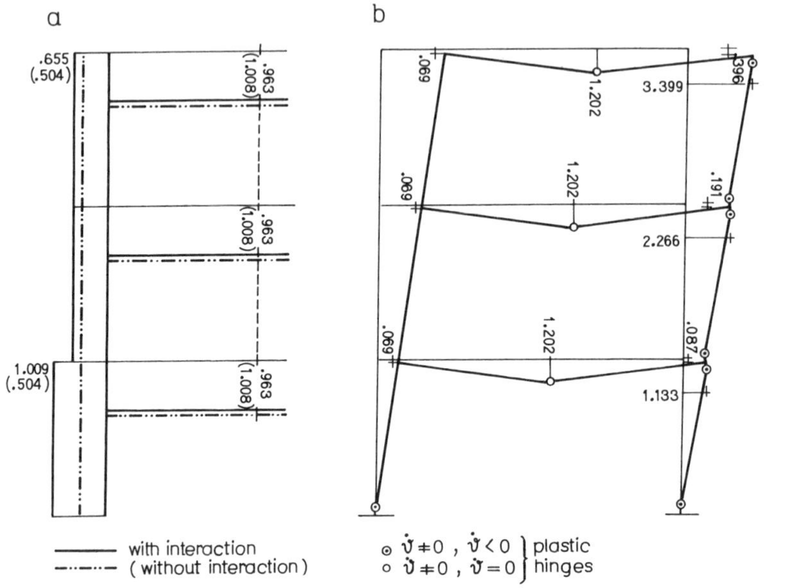

Figure 13.18. Plastic Collapse of Optimal Three-Story Frame in Case III: (a) Optimal Yield Moments; (b) Collapse Mechanism.

in a form desired by the user. As far as the pre- and postprocessors are concerned, the codes widely used in elastic FEM analysis can be easily modified to work with the additional data encountered in plasticity.

As mentioned in Section 10.6, since there are many good commercially available LP solvers for mainframe computers, it is not a good idea to use a self-made routine. The software supplied by IBM, ICL, or KENTRON uses sophisticated versions of the simplex algorithm that allow one to solve very large LP problems. As a rule, only nonzero elements of the simplex table are stored, a special product form is used to obtain a new vertex of the admissible domain, side constraints of the type $x_i^- \leqslant x_i \leqslant x_i^+$ are treated in a special way, etc.

Many of those memory-saving techniques can be implemented on microcomputers as well. One has to be aware, however, that some of them are quite time-consuming. The ease of man-machine interaction that characterizes modern small systems suggests wider application of a technique called *active-set strategy.* We used this method to solve some examples in this chapter. Based upon the elastic solution, experience, or even intuition, we can predict which yield locus faces are likely to be active at plastic collapse. Including only the inequalities that correspond to such faces into the set of constraints, we can reduce considerably the dimensions of problems (13.10), (13.32), or (13.44). Of course, after the solution has been obtained it is necessary to verify that it satisfies all the constraints. If some of them are violated, then they must be included in the active set and the problem must be solved once again. Such an iterative procedure could be performed automatically, but it is better to allow the user to interfere.

Finally, we remark on the optimal design of flexural systems. When considering the interaction between axial force and bending moment, we must assume a certain ratio β between the yield force and the yield moment before solving the problem. The true value of β becomes known after the cross sections of the structure are chosen. If it differs considerably from the assumed value, then we must solve the optimization problem again. It might be tempting at first glance to overcome this inconvenience by taking the yield force and the yield moment of the cross section as independent design variables. Remember, however, that β defines the orientation of the yield locus faces in the space of generalized stresses. Thus, a variable β means a variable N that destroys the linearity of constraints in problems (13.10), (13.32), and (13.44). Therefore, the ratio m^0/n^0 must be kept constant.

14

Optimization of Shells and Plates

14.1. Introduction

Shells are complicated mechanical objects: the set of differential equations describing their behavior is of order eight in spatial variables. The engineer's intuitive abilities for shell optimization problems are much weaker than for trusses, frames, and other bar structures. This is especially true with respect to the optimal design problem: skillful engineers can intuitively predict the layout of minimum-weight trusses with an error of at most 3% in terms of the objective function; but they make errors from 20 to 100% in their attempts to predict intuitively either the optimal shape of a shell's mid-surface, or the distribution of shell thickness, or both. (An error of 100% refers to the case when researchers do not notice the vanishing of their objective functions; see examples of such "erosion" in Volume 1, p. 290 for ill-posed plates and here for shells.)

The two types of problems in shell optimization are:

1. For a given geometry of the mid-surface when only the thickness of shell or the reinforcement is to be optimized.
2. When the shape of the mid-surface is to be optimized subject to constraints on the distribution of thickness or reinforcement (or both).

Case 2 is generally much more difficult than case 1, although chronologically case 2 was considered earlier. If we omit the related problems of shells of uniform strength (stated by Laplace), the first solution of an optimization problem for arches that was of interest to practical engineers was published by Jorgensen (1915). He considered an arch dam not as a shell but as a system of independent arches subjected to the hydrostatic pressure p appropriate to the water depth at the level of the arch. For a given span $2l$ of the arch, the radius r of each arch or the angle 2ϕ is the design variable defining the optimal structure of minimal volume (Fig. 14.1a):

$$V = 2\phi r F \to \min. \tag{14.1}$$

After substituting $r = l/(2\sin\phi)$ and $F(\text{area}) = pr/\sigma_c$ (where σ_c is the yield stress of material in compression; see Volume 1, p. 205), we have

$$V = \frac{pl^2}{2\sigma_c}\frac{\phi}{\sin^2\phi}. \tag{14.2}$$

Stationary values of V occur at roots of the equation

$$2\phi - \tan\phi = 0. \tag{14.3}$$

The nontrivial solution $2\phi_1 = 133°34'$ for arches of minimal volume until now has been considered as fundamental in the field of engineering optimization. The last formula gives the optimal value 2ϕ only for arches of constant radius, but a similar expression can be found for a spherical shell (see Exercise 14.1). Figure 14.1b gives the graphical interpretation of the optimal solution for this problem.

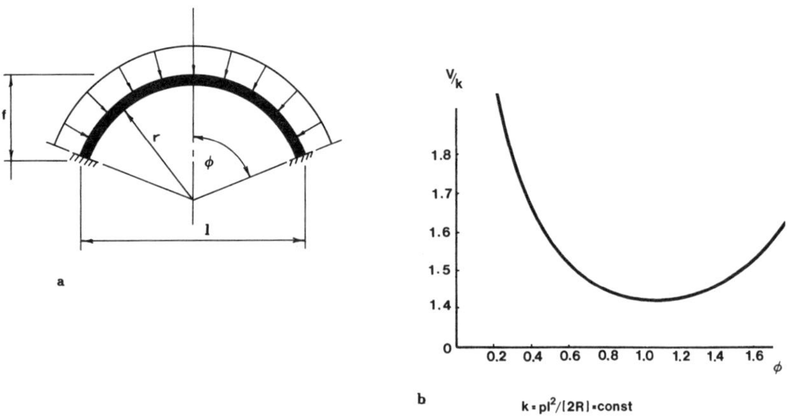

Figure 14.1. Cylindrical Optimal Shell with Uniform Thickness (Jorgensen's problem): (a) Scheme of Optimal Horizontal Arch. (b) Volume of the Arch as a Function of Angle ϕ.

But $\phi_1 = 66°47'$ is not the unique solution of the problem: the general solution will give us many values of ϕ. For instance, for large integers k, all values $\phi_k = (2k+1)\pi/2$ are approximate solutions of (14.3). This evidently indicates that the problem has many extrema. Direct verification shows that the problem has another source of multiextremality: under the same conditions, we shall have the same shell, but turned inside out, if we do not exclude this case by a special condition—say considerations of buckling or brittle fracture failure.

Some authors assume that the optimal forms of arches should follow the pressure curves, either precisely (for one loading) or approximately (for alternative loadings, see Volume 1, p. 98, when the bending moments cannot be excluded but can be minimized by a proper selection of the arch midline); see Kiselev (1953) or Khuberyan (1956). (The pressure curve in compression is the same as the chain line in tension.) Khuberyan has studied pressure surfaces that can be constructed in various ways (in contrast with pressure curves that are usually unique). His optimization-based research, starting from the idea of pressure surfaces, led the author to a more complex and reliable scheme of arch-cantilevers (1956).

The concept of discretization has always been of interest to engineers; in particular, purely analytical approaches are efficient only for axisymmetric shells and plates. But ordinary approximate approaches were used usually only on the last stage of structural analysis and design. However, during the last two decades model discretization has been most often used in the initial stages of the process. The realization of these procedures is based on the use of finite difference and finite element methods.

Optimization problems for axisymmetrical shells and plates were analyticaly solved in the 1960s by Prager, Shield, Save, and others [see Volume 1 and Save (1981)]. We shall not discuss this field except that (1) we illustrate the general approach to these problems, using the Pontryagin maximum principle and dynamic programming; (2) we discuss some problems for axisymmetric shells with an unspecified mid-surface. We study both ill-posed or degenerate cases arising in axisymmetric problems and regular problems arising in the general nonaxisymmetric case. The latter are our primary concern. These problems began with Ramaswamy (1958), who showed that for a definite class of shells (particularly textile shells, which have no shear stresses), one may find a mid-surface that guarantees the realization of these stress states under at least one loading. The search for this form was based on the solution of the classical boundary value Dirichlet problem for Poisson's equation.

The problem of constructing a proper optimal model for shells was not exhausted, and, as we shall see, this procedure sometimes fails. Now we have no definition of the physically and logically correct model of shell optimization. Physical properties of the shell material, admission or prohibition of plastic strains, or even the ways of load application are not as important as geometrical conditions or special requirements stated about the shape of the shell: the degeneration of ill-posed problems for optimized shells often happens independently of plastic strains or of the method of loading.

The important part of the optimization procedure is the selection of a proper physical model. If the model is not adequate, the optimization makes no sense. Plasticity theory allows for plastic strains and, in general, gives a reduction of cost unless plastic behavior is not permitted for the considered class of structures. Elasticity theory often leads researchers to very inefficient designs, even if optimization is used. If we require the wing of an aircraft to remain elastic in a crash, we cannot provide designers with an efficient procedure, even when the rest of the design is excellent.

Numerical methods for shell optimization began with Wolfensberger (1964), who included the optimal reinforcement of concrete slabs for specified loading and shape of the plan view on the horizontal plane. To find the minimal reinforcement, Gvozdev's statical approach to limit design was used in combination with finite difference discretization. The volume of steel reinforcement was the objective function, and the constraints relating to reinforcement and bending moments were linear inequalities. They were based on the Johansen plasticity criterion. The resulting LP problem was resolved with the use of a standard code. Later, Reitman (1967) used the same approach for shallow shells, but the Johansen plasticity criterion was replaced by the Karpenko–Morley (1966) conditions. Simultaneously, three other plasticity criteria [by Mises, Balandin–Geniev–Kissyuk (1965), and Reitman (1967)] were applied for shells of metal, reinforced concrete, and plain concrete, respectively. See the discussion in Reitman and Yarin (1974). Later problems with unknown mid-surfaces were resolved by Reitman (1983, 1985) for plastic structures and by Plaut, Johnson, and Parbery (1984) for elastic shells. Although the finite element method has often been used in shell optimization [and we discuss its applications by Prasad and Haftka (1979)], it is subject to some difficulties and cannot fully replace finite differences or polynomial discretization.

At the same time, the range of problems was widened to include more sophisticated problems, such as dynamic loading, "standardization" of structural elements, and optimal "stocking" in a warehouse, as proposed by Yarin for optimal continuous beams; see Section 14.4. Brandt (1984) gives the fullest list of references in optimal design up to 1980. More recent developments are covered by Yu. Lepik (1982) (analytical methods for dynamic problems) and by Kruzelecki and Zyczkowski (1985) (a wide range of problems with emphasis on buckling and dynamic constraints). Some problems of shell optimization are covered by Malkov and Ugodchikov (1981), Haug (1984), Haug and Cea (1981), Massonnet and Save (1963), and in Soares (1986) [see the reference to Plaut (1986)]. Raskazov and Dekhtyar' (1978) considered problems of optimal limit design of shells, and Jakubowska and Lesniak (1982) studied shape optimization of double-curved elastic shells.

14.2. Fundamentals of the Engineering Theory of Shells

14.2.1. Statics of Double Curvature Shells

The theory of shells allows different approaches, depending on the needs of the potential audience. We do not use tensors and differential geometry, but give instead an "engineering version" of shell theory based on Flügge (1960). We cannot name all the pioneers of this field, but we mention Donnell, Mushtari, Vlasov, Pucher, and Marguerre, whose independent contributions were especially significant in the 1930s when exchange of technical information between countries was extremely difficult.

14.2.1.1. Membrane Theory. Let $z = z(x, y)$ be the equation of the mid-surface of the shell as a function of the plane Cartesian coordinates x, y, and let $q = q(x, y)$ be the load per unit x, y area in the z-direction. The lines $x = \text{const}$ and $y = \text{const}$ on the shell mid-surface will not meet at right angles ($\beta \neq 90°$ as shown in Fig. 14.2). The angles ϕ and ψ illustrate this difference too. It seems reasonable to describe the membrane stresses in a shell by skew forces $\tilde{N}_x$, $\tilde{N}_y$, and $\tilde{N}_{xy} = \tilde{N}_{yx}$ (Fig. 14.3). The skew forces are forces per unit length of the line elements through which they are transmitted. The projections of forces $\bar{N}_x$, $\bar{N}_y$, and $\bar{N}_{xy}$ are convenient for constructing the equilibrium equations. In the general theory of membrane shells, $\bar{N}_x$, $\bar{N}_y$, and $\bar{N}_{xy}$ are used as intermediate values in transforming to the skew internal forces

$$\tilde{N}_x = \bar{N}_x\sqrt{\frac{1 - z_{,x}^2}{1 + z_{,y}^2}}, \quad \tilde{N}_y = \bar{N}_y\sqrt{\frac{1 + z_{,y}^2}{1 + z_{,x}^2}}, \quad \tilde{N}_{xy} = \bar{N}_{xy}. \tag{14.4}$$

Here $z_{,x}$ and $z_{,y}$ are the tangents of slope angles ϕ and ψ, and the comma denotes the partial derivative with respect to the independent variables.

The angle between the two skew axes is defined by the formula (see Fig. 14.3)

$$\beta = \sin^{-1}\sqrt{\frac{1 + z_{,x}^2 + z_{,y}^2}{(1 + z_{,x}^2)(1 + z_{,y}^2)}}. \tag{14.5}$$

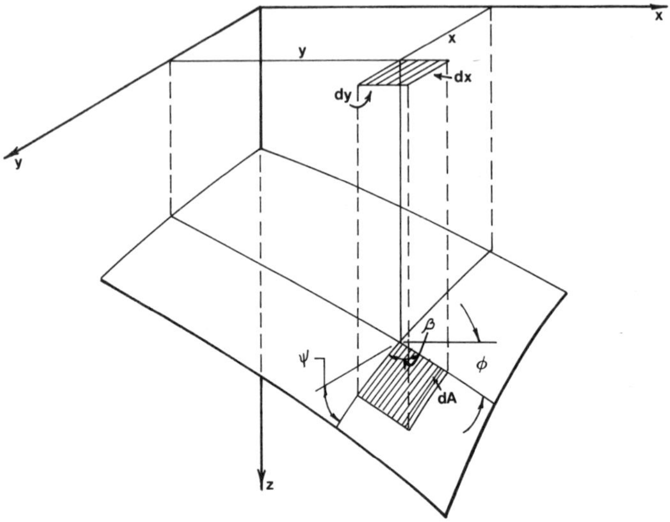

Figure 14.2. Shell Element and Coordinate axes.

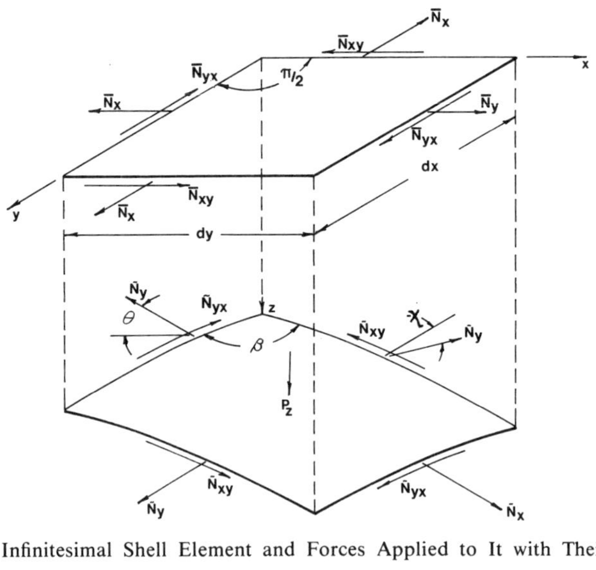

Figure 14.3. Infinitesimal Shell Element and Forces Applied to It with Their Horizontal Projections.

The statical equilibrium condition of a small shell element (not necessarily shallow!) in the z-direction leads to the equation

$$\bar{N}_x z_{,xx} + 2\bar{N}_{xy} z_{,xy} + \bar{N}_y z_{,yy} + q = 0. \tag{14.6}$$

Meanwhile the equilibrium conditions in the x- and y-directions give

$$\bar{N}_{x,x} + \bar{N}_{xy,y} = 0, \qquad \bar{N}_{xy,x} + \bar{N}_{y,y} = 0. \tag{14.7}$$

We have assumed that only the vertical component of the distributed load exists ($p_z = q\,dx\,dy$), whereas horizontal components of the load vanish. If we introduce Airy's function $F(x, y)$ and assume

$$\bar{N}_x = F_{,yy}, \quad \bar{N}_y = F_{,xx}, \quad \bar{N}_{xy} = -F_{,xy}, \tag{14.8}$$

Eqs. (14.7) are satisfied identically and the substitution in (14.6) gives the nice equation proposed by Pucher [see Flügge (1960)]:

$$\frac{\partial^2 F}{\partial x^2}\frac{\partial^2 z}{\partial y^2} - 2\frac{\partial^2 F}{\partial x\,\partial y}\frac{\partial^2 z}{\partial x\,\partial y} + \frac{\partial^2 F}{\partial y^2}\frac{\partial^2 z}{\partial x^2} + q = 0 \tag{14.9}$$

or $\mathrm{PU}(F, z) + q = 0$, where

$$\mathrm{PU} = F_{,xx} z_{,yy} - 2F_{,xy} z_{,xy} + F_{,yy} z_{,xx} \tag{14.10}$$

is Pucher's differential operator. The skew forces are the actual forces in membrane shells, but it is not convenient to use them directly in the plasticity criteria. Therefore, we have to convert the forces into Cartesian components (simultaneously orienting them along the principal directions of stress):

$$N_1 = \frac{A_x + A_y}{2\sin\beta}, \qquad N_2 = \frac{A_x - A_y}{2\sin\beta}, \tag{14.11}$$

where A_x and A_y are new auxiliary variables defined by

$$\begin{aligned} A_x &= \tilde{N}_x + \tilde{N}_y + 2\tilde{N}_{xy}\cos\beta, \\ A_y &= \sqrt{A_x^2 - 4\sin^2\beta(\tilde{N}_x\tilde{N}_y - \tilde{N}_{xy}^2)}. \end{aligned} \tag{14.12}$$

We will not derive these equations, but we will discuss them. The concept of shallowness is not used in this approach. Pucher's differential equation governs the equilibrium of the internal force projections on all three axes, but the forces are not subject to ordinary physical definitions: for shells of arbitrary rise, these forces are skew to each other and not identical to their projections on the x-, y-, and z-axes. Both properties are important for interpreting the results.

If a shell is shallow, the derivatives $z_{,xx}$, $z_{,yy}$, and $z_{,xy}$ are approximately equal to the curvatures and twist k_x, k_y, and k_{xy}, respectively. These values have rigorous geometrical interpretations: k_x and k_y are the direct curvatures of the shell mid-surface in the x- and y-directions, and k_{xy} is the twist curvature. Pucher's equation takes the simple form

$$k_x N_x + 2k_{xy} N_{xy} + k_y N_y + q = 0 \tag{14.13}$$

along with the two equilibrium equations in the horizontal plane (14.7) but now in terms of actual forces. Generally speaking, the form (14.9) of the equilibrium equation is convenient if the shape of the mid-surface is also an unknown; otherwise another form of the equilibrium equations is preferable (see arch dams discussed later).

If the mid-surface of a shell is specified, then the function $z(x, y)$ and its derivatives are known. The system defined by the Pucher equation is then apparently statically determinate (one equation in one unknown function F). But the problem could be redundant because of the support conditions. These facts will be illustrated with examples of optimization problems.

The inverse approach occurs when the stress state in a shell is given, but the mid-surface corresponding to this stress state is desired. Pucher's equation (14.4) may be simplified, and we can obtain the Dirichlet boundary value problem. Specifically, if before any optimization is attempted, the stressed shell admits only homogeneous stress states (for instance, $N_x = N_y = 1/a = \text{const}$, $N_{xy} = 0$), the set of equilibrium equations may be reduced to the Poisson partial differential equation governing the function $z(x, y)$:

$$z_{,xx} + z_{,yy} = -aq. \tag{14.14}$$

Since Eq. (14.14) has no free design parameters to use in an optimization procedure, the only remaining possibility comes from the variation of boundary conditions imposed on the full boundary value problem. This is certainly an inverse problem of shell theory, but it is not always the optimization problem: it is instead a problem about the uniform strength shell [Ramaswamy (1958) was the first contribution]. This problem has its beginning in the problem of the drop shape initiated by Laplace [see the complete references in Brandt (1984)]. Equation (14.14) leads also to the analogy between optimal structural design and soap films: as early as the 1930s researchers produced optimal shells by the study of soap films spanned on proper closed or open space contours. Changing the pressure inside the film allowed them to control the shell rise.

In the highly degenerate but still realistic case of the plane arch, the equation

$$\frac{d^2z}{dx^2} = -aq \tag{14.15}$$

can be solved immediately:

$$z = -aqx^2 + C_1x + C_2. \tag{14.16}$$

The solution defines a pressure curve under a homogeneous load, and the curve is a parabola. [Alternatively, a suspended chain may also be considered if the structure is under tension; see Kiselev (1953)]. The two constants of integration in this formula may be determined by two boundary conditions. But they also may be considered as design parameters and defined by minimizing an objective function.

14.2.1.2. Bending of Shells of Double Curvature. If a shell can sustain bending and torsional moments, this property contributes to the strength of the shell, but the scale of this effect depends on some other properties of the shell. The simplest way is to include the bending term in Pucher's equation. We shall do this only for a shallow shell and will identify the bending moments of this shell with the moments in a plane plate. We introduce Lagrange's operator

$$\mathrm{LA}(\mathbf{M}) = M_{x,xx} + 2M_{xy,xy} + M_{y,yy}, \tag{14.17}$$

where M_x, M_y, and M_{xy} are, respectively, bending and twisting moments forming a 3-tuple $\mathbf{M}$. Sometimes it is convenient to use a shortened notation, using the differential operators LA and PU(F, z):

$$\mathrm{LA}(\mathbf{M}) + \mathrm{PU}(F, z) + q = 0. \tag{14.18}$$

If $z = 0$ identically, that is, for the flat plate, the equilibrium equation takes its simplest form:

$$\mathrm{LA}(\mathbf{M}) + q = 0. \tag{14.19}$$

This equation of plate equilibrium has been used many times in optimization problems. If we use it for a plate, the operator LA in combination with Hooke's law gives the biharmonic operator and Sophy Germain's equation:

$$\Delta\Delta w = q/D, \tag{14.20}$$

where Δ is Laplace's operator and D is the bending stiffness:

$$\Delta = \frac{\partial^2}{\partial x^2} + \frac{\partial^2}{\partial y^2}, \qquad D = \frac{EH^3}{12(1-\nu^2)}. \tag{14.21}$$

Here E is Young's modulus, ν is Poisson's ratio, and H is the plate thickness. The material of the shell and plate has been assumed to be isotropic, but all the statical equations are valid also for anisotropic shells and plates, and the restatement of Hooke's law is obvious. In Eq. (14.20) we have assumed that H is constant; for variable H see the corrections in the sequel.

14.2.2. Plasticity Conditions

Plasticity criteria or conditions are used for estimating the stress states in shells. The main criteria by Mises, Tresca, and Johansen and discussed in Volume 1 for axisymmetrical plates and cylindrical shells. Here we consider criteria for arbitary forms of shells made of the same materials (metals and concrete). Currently, limit design may be used in some countries and must be used in many others, but the degree to which plastic strains are permitted in the normal life of a structure is sometimes qualified by other considerations. However the choice of a plasticity criterion is sometimes arbitrary, and we may choose it based on technical knowledge in any of several fields: manufacturing engineering, civil engineering, aircraft, shipbuilding, etc. So any of the criteria presented here may be replaced by any other criteria admissible on mechanical grounds.

14.2.2.1. Metals. Many metals can undergo large plastic strains without a corresponding increase in stress. Among possible yield critiera for such ductile metals is the "perfectly plastic" von Mises criterion: stress states $\mathbf{N}$ that cause equality to hold in (14.22) permit plastic strains to occur, whereas inequality means that no further plastic straining can happen at that stress:

$$N_x^2 - N_x N_y + N_x^2 + 3N_{xy}^2 \leqslant H^2\sigma_0^2, \tag{14.22}$$

where H is the thickness of the shell and σ_0 is the yield stress in tension and compression. This criterion is formulated by using the in-plane forces, but after dividing by H^2 it can be written in terms of stresses, since $N/H = \sigma$. In the space of the principal values N_1 and N_2, condition (14.22) is defined as an ellipse with the slope angle of the major axis equal to $\pi/4$ (see Fig. 6.24, p. 254, Volume 1). A Mises condition can be written for the bending stress state and used for the design of a metal plate; a section becomes plastic when the inequality

$$M_x^2 - M_x M_y + M_y^2 + 3M_{xy}^2 \leq H^4\sigma_0^2/4 \tag{14.23}$$

becomes an equality. The symbol H again defines the thickness of the solid shell, and the inequality is evident for plates in which only bending effects occur. But the attempt to allow for both bending and in-plane components of the stress state meets serious difficulties. Many proposals are based on assumptions about the values of the moment arm or lever of the internal couple, but we mention only a three-layer shell (often called *sandwich shell*): the thicknesses h and h^* of two extreme layers of material are small compared with the total thickness H. In this case, the thickness of the middle layer is $H - (h + h^*)$. The internal layer bears only shear forces, and the bending and twisting moments are borne only by the external layers. In such problems we usually have to determine optimal distributions of h and h^* (or only h, when both layers have the same thickness). Such three-layer plates and shells have been considered by many authors, but we must admit that requiring continuous change of the layer thickness is a hard manufacturing problem.

As we mentioned, it is often useful to replace the Mises condition by the Tresca condition; we shall see examples of this. Another condition used in the first shell optimization problems is Hill's plasticity condition: see Mikeladze (1969), who covers work back to 1950. This criterion is used for anisotropic plastic shells.

14.2.2.2. Reinforced Concrete. The applicability of plastic design to reinforced concrete strongly depends on the reinforcement ratio and the properties of the stress state, as will be discussed in detail in Volume 3. In plates and shells, reinforcement ratios are, as a rule, small enough to ensure collapse by plastification of the steel before the maximal strain capacity of the concrete is reached.

But in any case, von Mises' plasticity criterion is not admissible for concrete, since the material possesses asymmetric behavior in tension and in compression. Moreover, experts in reinforced concrete think that only plastic strains in the steel reinforcement may be permitted in many categories of concrete structures (for instance, even small plastic strains in the concrete are accompanied by microcracks that are not dangerous for the immediate strength of the structure, but lead to quick corrosion of the steel reinforcement). Another objection to plastic strains in concrete is based on the brittle behavior of plain concrete and its inability to resist reliably any extension. Only plastic strains in the reinforcement itself can be permitted by the designer. These considerations were used by Karpenko and Morley in 1966 to propose independently, and in the same notation (!), the following plasticity condition:

$$\begin{aligned}(\pm M_{xy} - \delta N_{xy})^2 - (\pm M_x - \delta N_x - \delta H\mu_x\sigma_0)(\pm M_y - \delta N_y - \delta H\mu_y\sigma_0) \leqslant 0,\\ \pm M_x - \delta N_x - \delta H\mu_x\sigma_0 \leqslant 0,\\ \pm M_y - \delta N_y - \delta H\mu_y\sigma_0 \leqslant 0,\end{aligned} \tag{14.24}$$

where 2δ is the given moment arm of the internal couple considered constant in the whole shell, H is the thickness of the shell section, and μ_x and μ_y define the reinforcement factors in the shell in the x- and y-directions, respectively, or ratios of the reinforcement section area to the total section area; in a more general case δ could be considered variable and subject to iteration. Graphically, this criterion is presented in the space of three arguments by two conical surfaces. Their trace in the axisymmetrical case is a rectangle (see pp. 207–208, Volume 1, where the case $M_{xy} = 0$ and $N_{xy} = 0$ is discussed).

If the structure cannot sustain membrane forces or if those forces are excluded for some other reasons, the Karpenko–Morley criterion is reduced to the form

$$M_{xy}^2 - (M_x - M_{0x})(M_y - M_{0y}) \leqslant 0, \tag{14.25a}$$

where

$$M_x - M_{0x} \leqslant 0, \qquad M_y - M_{0y} \leqslant 0 \tag{14.25b}$$

and

$$M_{0x} = \delta H\mu_x\sigma_0, \qquad M_{0y} = \delta H\mu_y\sigma_0. \tag{14.25c}$$

This plasticity condition is usually called *Johansen's condition*, based on his work in the early 1940s [see Johansen (1962)], though the author saw it for the first time in Massonnet and Save (1963). In all these sources the plasticity condition was formulated for the flat plate, but the curvature of the shell is not involved at all: it will influence only the equilibrium and compatibility equations.

14.2.2.3. Plain Concrete. Plain concrete can be considered as plastic only in compression. Many other requirements have been mentioned by different experts, who frequently failed to convince each other. The author decided to select the criterion by Geniev and Kissyuk (1965), because they contributed much to the problem. Since they based their plasticity criterion on experiments and on an earlier study by Balandin, in the following the criterion will be called BGK. Its form is

$$H = \frac{(\sigma_c - \sigma_t)I_1 + [(\sigma_c - \sigma_t)^2 I_1^2 + 4I_2\sigma_c\sigma_t k]^{1/2}}{2\sigma_c\sigma_t}, \tag{14.26}$$

where

$$I_1 = N_1 + N_2, \qquad I_2 = N_1^2 - N_1N_2 + N_2^2, \tag{14.27}$$

N_1 and N_2 are the principal values of the membrane forces,

$$k = 1 - \left(1 - \frac{3\sigma_s^2}{\sigma_c\sigma_t}\right)\left(1 - \frac{2(N_1^3 + N_2^3) - 3N_1N_2I_1}{2(N_1^2 - N_1N_2 + N_2^2)^{3/2}}\right), \tag{14.28}$$

and σ_t, σ_c, and σ_s are the yield limits of plain concrete in one-dimensional tension, compression, and pure shear, respectively. This criterion evidently shows that in the plane stress state under equal compression in the x- and y-directions the material is stronger than under one-dimensional loading; this opinion is not shared by all experts in this field. But this property of the criterion helps to construct efficient optimal stress distributions; that is particularly important in optimal design, and the author wants to believe that this property is correct.

Computing with this criterion is time-consuming, since it involves many arithmetic operations. So, we may use a simpler criterion, the modified BGK criterion. It uses only the physical property σ_c and has the form

$$N_x^2 + N_y^2 + 2N_{xy} - 1.576\sigma_c H(N_x + N_y) + 0.576\sigma_c^2 H^2 \leqslant 0. \tag{14.29}$$

This condition defines a circle with radius $0.788\sigma_c$ or $0.788\sigma_c H$ in the plane of principal stresses σ_1 and σ_2 or forces N_1 and N_2, respectively. The criterion describes only stresses in the positive quadrant of the N_x-N_y plane, and all other cases are eliminated by the additional condition $N_xN_y \geqslant N_{xy}^2$, which enforces the nonnegativity of the smallest principal stress. At first glance, this condition would seem to rule out unnecessarily possible candidates for the optimal solution because it does not permit forces of small magnitude. However, the optimal solution does not occur for forces near zero, so imposition of this constraint is not unduly restrictive. In the formulation of these criteria for concrete, compressive force components are considered as positive, whereas for all other criteria they were assumed negative.

The five plasticity criteria discussed, however different they are, cover a wide range of the shells designed optimally, although many other formulations (often different only in notation) are presented in other publications. The criteria are used here, sometimes in simplified forms, to achieve the goals relating to the desired mathematical models. For instance, if we are satisfied by a linearized approach, an inscribed polygon will replace the limit surface in the space of forces and give a conservative estimate of the limit design.

14.2.3. Compatibility Conditions

We shall consider perfectly plastic and elastic physical laws of shell deformations. In the first approach, limit design theory will be used [Gvozdev (1936)]. In both formulations the kinematical compatibility conditions will be used. For instance, for a slab they are

$$W_{,xx} = \kappa_x, \quad W_{,yy} = \kappa_y, \quad W_{,xy} = \kappa_{xy}, \tag{14.30}$$

where $W = W(x, y)$ is the lateral deflection of the slab, and κ_x, κ_y, and κ_{xy} are bending curvatures in the x- and y-directions and the twisting of the slab, respectively. In the process of elastic optimization, these compatibility conditions must be accounted for explicitly, whereas in the limit design, due to Gvozdev's statical theorem, the conditions are satisfied automatically if the material of shell admits unlimited deformations without cracks or destruction. The shell design, as obtained by the statical approach, will be safe (this means that the shell will not fail under the postulated conditions), although the variation in value of the objective function from the optimal design may be significant.

14.2.4. Hooke's Law for Shells

The forces and moments in shells **N** and **M** are connected with the strains by the formulas

$$N_x = \frac{EH}{1 - \nu^2}(\varepsilon_x + \nu\varepsilon_y),$$

$$N_y = \frac{EH}{1 - \nu^2}(\varepsilon_y + \nu\varepsilon_x),$$

$$N_{xy} = N_{yx} = \frac{EH}{2(1+\nu)}\,\varepsilon_{xy},$$
$$M_x = \frac{EH^3}{12(1-\nu^2)}(\kappa_x + \nu\kappa_y),$$
$$M_y = \frac{EH^3}{12(1-\nu^2)}(\kappa_y + \nu\kappa_x),$$
$$M_{xy} = M_{yx} = \frac{EH^3}{12(1+\nu)}\,\kappa_{xy}, \tag{14.31}$$

where the κ's and ε's are bending and extensional strains in the shallow shells. The existence of in-plane components of displacements u and v in the x- and y-directions, respectively, has been tacitly ignored, but they are easily included [see Novozhilov (1964)].

The theory of bending of plates and shells usually is developed for constant thickness H and moment arm 2δ. Otherwise, the substitution in all differential equations produces terms with derivatives of H and δ with respect to x and y. The first application of the theory for plates to the optimal design of a vibrating rectangular platelike structure belongs to Armand (1971). Few researchers have studied this equation for optimization, because it is highly nonlinear and requires a meticulous interpretation of the results. This is no reason to ignore the terms containing the derivatives of H; moreover, these terms tend to infinity if the problem has no constraints (see Sections 14.4 and 14.6). Later, Olhoff (1973) showed that the "pillow-like" solutions with continuous variation of thickness were caused by local extrema (see Section 14.6).

14.2.5. Preliminary Remarks on the Statement of Problems

Due to great differences between various approaches to shell optimization problems, we mention initially only two classical objective functions:

$$C = \int_S H\,dS = \int_{S_p} H\sqrt{1 + z_x^2 + z_y^2}\,dS_p \to \min, \tag{14.32}$$

with S is the mid-surface area, S_p is the area of its projection on the plan view, and the integrations are taken over all of these areas. In shallow shells where the geometry of the mid-surface is identical to that of the plane plate, we obtain the simpler equation

$$C = \int_{S_p} H(x, y)\,dS_p \to \min. \tag{14.33}$$

We add one more objective function for reinforced shells:

$$C = \int_{S_p} (\mu_x + \mu_y) H \sqrt{1 + z_{,x}^2 + z_{,y}^2}\, dS_p \to \min, \qquad (14.34)$$

where μ_x and μ_y are the reinforcement factors, the integration is made over the whole mid-surface, and the total volume of reinforcement is minimized.

We discuss some other optimality criteria in connection with the different problem statements. (They include the concepts of scale economy and optimal stocking; see Section 14.6.) From the point of view of manufacturing, we must write the optimal sequence of material orderings by their adjustment to the optimal design: (1) the steel reinforcement layout in the concrete structures; (2) shape of solid structures of concrete; (3) shape of the cast metal structures; (4) shape of the forged metal elements; (5) shape of the rolled thin-walled structural members. We will not consider optimization problems for composite materials, prestressed structures, and chemical, heating, cooling, or radioactive optimal processing of material elements aiming to improve their mechanical properties.

All derivations have been made in Cartesian coordinates, although they may be developed in polar coordinates as well. In the next section, we use the corresponding theory of shells in polar coordinates without explicit derivation based on the change of coordinates:

$$r = \sqrt{x^2 + y^2}, \qquad \phi = \tan^{-1}(y/x).$$

Similar transformations will be carried out for the differential and integral operators, but we shall not repeat them, since an independent survey of the equation of the theory of axisymmetric membrane shells is given in Volume 1, p. 252. The contribution of the moment stress state may be studied in detail in Novozhilov (1964) and Flügge (1960).

14.3. Axisymmetric Shells and Plates

14.3.1. Cylindrical Plastic Shell

The problem of optimization of cylindrical shells was considered in Volume 1 by an analytical method. We consider a similar problem to describe a general approach to the optimization of the cylindrical shells on the basis of Pontryagin's maximum principle (see Section 14.7 for a brief statement of the principle). The shell is in equilibrium if the following equation is satisfied:

$$\frac{d^2M_x}{dx^2} + \frac{N_\theta}{r} + p = 0. \tag{14.35}$$

Here M_x is the longitudinal bending moment, N_θ is the circumferential normal force, r is the radius of the middle surface, and $p(x)$ is the given internal pressure (Fig. 14.4). The x-axis is directed along the shell axis. Sometimes this version of the theory of shells is called *semimomentless*, since it ignores circumferential bending moments but accounts for longitudinal ones. One end of the shell is clamped while the other is free. The objective function $C(L)$ reflects the minimum condition for total reinforcement:

$$C(L) = \int_0^L (\mu_x + \mu_\theta) r\, dx \rightarrow \min, \tag{14.36a}$$

or, by expressing the reinforcement factors in terms of generalized forces M_x and N_θ [see Karpenko (1966) and Morley (1966)],

$$C(L) = \int_0^L (M_x + N_\theta \delta)\, dx, \tag{14.36b}$$

where δ is the half-arm of the internal forces (see Karpenko–Morley's plasticity criterion in Section 14.2). (In the more general case of loading, we must use the absolute value because the reinforcement ratios μ_x and μ_θ

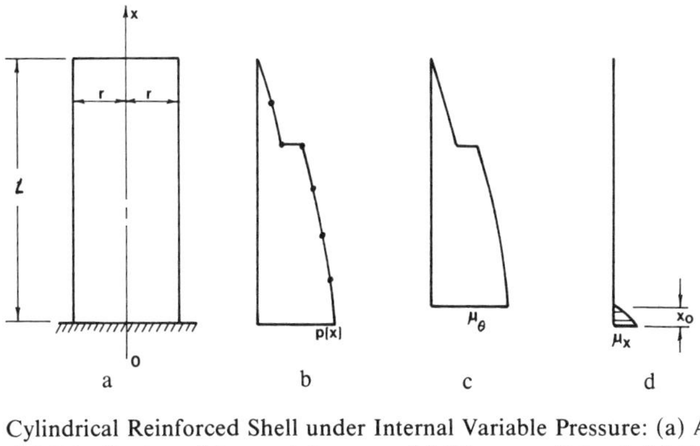

Figure 14.4. Cylindrical Reinforced Shell under Internal Variable Pressure: (a) Axial Section of the Shell. (b) Distribution of Internal Pressure. (c) Diagram of the Axial Reinforcement Factor. (d) Diagram of Circumferential Reinforcement Factor.

are nonnegative but M_x and N_θ can have positive and negative values. Such a cantilever-like behavior is common for the shallow-water reservoir walls. Now $C(x)$ may be considered as a design variable, meaning the reinforcement of the shell length x, not L. Introducing a new variable $Q = M_x'$ enables us to rewrite (14.35) as a linear system of second order:

$$\frac{dQ}{dx} = p - \frac{N_\theta}{r}, \qquad \frac{dM_x}{dx} = Q. \tag{14.37}$$

The generalized forces M_x and Q play the role of state variables, and N_θ is the control function. If the shell is in a zero-moment stress state, it means that the structure sustains the pressure solely through the circumferential forces. Then $0 \leqslant N_\theta \leqslant pr$, and the Hamiltonian of the system is

$$H = \psi_0(M_x + N_\theta \delta) + \psi_1 Q + \psi_2(p - N_\theta/r), \tag{14.38}$$

where δ is the given half-arm of the internal moment. Using the expression for the Hamiltonian, we construct the auxiliary system governing the dual variables (ψ_0, ψ_1, ψ_2):

$$\frac{d\psi_0}{dx} = -\frac{\partial H}{\partial C} = 0, \tag{14.39}$$

$$\frac{\partial \psi_1}{dx} = -\frac{\partial H}{\partial M_x} = \psi_0, \tag{14.40}$$

$$\frac{d\psi_2}{dx} = -\frac{\partial H}{\partial Q} = -\psi_1, \tag{14.41}$$

or, more explicitly,

$$\psi_0' = 0, \quad \psi_1' = \psi_0, \quad \psi_2' = \psi_1. \tag{14.42}$$

From (14.42) it follows that $\psi_1'' = \psi_2$; this condition is identical to the well-known strain compatibility equation $u_r'' = \varepsilon_\theta / r$, where $u_r = \psi_1$ is the radial displacement and $\varepsilon_\theta = \psi_2 r$ is the circumferential deformation of the shell. Solving this system with the boundary conditions

$$\psi_0(0) = -1, \qquad \psi_1(L) = \psi_2(L) = 0, \tag{14.43}$$

we get

$$\text{for } (L - x) - 2\delta r \leqslant 0, \text{ the solution } N_\theta = 0,$$
$$\text{for } (L - x) - 2\delta r > 0, \text{ the solution}$$
$$N_\theta = pr. \tag{14.44}$$

We know from the theory of optimal systems that the solution of this problem must have only one switchpoint at which N_θ changes its analytical expression; the number of such points is one less than the order of the system. Physically, the switchpoint divides the membrane and moment fields of the shell behavior, respectively. Formulas (14.43) define the stress state in the optimal shell: almost all of the shell is momentless, and only at a boundary layer is the cantilever element more efficient than a ring; the load is supported by the clamped end. The optimal solution does exist because the moment arm 2δ of the internal couple is assumed constant.

One can calculate that the role of the boundary layer is limited and contributes to the bearing capacity of the structure rather moderately [not greater than 2%, as calculations show; see Reitman and Shapiro (1976)]. But this conclusion was made under ordinary conditions for shell tanks, pressure vessels, and silo design; it is obviously untrue for very short shells for which L is greater than the coordinate of the switchpoint x_0. It seems that this is true not only for cylindrical shells but also for shells of general mid-surface.

Moreover, the solution described is true only when the reinforcement is bounded. If there is no such constraint (or a proper thickness constraint, when the shell is assumed solid), we obtain a flange in the form of Dirac's delta: the function of thickness whose value is infinity, but the integral of this function is finite. This solution was obtained by Shield (1960) and described in Volume 1. We mention this solution again to emphasize that the theory of optimal control and its standard extensions do not produce such singularities; Pontryagin's maximum principle assumes that the control function is bounded and has a finite number of bounded discontinuities [see, for instance, Gelfand and Fomin (1972)]. But Shield's degenerate solution is sound: the total amount of material is finite, and the use of flanges on pressured pipes was well known by design engineers long ago. Some mathematical aspects of this problem will be discussed in Section 14.6.

14.3.2. Shallow Elastic Shells

The problem of optimization of the mid-surface for a given uniform thickness of an axisymmetric shell has been considered by Plaut, Johnson, and Parbery (1984). The radius of the supporting contour and the total area of the surface are also specified, which means the volume of material V is also given. The last assumption for an axisymmetric shell leads to the constraint

$$V = 2H\pi \int_0^{r_s} \sqrt{1 + \left(\frac{dz}{dr}\right)^2}\, r\, dr, \qquad (14.45a)$$

where V is given and we must find the maximum natural frequency together with the corresponding functions w and z, the displacement and the shape of the mid-surface, respectively; r now stands for the radial coordinate. For shallow shells, when $\partial z/\partial r \ll 1$, this condition may be approximately rewritten in the form

$$\frac{V}{H} = S_s = \pi r_s^2 + \frac{\pi}{2}\int_0^{r_s} \left(\frac{dz}{dr}\right)^2 r\,dr, \tag{14.45b}$$

which is used not for a simpler integration (the integration is numerical in any case) but to distinguish between the integrated difference of volumes between the shell and the flat plate. For plates, the second integral term vanishes, and the rise of the shell will be measured by the parameter λ, which is defined by the ratio

$$\lambda^4 = 48(1-\nu^2)\frac{S - \pi a^2}{\pi H^2}. \tag{14.46}$$

So λ measures the average value of the rise of a shell, although it permits any value for the local slope. (Compare with the parameter α, which measures the slope in Section 14.5 and does not take into account the average rise.)

If we neglect the in-plane inertia, we obtain the Marguerre–Vlasov equations in polar coordinates:

$$\begin{aligned} D\Delta\Delta W &= \Omega_1(r, z, F, W), \\ \Delta\Delta F &= \Omega_2(r, z, F, W), \end{aligned} \tag{14.47}$$

where Ω_1, Ω_2 are differential operators that depend on which version of shell theory is used. Here

$$D = \frac{EH^3}{12(1-\nu^2)}$$

and

$$\Delta = \frac{\partial^2}{\partial r^2} + \frac{1}{r}\frac{\partial}{\partial r} + \frac{1}{r^2}\frac{\partial^2}{\partial \theta^2} \tag{14.48}$$

are the plate stiffness and Laplace's operator, respectively, r and θ are polar coordinates of a point of the shell plan view, ν is Poisson's ratio, and E is Young's modulus.

These equations and boundary conditions together with maximization of the fundamental natural frequency ω define the problem of control theory for systems with distributed parameters. The unknowns are the frequency ω and the functions $z(r)$, $W(r, \theta)$, and $F(r, \theta)$. If $z(r)$ is specified iteratively, the corresponding direct problem consists in finding the maximal lowest eigenvalue and eigenfunctions for Eqs. (14.42). The unknown functions W and F are written in the form

$$W = \sum_{n=0}^{\infty} \beta_n(r) \cos n\theta \cos \omega_n t, \tag{14.49}$$

$$F = \sum_{n=0}^{\infty} \xi_n(r) \cos n\theta \cos \omega_n t, \tag{14.50}$$

where ω_n are the natural frequencies of free vibration and β_n and ξ_n are the corresponding radial mode shapes. At $n = 0$, we obtain an axisymmetric vibration, with $\xi_0(r)$ and $\beta_0(r)$ as its mode functions. The corresponding lowest frequency ω_0 can be found for any given design $H(r)$ by classical Rayleigh quotient methods. The problem is now to find the design function $H(r)$ that maximizes the fundamental vibration frequency for a given value of the surface parameter λ. In some cases the optimal design is associated with a double frequency, and the method allows such a possibility. We omit the detailed description of the iterative process in which each new function $g(r) = dz/dr$ is found as a given functional of already tested curves. The procedure is based on an optimality condition and on the shooting method; a spherical shape is selected to initiate the process. Typical results for optimal shells are presented in Fig. 14.5a, b. The optimal meridians of shells are obviously different from the spherical shape, so, for better substantiation of results, the authors used various radii of the initial spheres and came to the same results.

The clamped edge of an optimal mid-surface is always normal to the supporting cylinder: this fact may be proved theoretically by using the Weierstrass-Erdmann conditions (as in the optimal cylindrical shell in Section 14.5.4). However, a simply supported edge is generally inclined with respect to the horizontal plane. If the slope on the boundary of the shell vanishes, it means that the material is highly efficient in bending. This effect is only in shells that transmit moments, since for membrane shells we cannot find an equilibrium state for vanishing slopes in the planes normal to an edge. For axisymmetric shells this fact is illustrated by the results of dynamic programming (see the next section).

As for convergence of the process to the theoretical exact solution, the quotation from Plaut, Johnson, and Parbery (1984) sounds discouraging:

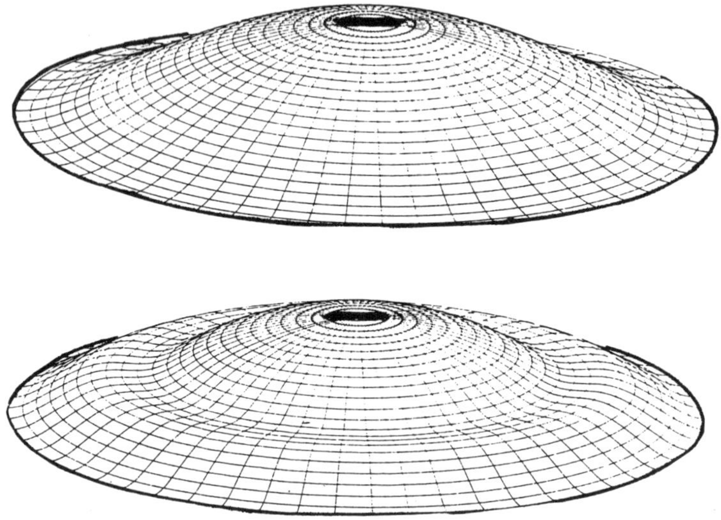

Figure 14.5. Optimal Shapes of Constant Thickness Axisymmetric Shells: (a) Simply Supported at the Edge. (b) Clamped at the Edge.

"We cannot guarantee that the solution procedure leads to the global optimum. In a few cases,we chose different initial designs and arrived at the same final design." This point deserves more study, since in many cases the attraction domain of local extrema is much larger than the attraction domain of the global extrema. Moreover, in some problems for optimal bar structures the attraction domain of global extrema may vanish and degenerate into a point of the design space with vanishing attraction domain [see Kirsch (1986)]. There is no reason to think that this possibility does not occur in shell optimization problems. Moreover, each problem has to be considered as multiextremal unless the opposite statement is proved.

14.3.3. Axisymmetric Membrane Shell

14.3.3.1. Dome with a Specified Mid-Surface. The problem of the non-shallow membrane dome is statically determinate: if we know the load that must be transmitted to the lower part of the shell, it is easy to find the distribution of the dome material and/or the reinforcement over the meridian of the shell mid-surface (see Volume 1). We must use the universal approach based on Pontryagin's maximum principle or on some other rational procedures. Due to statical determinacy of this problem, the solution described here may be equally adequate for both plastic and elastic properties of the shell material (compare with trusses in Volume 1). The same problem within the framework of the theory of the bending of shells is ill-posed (see Section 14.6).

14.3.3.2. Dome with an Unspecified Mid-Surface. This problem is also a problem of the theory of optimal systems that can be resolved by using Pontryagin's maximum principle. But remember that, contrary to the problem with a specified mid-surface, this problem is multiextremal. It is easy to argue with anyone who objects to this statement: if one says that he knows the optimal solution in the shape of the meridian function $z(r)$, it is easy to disprove it: at least for von Mises' plasticity criterion, if one replaces $z(r)$ by $-z(r)$, the volume of the new shell will be the same. (Buckling is not allowed in this approach.)

In fact, the problem has many local extrema, even an infinity of them, and this must be allowed in the choice of the solution algorithm. We use the procedure based on dynamic programming [Kuznetsov *et al.* (1971)]. In the first stage of solution, we divide the radius of the supporting edge of a shell into finite, not necessary equal, segments. Then we fix the coordinates of the top of the shell at $x = 0$, $z = 0$. The problem is to find the meridian of the shell under which the volume of reinforcement in the directions r and ϕ is minimal. Within the framework of each step, the proper "belt" of a shell is assumed to be conical. In accordance with the theory of momentless shells [see Novozhilov (1964) and Volume 1], meridional and circumferential forces in the directions N_ϕ and N_θ are defined by the formulas

$$N_\phi^i = \frac{Q_x}{2r_i \sin \phi_i} \tag{14.51}$$

and

$$N_\theta^i = -\frac{q_i r_i \cos^2 \phi_i}{\sin \phi_i} - Q_{i-1} \frac{\sin(\phi_i - \phi_{i-1})}{2 \sin \phi_{i-1}(r_i - r_{i-1})}, \tag{14.52}$$

where Q_i is the vertical lateral shear force on the cylindrical surface with radius r_i, and q_i is the intensity of the load on the ith belt level (see Fig. 14.6). So it follows that

$$Q_i = Q_{i-1} - q_i(r_i^2 - r_{i-1}^2). \tag{14.53}$$

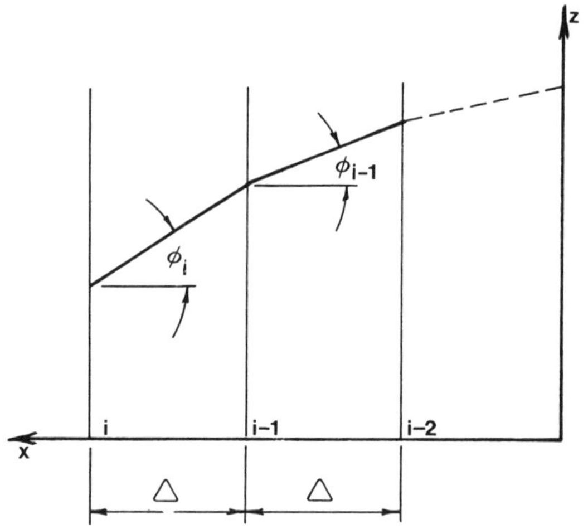

Figure 14.6. The Elemental Belt of an Optimal Axisymmetric Membrane Shell.

Thus the problem is reduced to searching for the values of $\phi_0, \phi_1, \ldots, \phi_n$, where n is the number of the last belt. Using (12.49) and (14.50) as functional equations, we obtain

$$\begin{aligned} N_\phi^i &= N_\phi^i(\phi_{i-1}, \phi_i), \\ N_\theta^i &= N_\theta^i(\phi_{i-1}, \phi_i). \end{aligned} \tag{14.54}$$

Keeping these formulas in mind and specifying H, we may find the optimal values of μ_1 and μ_2 from the BGK plasticity condition (Section 14.2). The reinforcement factors μ_1 and μ_2 may be found, based on the local cost of each belt:

$$C_i = [C_c + (\mu_1 + \mu_2)C_r]Hr_i\sqrt{(z_i - z_{i-1})^2 + (r_i - r_{i-1})^2}. \tag{14.55}$$

After some algebra, taking into account that there are no shear forces and that θ and ϕ are principal directions, we obtain the plasticity criterion:

$$\begin{aligned} &N_1^2 - N_1N_2 + N_2^2 - H(\sigma_c - \sigma_t)(N_1 + N_2) \\ &\quad - \mu_1\sigma_r h(2N_1 - N_2) - \mu_2\sigma_0 H(2N_2 - N_1) \\ &\quad + (\mu_1^2 - \mu_1\mu_2 + \mu_2^2)H^2\sigma_a^2 + H(\sigma_c - \sigma_t)(\mu_1 + \mu_2)\sigma_0 \leqslant \sigma_c\sigma_t, \end{aligned} \tag{14.56}$$

where σ_0 is the yield stress for the steel reinforcement. In this case, both possibilities of failure are allowed for: the failure of the reinforcement and the failure of plain concrete. It requires some tacit additional assumptions, but we shall not dicuss them.

If we introduce Bellman's function (see Fig. 14.6)

$$f_i(\phi_i) = \min_{\{\phi_0, \phi_1, \ldots, \phi_{i-1}\}} \sum_{j=1}^{i} g_i(\phi_{j-1}, \phi_j), \tag{14.57}$$

we obtain Bellman's equation

$$f_i(\phi_i) = \min_{\{\phi_{i-1}\}} [g_i(\phi_i, \phi_{i-1}) + f_{i-1}(\phi_{i-1})]. \tag{14.58}$$

This equation states that we must move from the top of the shell (where $\phi = 0$) down to the edge of the shell and to remember optimal values of ϕ_{i-1} for each ϕ_i. So, the values of ϕ_i are the main variables of the dynamic programming procedure. The mere enumeration used here helps avoid complications related to the local extrema. The process converges to the exact solution of optimal system theory, when the number of steps n tends to infinity. This dynamic programming problem is unidimensional; any attempt to take into account two variables leads to the two-dimensional problem, and the "dimension curse" appears immediately.

The results of the calculations are presented by Reitman and Shapiro (1976) and by Kusnetsov *et al.* (1971). They are also partially shown in Fig.

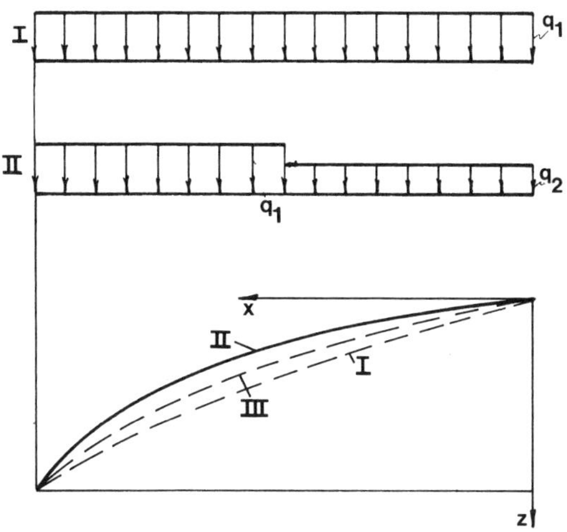

Figure 14.7. Optimization Results for an Axisymmetric Membrane Shell under Different Loads: (I) Is for a Uniformly Distributed Vertical Load q_1. (II) Is for Load with Different Intensity q_1 and q_2 on Left and Right Parts of the Radius, Respectively. (III) Is for Both Loads.

14.7. A rare opportunity to optimize the structure for two different loads is also presented. The cost of the support edge has also been considered and has strongly influenced the optimal solution. The approach may also be extended to the case when not only the external load but the structure's own weight is allowed for. But this improvement immediately increases the dimension of the problem from one to two. Interesting results relate to the influence of the cost of the supporting edge: the optimal shell will be steeper at the edge compared with the shell with free support when the horizontal projection of the reaction causes no change of the objective function.

14.4. Shells with Given Mid-Surface

14.4.1. Plastic Reinforced-Concrete Shallow Shells

14.4.1.1. Statement of the Problem. Reinforcement is the most easily variable property of any composite or concrete structure. Its variation does not require any change of the falsework, forms, or thickness H of the structure. The plasticity criterion is selected in the Karpenko–Morley form (see Section 14.2) in our approach. It includes the membrane forces **N** and moments **M**, which each have three scalar components identified by the subscripts x, y, and xy. The plasticity criterion (which initially was represented as a conical surface in the space of three force components) may be linearized by an inscribed truncated pyramid in the form

$$M_{xy} + \delta N_{xy} - M_x - \delta N_x - \delta H \mu_x \sigma_0 \leqslant 0, \tag{14.59}$$

$$M_{xy} + \delta N_{xy} - M_y - \delta N_y - \delta H \mu_y \sigma_0 \leqslant 0, \tag{14.60}$$

where 2δ is the moment arm of the internal moment; these inequalities form the plasticity criterion.

One may consider a shallow shell the mid-surface of which is given by the equation

$$z = 0.5 f(x^2 + y^2)/A^2, \tag{14.61}$$

where f is the rise of the shell and $2A$ is a side of the square domain of the shell plan view. By the assumption of shallowness, it follows that $f < A$.

The optimality criterion is the minimum of reinforcement of the shallow shell:

$$C = \int_S (\mu_x + \mu_y)\, dS \approx \sum_{i=1}^{K_1} d_i(\mu_x^i + \mu_y^i) \to \min, \tag{14.62}$$

where S is the domain in the xy-plane contained in the shell plan view where reinforcement is found and, in the discretization, K_1 is the total number of cells of unit area into which the plan view is divided, and d_i is an area parameter equal to 1 if cell i belongs wholly to S, 0 if $i \notin S$, and between 0 and 1 if i is cut by the boundary of S. Also the shallowness assumption has been used to neglect squares of the mid-surface, compared with 1. Figure 14.8 shows a 49-cell example; in a later numerical calculation we take $K_1 = 1600$. As a result of shallowness, the equilibrium equations can be expressed as

$$\begin{gathered} KN_x + KN_y - M_{x,xx} - 2M_{xy,xy} - M_{y,yy} - q = 0, \\ N_{x,x} + N_{xy,y} = 0, \qquad N_{y,y} + N_{xy,x} = 0, \end{gathered} \tag{14.63}$$

where the curvature K of the mid-surface is determined by the formula

$$K = K_x = K_y = z_{,xx} = z_{,yy} = f/A^2. \tag{14.64}$$

We introduce dimensionless forces and moments denoted by n and m:

$$\begin{gathered} n_x = \frac{16N_x\delta}{qA^2}, \quad n_y = \frac{16N_y\delta}{qA^2}, \quad n_{xy} = \frac{16N_{xy}\delta}{qA^2}, \\ m_x = \frac{16M_x}{qA^2}, \quad m_y = \frac{16M_y}{qA^2}, \quad m_{xy} = \frac{16M_{xy}}{qA^2}. \end{gathered} \tag{14.65}$$

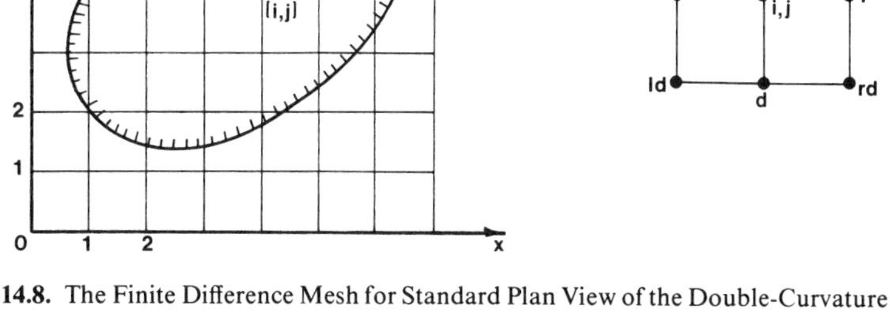

Figure 14.8. The Finite Difference Mesh for Standard Plan View of the Double-Curvature Shell.

The equilibrium equations become

$$n_{x,x} + n_{xy,y} = 0, \qquad n_{y,y} + n_{xy,x} = 0,$$
$$-\frac{k}{16}(n_x + n_y) + m_{x,xx} + 2m_{xy,xy} + m_{y,yy} + 1 = 0. \tag{14.66}$$

Using the finite difference mesh (see Fig. 14.9), we replace the previous equations by

$$n_x^{\rm r} - n_x^{\rm l} + n_{xy}^{\rm u} - n_{xy}^{\rm d} = 0, \qquad n_y^{\rm u} - n_y^{\rm d} + n_{xy}^{\rm r} - n_{xy}^{\rm l} = 0,$$
$$\frac{-k}{16}(n_x^0 + n_y^0) + m_x^{\rm r} - 2m_x^0 + m_x^{\rm l} + 0.5(m_{xy}^{\rm rd} + m_{xy}^{\rm lu} - m_{xy}^{\rm ru} - m_{xy}^{\rm ld})$$
$$+ m_y^{\rm u} - 2m_y^0 + m_y^{\rm d} + 1 = 0. \tag{14.67}$$

(See Fig. 14.8b, where ru = right up, rd = right down, lu = left up, ld = left down.) In addition to the equilibrium equations, statically admissible fields of internal forces must satisfy the boundary conditions

$$m_\eta = 0, \qquad n_\eta = 0, \tag{14.68}$$

where the subscript η stands for the unit normal to the boundary of the shell. These restrictions are valid on a simply supported edge; on a clamped edge, there are no restrictions on force or moment. This means that the shell is supported by the contour, which permits any values of all other components of the forces.

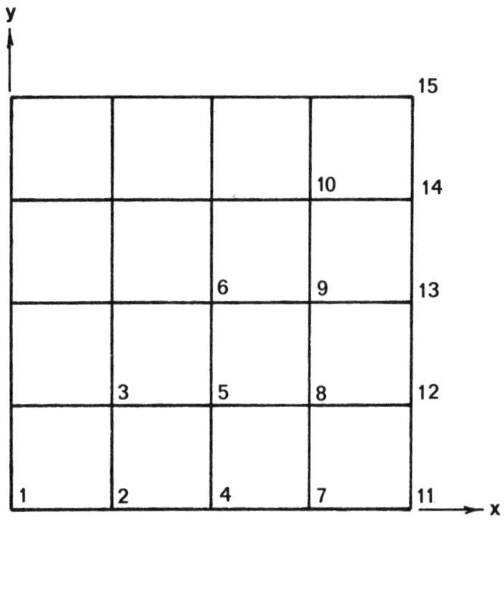

Figure 14.9. The Finite Difference Approximation of the Equilibrium Equation for the Double-Curvature Shell: Due to Symmetry, Only the Finite Difference Mesh for One-Eighth of the Shell is Provided with Node Numbers.

The procedure of transforming the problem to a dimensionless form has been performed for all other problems as well, but we shall not mention it any more due to its routine character.

The Karpenko–Morley condition (14.59)–(14.60) can be written in dimensionless form as

$$\begin{aligned} m_{xy} + n_{xy} - m_x - n_x - m_{0x} &\leqslant 0, \\ m_{xy} + n_{xy} - m_y - n_y - m_{0y} &\leqslant 0, \end{aligned} \tag{14.69}$$

where

$$m_{0x} = \frac{16M_{0x}}{qA^2}, \qquad m_{0y} = \frac{16M_{0y}}{qA^2}. \tag{14.70}$$

Satisfaction of the additional inequalities in the full Karpenko–Morley condition (14.24) follows naturally.

14.4.1.2. Method of Solution. Now Eqs. (14.62) (for the whole system) as well as (14.66) and (14.69) (for each internal node of the finite difference mesh) and (14.68) (for each point on the edge of the shell) form a LP problem where the unknown variables are the nonnegative values of μ_x^i and μ_y^i at each node i of the mesh. Meanwhile, the forces n^i and moments m^i are unknown 3-tuples that are also indexed by the nodal number i. These values may be positive or negative. The solution may be obtained with the help of any computer, but the standard formulation of the LP problems requires all variables to be nonnegative. To reduce the problem to the standard LP problem, we may need to replace m and n by new variables n' and m', where $n' = n + a$ and $m' = m + a$. The value of a may be defined in different ways:

1. We may give a any value greater than any absolute value of negative components of m and n.
2. It may be included in the set of unknown variables.
3. Each variable (for instance, n_y^6) could be replaced by the difference of two nonnegative variables (for instance, $n_y^6 = n_y^{6+} - n_y^{6-}$).

Although the last method is recommended by some textbooks on LP, it is not the best since it greatly increases the number of variables, whereas the first two are almost free of this drawback.

Since only the plasticity condition is nonlinear (all other conditions are linear initially), we may use convex programming as well (the Karpenko–Morley condition is conical and hence convex) by the use of standard codes for convex programming. The problem of comparison between the two approaches will be discussed later.

14.4.1.3. Numerical Examples and Discussion. As the first nonaxisymmetric problem, we consider a reinforced shell of double curvature made of concrete, which follows the linearized Karpenko–Morley plasticity condition. If we assume that the shell is supported by diaphragms that are perfectly stiff in their planes and perfectly flexible out of their planes, the boundary conditions of the optimization problem are

$$\begin{aligned} M_x = N_x = 0, \qquad & \text{at } x = \pm A, \\ M_y = N_y = 0 \qquad & \text{at } y = \pm A. \end{aligned} \tag{14.71}$$

The case of reinforced slab optimization follows from this theory, if we assume $K = \infty$ and $N_x = N_y = N_{xy} = 0$. [The first description of this case was given by Wolfensberger (1964).] The figures give the layout of the reinforcement for the optimal slab (Fig. 14.10) and shell (Fig. 14.11). The dimensionless optimal data concerning the reinforcement of the slabs and shell are given in Tables 14.1 and 14.2, respectively, where the indexed points correspond to the nodes of Fig. 14.9.

As follows from Table 14.1 (for the slab), optimal reinforcement is condensed at the two symmetry axes, vertical and horizontal, that form something like two supporting beams of variable reinforcement. The twisting moments vanish over the whole area of the slab. The total reinforcement equals $33.7qA^4/\delta\sigma_0$. A similar (nonnumerical!) solution was obtained by Rozvany (1966), who used quite a different approach, based on analytical

Table 14.1. Numerical Results for Optimization of the Reinforced Square Slab

Point	m_x	m_y	m_{xy}	m_{0x}	m_{0y}
1	16	16	0	16	16
2	15.75	0	0	15.75	0
3	1.5	1.5	0	1.5	1.5
4	11.5	0	0	11.5	0
5	2.5	0	0	2.5	0
6	0.75	0.75	0	0.75	0.75
7	6.25	0	0	6.25	0
8	1.75	0	0	1.75	0
9	1	0	0	1	0
10	0.25	0	0	0.25	0
11	0	0	0	0	0
12	0	0	0	0	0
13	0	0	0	0	0
14	0	0	0	0	0
15	0	0	0	0	0

Table 14.2. Numerical Results for Optimization of the Reinforced Double-Curvature Shell

Point	n_x	n_y	n_{xy}	m_x	m_y	m_{xy}	m_{0x}	m_{0y}
1	0.12	0.12	0	0.06	0.06	0	0	0
2	0.24	0.27	0	0	0.05	0	0	0
3	0.3	0.3	0.15	0	0	0.15	0	0
4	0.11	0.26	0	0.05	0.13	0	0	0
5	0.20	0.32	0.07	0.06	0	0	0	0
6	0.71	0.21	13	0.04	0.04	0	0	0
7	0.09	0.31	0	0	0.15	0	0	0
8	0.17	0.43	0.29	0.20	0	0	0	0
9	0.2	0.29	0.2	0	0.05	0	0	0
10	0.21	0.21	0.25	0	0	0	0	0
11	0	0.53	0	0	0.05	0	0	0
12	0	0.57	0.9	0	0	0	0	0.05
13	0	0.46	0.35	0	0.03	0	0	0.17
14	0	0.27	0.49	0	0	0	0.11	0.24
15	0	0	0.76	0	0	0.53	0.91	0.91

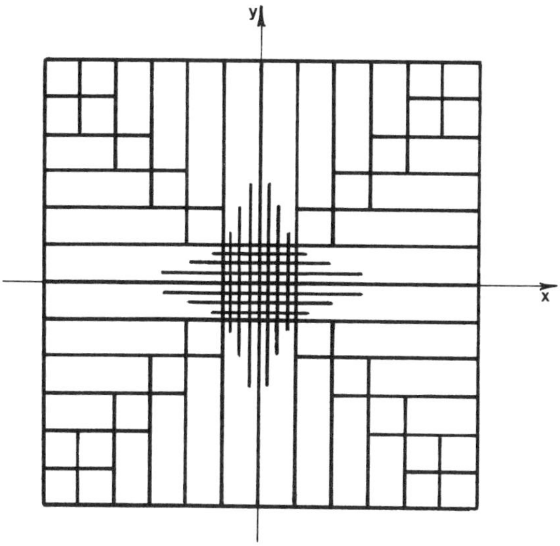

Figure 14.10. Optimal Layout of Reinforcement in a Plane Slab.

considerations, and whose derivation also led to the "cross" of reinforcement as shown in Fig. 14.10. Rozvany has also shown that the amount of reinforcement can be reduced further if it is assumed to be parallel to the diagonals rather than to the sides of the slab. The rest of the slab reinforcement facilitates the transfer of loads to the crossbeams and to the edges of the slabs.

A quite different pattern of reinforcement is presented in Table 14.2. The optimal reinforcement is developed in the corners of the structure. These results are indirectly related to the singular fields of forces generated by the elastic solution of the problem. It reminds us of the "boundary layer effect" occurring for almost all kinds of moment-resistive shells (see Volume 1, Chapter 6). The total amount of reinforcement in the shell is much less than in the slab: for the case considered, it is $0.68qA^4/\delta\sigma_0$—a tremendous saving (see Fig. 14.11). Figure 14.12 illustrates the change of the optimal reinforcement as a function of the dimensionless height of the shell, f/δ. At some point in the growth of this ratio, there is a jump from the scheme of Fig. 14.10 to the scheme of Fig. 14.11, and the graph in Fig. 14.12 (the total amount of reinforcement) behaves continuously. Attempts have been made to find a range of heights at which a weighted combination of the two layouts occurs. However, this way was not productive; the transition appears as a catastrophe, using Rene Thom's terminology, since all intermediate designs are unstable. Using a finer mesh does not necessarily show better convergence and tends to confirm the existence of a numerical

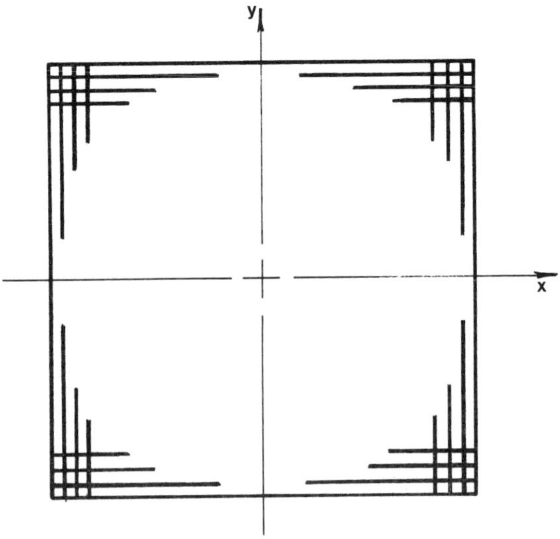

Figure 14.11. Optimal Layout of Reinforcement in a Double-Curvature Rectangular Shell.

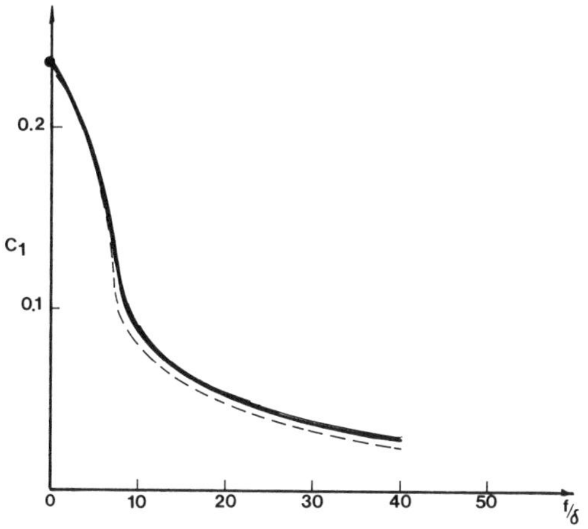

Figure 14.12. Total Shell Reinforcement C as a Function of Nondimensional Rise: Solid Line Shows the Results for Linear Model; Dashed Line Is for Nonlinear Model.

instability. The convergence of the process is not monotonic: sometimes the solution with a finer mesh led to a greater (within 2%) value of the objective function.

The optimal reinforcement of a square slab for assigned limit load allowed the reduction of reinforcement by 32.5%, as compared with uniformly reinforced slabs. Meanwhile, the optimal reinforcement layout for the shell was 5.2 times as small. We emphasize that all the examples were considered under the assumption that the arm of the internal moment 2δ was constant. If the value is variable, it makes the problem more difficult. However, convexity is maintained, since the growth of the bending moment at a given thickness of the shell causes growth of the compressed zone and reduces the arm of the internal moment.

We have discussed the results for linearized problems, but appropriate calculations were made also for nonlinear problems by using the original Karpenko-Morley criterion. Using the standard code SUMT gives numerical results that are qualitatively the same as the LP results. As for quantitative results, they are identical for slabs and very shallow shells. This fact is well known in theory: the optimal reinforcement of slabs along the prescribed directions leads to zero twisting moments [see Rozvany (1976)]. But when the rise of the shell grows, the role of twisting moments in the optimal solution becomes more and more significant. And the nonlinear problem gives a smaller optimal value of the objective function, since linearization cuts off significant parts of the admissible domain in the moment space. The numerical calculations made with the help of the SUMT standard code support this statement. (For details of this phenomenon as applied to another problem, see Section 14.3.)

In this problem the reinforcement directions have been given in advance. But it is more efficient to reinforce the square slab along the directions parallel to the diagonals [see Rozvany (1976)]. For composite shells many solutions have been published when the area of reinforcement is given but the common angle of its alignment is to be found. From the computational point of view, this problem is oversimplified; although it is very important for practice, it is beyond our scope. The problem of simultaneous determination of reinforcement and alignment of the reinforcement is much more difficult, and we do not know of any published solutions. Contrary to the problem considered here, that one is probably multiextremal. The role of this complication and a method of dealing with it will be discussed in Section 14.5.

Here the finite difference discretization was used, but we may use the finite element discretization as well. The comparison of results for cylindrical shells and slabs using von Mises', Tresca's, Johansen's, and Karpenko-Morley's plasticity criteria can be found in Čyras, Borkauskas, and Karkauskas (1974).

14.4.2. Plastic Shells of Plain Concrete

14.4.2.1. Formulation of the Problem. The optimization problem for a homogeneous material must be approached carefully because of different forms of degeneration (see Section 14.6.3 and p. 290, Volume 1). Now the shell is considered as a nonshallow membrane with a given middle surface and variable thickness. Loads allowed may include components besides p_z (treated earlier). It becomes inconvenient to use Pucher's equation and other similar equilibrium forms, and another form of the equations will be applied.

By Gvozdev's statical theorem, the structures will not collapse if the equilibrium equations and plasticity criterion are satisfied. We shall use the equilibrium equations in the form

$$\frac{\partial}{\partial x}\left(\tilde{N}_x \frac{\cos\phi}{\cos\psi}\right) + \frac{\partial \tilde{N}_{xy}}{\partial y} + p\tan\phi = 0,$$

$$\frac{\partial}{\partial y}\left(\tilde{N}_y \frac{\cos\psi}{\cos\phi}\right) + \frac{\partial \tilde{N}_{xy}}{\partial x} + p\tan\psi - \frac{\gamma H}{\cos\phi\cos\psi} = 0, \tag{14.72}$$

$$\frac{\partial}{\partial x}\left(\tilde{N}_x \frac{\sin\phi}{\cos\psi}\right) + \frac{\partial}{\partial y}\left(\tilde{N}_y \frac{\sin\phi}{\sin\psi}\right) + \frac{\partial}{\partial y}(\tilde{N}_{xy}\tan\phi) + \frac{\partial}{\partial x}(\tilde{N}_{xy}\tan\psi) - p = 0,$$

where the 3-tuple $\tilde{\mathbf{N}}$ is skew forces, γ is the unit weight of the shell material, ϕ and ψ are the slope angles between tangents to the surface at the point considered in planes xOz and yOz and the axes x and y, respectively, and p is the pressure as a function of the vertical coordinate z.

Besides these equations, the admissible values of the 3-vector function $\bar{\mathbf{N}}(x, y)$ must satisfy the boundary conditions

$$\bar{N}_x \cos\chi_0 + \bar{N}_{xy} \sin\chi_0 = X, \qquad \bar{N}_{xy} \sin\chi_0 + \bar{N}_y \cos\chi_0 = Y, \tag{14.73}$$

where χ_0 is the angle between the normal to contours of the shell in the xy-plane and the x-axis, the 3-tuple $\bar{\mathbf{N}}$ is the force projections, and X and Y are the components of the external load in the x- and y-directions, respectively.

14.4.2.2. Solution of the Problem for an Arch Dam. One of the simplest ways to obtain the solution of the problem is by using mathematical programming: the problem is convex, and this guarantees unique results. In this case, the boundary condition is reduced to $\tilde{N}_y = \tilde{N}_{xy} = 0$ on the upper free side of the dam, whereas on other sides there are no boundary conditions. This is equivalent to the assumption that the walls of the canyon may carry any load, and failure is probable only in the dam. Contrary to the general formulation of the problem (see Section 14.2), we assume that the z-axis is aligned along the canyon, but not vertical, to preserve the z-axis as the direction of main loading.

We must state the optimality criterion, which in this case is minimization of the volume function

$$C = \int_S H\sqrt{1 + \tan^2\phi + \tan^2\psi}\, dS \tag{14.74}$$

subject to the BGK plasticity condition (see Section 14.2). The problem also is one of optimal control theory for systems with distributed parameters. This system is nonlinear only in the plasticity condition and the integrand (14.74), whereas all other equations are linear. The system may be discretized with the help of finite differences. This leads to a CP problem, since all the equations and inequalities are convex (the plasticity condition and the objective function) or linear (all other conditions of the problem).

The design of one of the versions of the Inguri dam was considered as a numerical example (Fig. 14.13). Due to the symmetry of the Inguri canyon section, only a half of the dam is specified. The geometry of the dam is described by Table 14.3.

The height of the dam is 225 m. Because we are designing a dam, the stress state may not include tensile principal components of the membrane forces, since these forces can cause an increase in filtering water by the dam. (In this problem tensile forces are considered to be negative.) As a

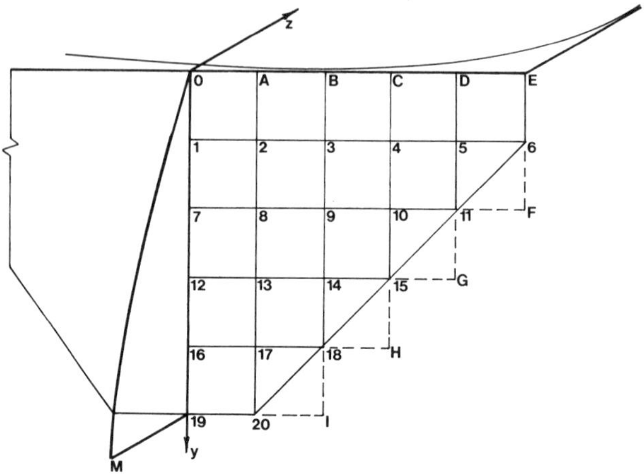

Figure 14.13. Finite Difference Mesh for Arch Dam Optimization.

Table 14.3. Geometrical Characteristics of the Arch Dam

y, m		$x = 0$	$x = 45$	$x = 90$	$x = 135$	$x = 180$	$x = 225$
	z	0	2.91	11.76	27.10	49.84	81.92
0	$\tan \phi$	0	0.13	0.266	0.418	0.600	0.840
	$\tan \psi$	−0.445	−0.440	−0.423	−0.392	−0.340	−0.253
	z	−18.76	−15.60	−5.87	11.05	36.57	73.51
45	$\tan \phi$	0	0.142	0.292	0.464	0.681	0.986
	$\tan \psi$	−0.387	−0.380	−0.358	−0.316	−0.243	−0.109
	z	−34.53	−30.98	−29.10	−0.90	28.69	73.70
90	$\tan \phi$	0	0.158	0.329	0.531	0.801	1.253
	$\tan \psi$	−0.309	−0.302	−0.269	−0.208	−0.094	0.152
	z	−46.09	−42.03	−29.41	−6.60	29.99	
135	$\tan \phi$	0	0.182	0.383	0.638	1.028	
	$\tan \psi$	0.197	−0.183	−0.135	−0.034	0.184	
	z	−51.26	−46.32	−30.78	−1.47		
180	$\tan \phi$	0	0.222	0.480	0.853		
	$\tan \psi$	0	0.011	0.101	0.318		
	z	−44.69	−37.94	−15.43			
225	$\tan \phi$	0	0.308	0.727			
	$\tan \psi$	0.365	0.431	0.705			

plasticity condition, the BGK modified condition has been used (see Section 14.2) in combination with the condition that rules out the negative (tensile) values of the forces:

$$\begin{aligned} -N_x + N_{xy} &\leqslant 0, & -N_y + N_{xy} &\leqslant 0, \\ -N_x + N_y + 2N_{xy} &\leqslant \sigma_c, & N_x + N_{xy} &\leqslant \sigma_c H, \\ 2N_x \pm 2N_{xy} &\leqslant 3\sigma_c H, & 2N_y \pm 2N_{xy} &\leqslant 3\sigma_c H. \end{aligned} \tag{14.75}$$

As mentioned in Section 14.2, there is an essential difference between the skew forces in Eqs. (14.72) and the actual forces in (14.75). They are interrelated by substantially nonlinear equations [see (14.11) and (14.12)]. But if we identify N and $\tilde{N}$ in (14.75), all conditions of the problem become linear. So the LP problem is formulated, and the results of its solution (using a code of linear programming) are given in Table 14.4. A study of the table shows some unpredictable features. First of all, in the linear solution, due to allowance for the weight of the structure, a part of the material plays the role of a useful burden, which contributes to loading rather than to the strength of the dam. The membrane solution in the form of a compressive stress state may not be found without it. Using the weight

Table 14.4. Some Results of the Optimal Dam Design

Point	$\tilde{N}_x$ (T/cm)	$\tilde{N}_y$ (T/cm)	$\tilde{N}_{xy}$ (T/cm)	H (m)
1	170.83	17.84	0	6.0
2	173.96	14.58	0	6.3
3	173.62	14.33	3.01	8.2
4	175.12	20.53	6.80	10.6
5	170.20	14.28	0	6.1
6	173.97	6.59	0	6.6
7	275.51	48.76	0	13.7
8	278.99	29.63	0	9.8
9	282.13	22.83	16.14	11.4
10	243.23	36.29	36.29	13.5
11	194.67	33.62	33.62	8.9
12	358.07	53.94	0	11.9
13	303.66	43.43	20.74	11.8
14	298.73	34.08	20.65	12.0
15	305.54	36.79	36.79	13.4
16	273.76	58.80	0	8.8
17	330.93	46.50	8.54	11.8
18	335.38	27.77	27.77	14.2
19	240.47	47.36	0	7.6
20	225.60	30.46	0	7.6

of the dam material instead of its strength is not efficient and prompts us to be careful when constructing the mathematical model: the linearization cuts off some operative parts of the plasticity condition. This effect is responsible for a 22% increase in the optimal value of the objective function for the optimal design in the linear problem, in comparison with the appropriate value in the nonlinear problem. Although in the previous (linearized) problem the relative volume of the admissible domain in the force space that is cut off by linearization was of the same order, the detrimental effect of linearization is not so pronounced, since the cutoff parts of the admissible domain are not operative.

The results at points O, A–E are not given, since the forces N_y and N_{xy} are zero and N_x and H are small. The results for points F–H are omitted also, since the structure has massive supporting foundations there and is not modeled by shell theory. The distribution of thickness in the first upper arches of the dam shows that the thickness from the middle point of the arch to a shore is initially increasing and reaches a maximum at about one quarter of the span. But further out there is a segment of decreasing thickness. This trend is ignored by designers of the dam, although the same trend may be indirectly inferred from the elastic analysis of the structure. A significant deviation from well-established design rules is related to the bottom of the dam: the optimal solution recommends a decrease in the thickness for the two bottom arches of the dam. This effect was also noticed by the designers, but a decrease of thickness at the bottom of dam looks dangerous to designers and it did not occur in the actual design. The equality of some values $\tilde{N}_y$ and $\tilde{N}_{xy}$ (for instance, at point 18) is by no means accidental: it shows that the second relation of (14.75) for this point is satisfied as an equality.

In calculations of this example, the skew forces $\tilde{N}$ were identified with the actual forces N in the plasticity criterion. This is a serious source of error, which will be discussed later in the problem with variable geometry. This is equivalent to considering the shell as shallow in the plasticity condition, although it was treated as not shallow in the equilibrium equations. Slopes of the shell do not support the assumption of shallowness (see Table 14.3). Actually, the point $x = 180$ m and $y = 135$ m, where $\tan\phi = 1.028$, does not look precisely shallow, but any attempt to follow the routine procedure of Section 14.2, as will be done in Section 14.5, inevitably led to a nonlinear problem, since the equation relating forces N in an arbitrary coordinate frame to principal forces cannot be linear.

14.4.3. Slabs Attached to a Plastic Foundation

14.4.3.1. Formulation of the Problem. We consider a concrete slab supported by a soil foundation. The design is based on the theory of plastic slab design, and the constraints on the pressure q at the interface of soil and slab are allowed for. Suppose σ_f is the normal yield stress of the soil. The layout of reinforcement of the slab must be selected so that the total volume of reinforcement is minimized. Although the allowable foundations are limited predominantly by the displacement constraint, in some cases the strength design may be useful too. The necessary condition for the problem is defined by the condition

$$\int_S q\,dS \leqslant \int_S \sigma_f\,dS. \tag{14.76}$$

Now the equilibrium condition leads to the double inequality

$$0 \leqslant \frac{\partial^2 M_x}{\partial x^2} + 2\frac{\partial^2 M_{xy}}{\partial x\,\partial y} + \frac{\partial^2 M_y}{\partial y^2} + q \leqslant \sigma_f. \tag{14.77}$$

The left side represents the one-sided connection between the foundation and soil.

14.4.3.2. Results of an Example. The problem of a square foundation uses almost the same conditions as the problem of simply supported edges, except that the boundary conditions are

$$\begin{aligned} M_x &= 0, \qquad Q_x = \frac{\partial M_x}{\partial x} + \frac{\partial M_{xy}}{\partial y} = 0, \\ M_y &= 0, \qquad Q_y = \frac{\partial M_y}{\partial y} + \frac{\partial M_{xy}}{\partial x} = 0. \end{aligned} \tag{14.78}$$

The optimization results are obtained by using an LP algorithm, and the optimal layout of the reinforcement is shown in Fig. 14.14. The dashed line defines the contour of the optimal slab as a cross (compare with the results of edge-supported slabs that also exhibit the crosslike layout). In Reitman and Yarin (1974), a more sophisticated case is considered, where the soil is represented not by Winckler's model but by the Coulomb–Mohr plasticity condition, which describes soil behavior.

If we turn the model upside down, it leads to another engineering structure, the floor slab supporting a vertical stock load. In the "rational stock" problem the task is to optimize simultaneously the load distribution and the reinforcement distribution over the floor slab. [See Reitman and Yarin (1974).] The total load is specified and, as in the optimal foundation problem, is inside the admissible range. If the problem is stated without constraints, it leads to immediate degeneration: if all load is concentrated

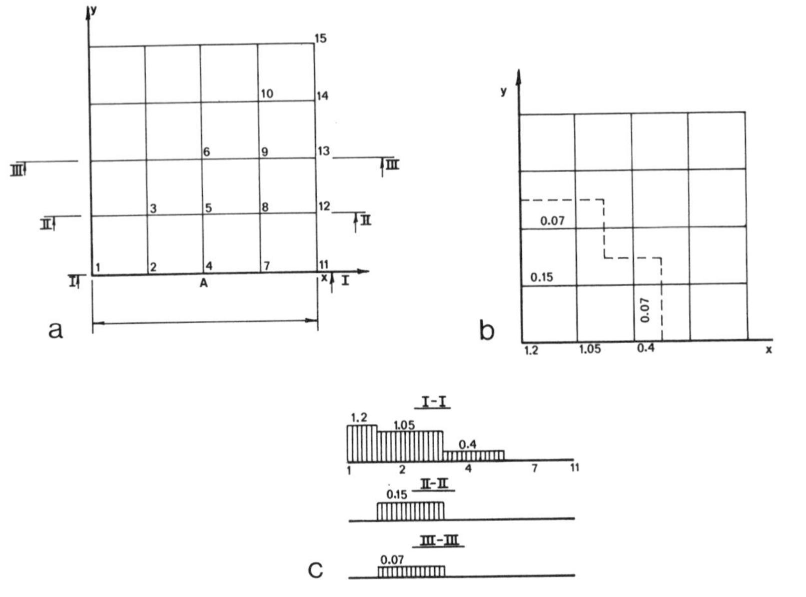

Figure 14.14. Optimal Layout of Reinforcement in a Foundation Slab: (a) Finite Difference Mesh for One Quarter of the Foundation Slab. (b) Reinforcement Factors μ_x and μ_y in the Optimal Slab. (c) Diagrams of the Reinforcement Factor μ_x in Three Sections.

on the columns, there will be no need for slabs. But at stock warehouses, there is usually an upper limit for the height of the pile; this constraint gives the load parameter, which corresponds to the ultimate compression for the foundation. The problem of optimal stock becomes identical to the foundation design problem. Besides the direction of loads, another difference is connected with the interpretation of results; in the foundation problem the part of the slab outside the dashed line in Fig. 14.14b may be removed and only the crosslike foundation be left, whereas for the floor this removal is impractical.

14.4.4. Elastic Metal Plates with Arbitrary Plan View

14.4.4.1. Statement of the Problem in Finite Elements. In all of the previous models, a finite difference discretization has been used. The character of the problem justified this, since the advantages of finite elements over finite differences appear only for rather sophisticated configurations of edges or shapes. The description of shells by finite elements is so cumbersome that authors of textbooks in finite element theory usually do not include these examples in their books. In the theory of plates one uses triangular or rectangular elements as in plane stress problems.

We consider a rectangular plate bending element and require the continuity of the vertical displacement for the two, three, or four finite elements to which the node belongs. For the plane stress state, this compatibility assumption enables us to construct the simplest theory. But for plates the compatibility condition must be augmented by another assumption, since the order of the set of equations is higher; for instance, we may require the continuity of the angular displacement θ across the one, two, three, or four joints of a finite element with other finite elements of the same assemblage.

Following Prasad and Haftka (1979), we describe the system PARS (program for analysis and revising of structures), which may be used for the optimization of elastic homogeneous plates in a finite element form. Equations (14.31) may be represented as $M = H^3[c]\kappa$, where the 3×3 matrix $[c]$ defines the linear relations between the moment and curvature 3-tuples. If we introduce the matrix $[A]$ by

$$\kappa = [A]M = H^{-3}[c]^{-1}M, \tag{14.79}$$

we can obtain an analytical expression for the derivative of $[A]$ with respect to H:

$$[\partial A/\partial H] = -3H^{-4}[c]^{-1}. \tag{14.80}$$

The matrix $[c]$ does not depend on H for each element of a grid or for the whole assemblage.

The moments within each element are defined by the shape functions b_i $(i = 1, 2, \ldots, 15)$:

$$m = [b]\iota, \tag{14.81}$$

where

$$[b] = \begin{bmatrix} b_1 & b_6 & b_4 & b_{14} & b_{13} & b_{10} \\ b_2 & b_5 & b_7 & b_{13} & b_{15} & b_{11} \\ b_3 & b_9 & b_8 & 0 & 0 & b_{14} + b_{15} \end{bmatrix}, \tag{14.82}$$

and

$$\iota = \{1, x, y, x^2, y^2, xy\}^{\mathrm{T}}. \tag{14.83}$$

We do not give the formulas for b_i, but merely remark that these functions do not depend on the design variables. This approach helps us to use the SUMT optimization method involving the constraints on stress, displacements, and natural frequencies of an elastic plate:

$$\|\sigma\| \leqslant \sigma_0, \quad w_1 \leqslant w \leqslant w_2, \quad \omega_1 \leqslant \omega \leqslant \omega_2. \tag{14.84}$$

These constraints are allowed for by means of different penalty functions.

14.4.4.2. Elastic Solid Plate Supported at Four Points. We consider the rectangular plate supported at four symmetrical points on principal diagonals at two-fifths of the distance between the plate's corner and its center (Fig. 14.15a). The plate is loaded by line loads along the central and boundary lines. The finite element model of the plate is shown in Fig. 14.15b. Because of symmetry, only one quarter of the plate is actually modeled.

14.4.4.3. Standardization of Finite Elements. Usually optimization of plates is associated (implicitly or explicitly) with the rate-specified standardization: the number t of unknown thickness parameters, or design parameters, is either equal to or less than the number n of finite elements. Here $S_1 = 1 - t/n$ may be considered as the "standardization level": If $t = n$ and $S_1 = 0$, the standardization level is minimal, and the thicknesses of all elements of the plate may be different; if $t = 1$ and n is large, $S_1 \approx 1$, the standardization level is large, and all elements of the plate are of the same

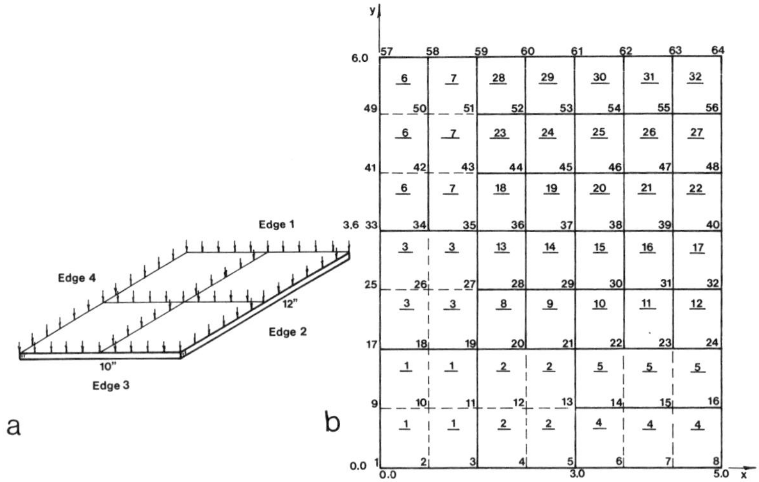

Figure 14.15. Optimization of Elastic Plate: (a) Line Loads on the Upper Surface. (b) Finite Element Mesh for One Quarter of the Plate.

thickness. The standardization problem, well studied for optimal bar systems [see Reitman and Shapiro (1976)], has been studied only indirectly for plates: in Prasad and Haftka (1979) the plate was divided into 49 elements, but they are partially collected in sets of cells with the same thickness. This set is defined by 32 parameters and gives the distribution of thicknesses shown in Fig. 14.16 and in Table 14.5.

For this case, $S_1 = 1 - \frac{32}{49} = \frac{17}{49}$, and the standardization level is not too high. The authors' plate thickness scheme illustrates rather chaotic distribution of thickness, but we can notice the formation of stiffeners of the type suggested in Megaref (1966) and Olhoff (1973); see Section 14.6. Certainly, and the authors emphasized this fact, there is no sense in discussing the convergence of this solution to an analytically exact solution as $n \to \infty$ because the problem for $n = \infty$ is ill-posed and the volume of material tends to zero according to the Megaref–Olhoff degeneration, although the convergence of satisfying the constraints has been fast. But for finite n and small t the problem makes physical sense and may be resolved, especially if $t \ll n$. However, at any n the structure that is the result of optimization ceases to

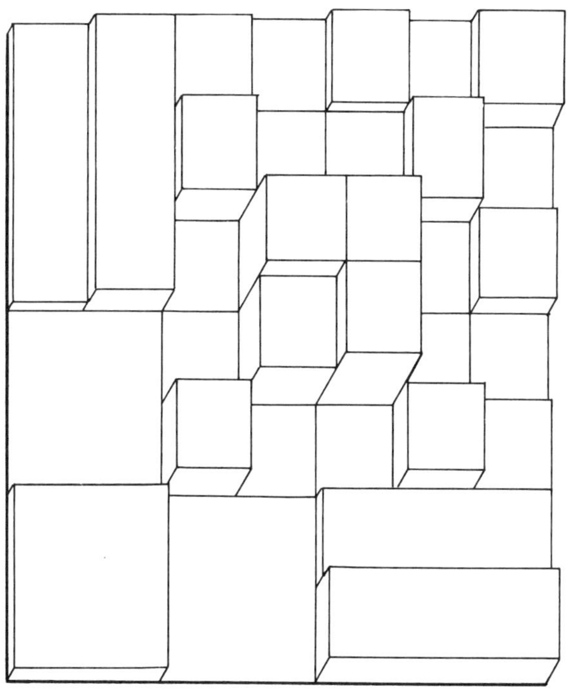

Figure 14.16. Optimal Layout of an Elastic Plate with Low Standardization Level [see Prasad and Haftka (1979)].

Table 14.5. Optimal Thickness of Rectangular Plate

Design variable	Element	H (in)	Deisgn variable	Element	H (in)
1	1–4	0.3760	17	34	0.0225
2	5–8	0.0432	18	35	0.0227
3	9–12	0.0531	19	36	0.8624
4	13–15	0.3802	20	37	0.8461
5	16–18	0.2089	21	38	0.0237
6	19–21	0.2321	22	39	0.2683
7	22–24	0.2837	23	40	0.7074
8	25	0.6172	24	41	0.0229
9	26	0.0217	25	42	0.0236
10	27	0.0224	26	43	0.6557
11	28	0.6545	27	44	0.0234
12	29	0.1878	28	45	0.1541
13	30	0.0220	29	46	0.0227
14	31	0.0754	30	47	0.2665
15	32	0.8725	31	48	0.0233
16	33	0.0200	32	49	0.2837

be covered by the elementary theory of plates. In particular, the theory is not applicable to plates with discontinuous variation of thickness (see Section 14.6 where we discuss the theoretical aspects of shell optimization).

14.5. Shells with Variable Mid-Surface

14.5.1. Plastic Metal Membrane Shells

Allowing variation of the mid-surface shape $z(x, y)$ is one of the most efficient tools for improving the design of shells, although it is not a universal tool, because we often have to apply many technological constraints. But frequently such variations of the mid-surface give designers more freedom to change the design than do thickness variations. In other words, independent changes of reinforcement, thickness, and mid-surface are equally pertinent, although their combination can lead to a very wide range of problems, including ill-posed and synergetic effects.

14.5.1.1. Statement of the Problem. For this category of shells we may select Pucher's equation (14.9) and von Mises' plasticity condition (14.22). After solving von Mises' condition with respect to H, we have

$$H = \frac{1}{\sigma_0} \int_S \sqrt{N_1^2 - N_1 N_2 + N_2^2}\, dS, \tag{14.85}$$

where σ_0 is the yield limit of the material. Here the actual principal internal forces N_1, N_2 are used, since they may be calculated for each value of the functions F and z. Substituting the actual forces into (14.85) followed by substituting H in the expression of the shell volume give an expression for the total volume of metal in the structure:

$$C = \frac{1}{\sigma_0} \int_S \sqrt{(N_1^2 - N_1 N_2 + N_2^2)(1 - z_{,x}^2 + z_{,y}^2)}\, dx\, dy. \tag{14.86}$$

The problem of shell optimization usually involves boundary conditions. If no force boundary conditions are present, it means that the supporting edge can sustain any forces without displacements.

Technological constraints for optimization of shells may be set on F and z or on their derivatives. For instance, sometimes an upper bound must be set on the slope of the mid-surface in the form

$$\sqrt{z_{,x}^2 + z_{,y}^2} \leqslant \tan \alpha, \tag{14.87}$$

where α is the limit angle of slope of the shell mid-surface based on considerations that are frequently far from mechanics or strength but are related to waterproofing or metal processing. Similar limitations are often connected with the maximal rise of the shell or its points:

$$z_1 \leqslant z \leqslant z_2. \tag{14.88}$$

The last constraint is necessary in many cases because optimal design often leads to shells that are too high. The problem of the theory of optimal control for the system with distributed parameters (or governed by partial differential equations) that results in this way has many extrema. Hence the method of solution has to be selected with due consideration of these obstacles. A polynomial representation of the unknown function was used, since this approach often helps to reduce the design variables to a reasonable number.

The polynomial representation of z and F can be approximated by sums

$$z = \sum_{n=0}^{m} \sum_{i=0}^{n} \alpha_{in} x^i y^{n-i}, \tag{14.89}$$

$$F = \sum_{k=1}^{s} \sum_{j=0}^{k} \beta_{jk} x^j y^{k-j}, \tag{14.90}$$

where α_{in} and β_{jk} are unknown coefficients found by solving the NLP problem. Substituting these finite sums (14.89) and (14.90) into the equilibrium equation (14.9) gives the following relations:

$$\sum_{n=2}^{m}\sum_{i=2}^{n} i(i-1)\alpha_{in}x^{i-2}y^{n-i}\sum_{k=2}^{s}\sum_{j=0}^{k-2}(k-j)(k-j-1)\beta_{jk}x^{j-1}y^{k-j-2}$$

$$-2\sum_{n=2}^{m}\sum_{i=i}^{n-1} i(n-i)\alpha_{in}x^{i-1}y^{n-i-1}\sum_{k=2}^{s}\sum_{j=1}^{k-1} j(k-j)\beta_{jk}x^{j-1}y^{k-j-1}$$

$$+\sum_{n=2}^{m}\sum_{i=0}^{n-2}(n-i)(n-i-1)\alpha_{in}x^{i}y^{n-i-2}\sum_{k=2}^{s}\sum_{j=2}^{k} j(j-1)\beta_{jk}x^{j-2}y^{k-j}+q=0, \tag{14.91}$$

where q is the vertical constant load. The equation must be satisfied for all admissible values of x and y; this means that we set equal to zero the coefficients of all possible degrees of the polynomials in x and y. Substituting (14.89)–(14.90) into the objective function (14.86) gives the reduced form of the objective function

$$C=\int_S H\,dx\,dy \approx \sum_{l=1}^{1600} d_l H \to \min, \tag{14.92}$$

where H is defined by formula (14.23), (14.26), or (14.29), d_l is the area of cell l of the standard domain, which has, as we assume, 1600 cells. These formulas define the MP problem with respect to the remaining values of α and β, after we have determined some of them from vanishing of the coefficients in (14.91).

When $F = D(x^2 + y^2)$, the middle surface of the shell is uniquely defined by a boundary value problem for Poisson's equation

$$\frac{\partial^2 z}{\partial x^2}+\frac{\partial^2 z}{\partial y^2}+q=0. \tag{14.93}$$

This case was considered by Ramaswamy (1958) and was a pioneering approach to the optimization of nonaxisymmetrical shells. However, we may use a more sophisticated approach in which the refined procedure permits us to drop an assumption about the stress state. Equation (14.93) is based implicitly on the hypothesis that in the optimal shell the shear force N_{xy} is zero; but the calculations show that the optimal shell sometimes has significant shear forces in the planes that are tangent the shell's mid-surface (see Section 14.4).

14.5.1.2. Method of Solution. Assume that

$$z = D(x^2 + y^2), \tag{14.94}$$

$$F = ax^4 + by^4 + cx^2y^2 + dx^2 + ey^2. \tag{14.95}$$

Therefore the shell mid-surface is a quadratic paraboloid or a parabola in each vertical section, and the stress state may have two axes of symmetry. Then Eq. (14.91) reduces to

$$b = a, \quad c = -6a, \quad d = -e - q/4D, \tag{14.96}$$

and the number of independent variables is now three: D, a, and e.

The design obtained must be safe, since the equilibrium equations and von Mises' plasticity criterion (14.22) are satisfied identically. So the deviation from the exact solution of the problem is now based on use of the objective function (14.92), and the equilibrium equations are satisfied rigorously. However, we must remember that the objective function is based on economical considerations that are, by their very sense, less precise than mechanical and geometrical data: the latter can be measured with up to six-digit precision, whereas in economics even one digit often looks unbelievable.

The program processes the information related to the problem in the form of the finite sum

$$C = 4 \sum_{i=1}^{20} \sum_{j=1}^{20} v_{ij} \to \min, \tag{14.97}$$

where

$$v_{ij} = \{[(12ax^2 + 2cy^2 + 2d)^2 - (12ax^2 + 2cy^2 + 2d)(12by^2 + 2cx^2 + 2e) + (12by^2 + 2cx^2 + 2e)^2 + 48(cxy)^2][1 + (2dx)^2 + (2dy)^2]\}^{1/2} \tag{14.98}$$

with x and y evaluated at the point (i, j) of a 40×40 square mesh similar to the mesh used in the last section. The program involves only three independent variables, but we may expect all kinds of tricks: multiextremality, singularity, multiadjoint admissible domain, and so on. In this case, we cannot trust the very concepts of gradient and sensitivity. We have selected random search with adaptation by "Agnesi curl." Specifically, at each step of the algorithm new values of variables T_k for assessment of the objective function were found from the relation

$$T_k = T_{k0} + \frac{T_k - T_{k0}}{1 + R_1(T_{k0} - TR_k)^2}, \tag{14.99}$$

where TR_k is a random number homogeneously distributed on the segment $[T1_k, T2_k]$, R_1 is the "skepticism parameter," and T_{k0} is the best value of T_k already found. If the second term in the denominator vanishes, then $T_k = T_{k0}$: adaptation is insignificant, since the algorithm is equivalent to a blind search. If the second term in the denominator is large, T_k is "attracted" to the so-far optimal value T_k. We do not discuss the numerical aspects of the random search, but we concentrate on the results.

14.5.1.3. Results of the Calculations and Discussion. The results are presented in Table 14.6 for different plan view [square, circle, rectangle (2:1), ellipse, and a complex plan without a title] in order to illustrate the universality of the approach (see Fig. 14.17). Moreover, the table contains results for shallow (slope below 18°) and nonshallow (slope between 18° and 35°) shells. Simultaneously, the role of the number of symmetry axes has been studied.

The areas for various plan views are also different so that, for better evaluation, not only dimensionless volumes but also relative volumes are given as well (as ratios between the volume and area of each figure). The plan figures were inscribed into the standard 40 × 40 mesh. Figures 14.18 and 14.19 show axonometric views of fields of normal forces and a mid-surface for some optimal shells. The force distribution is almost uniform,

Table 14.6. Amount of a Mises Material in Optimal Shells[a]

		Volume, V			Ratio volume/area		
			Nonshallow shell			Nonshallow shell	
Shape	Area	Shallow shell	2 axes	4 axes	Shallow shell	2 axes	4 axes
Square (2 × 2)	4.00	8.87	4.41	4.42	2.22	1.10	1.10
Circle ($r = 1$)	3.14	5.00	2.61	2.61	1.59	0.83	0.83
Rectangle (1 × 2)	2.00	3.14	1.74	1.36	1.57	0.87	0.68
Ellipse (1 × 2)	1.57	2.47	1.22	0.91	1.57	0.78	0.58
Complex contour (Fig. 14.17)	3.22	4.91	2.76	2.52	1.52	0.86	0.78

[a] From Reitman (1985).

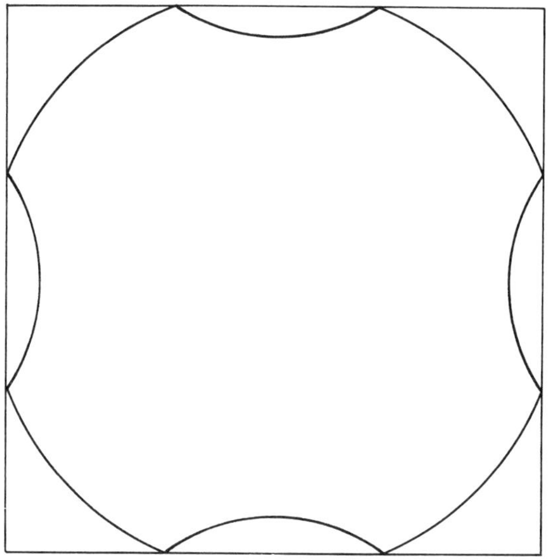

Figure 14.17. A Complex Plan View for Shell Optimization: Curved Line Shows the Contour of the Plate Inscribed in a 2×2 Square [Reitman (1983)].

with the variation not exceeding 18%. The distribution is almost symmetric with respect to the diagonals of the specified square plan view. These designs have been obtained for four axes of symmetry. But they were compared as well to the designs with two axes of symmetry (not nonshallow shells have been studied). For this case,

$$z = D_1 x^2 + D_2 y^2.$$

The results show without any doubt that for plans with four or more axes

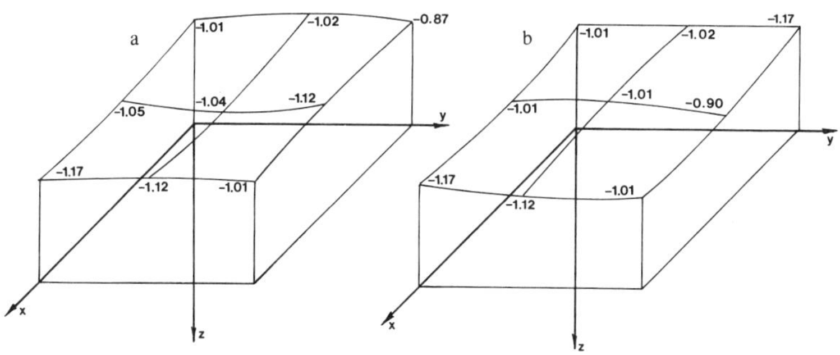

Figure 14.18. Force Distribution in the Optimal Plastic Metal Shell: (a) Component N_x. (b) Component N_y [Reitman (1985)].

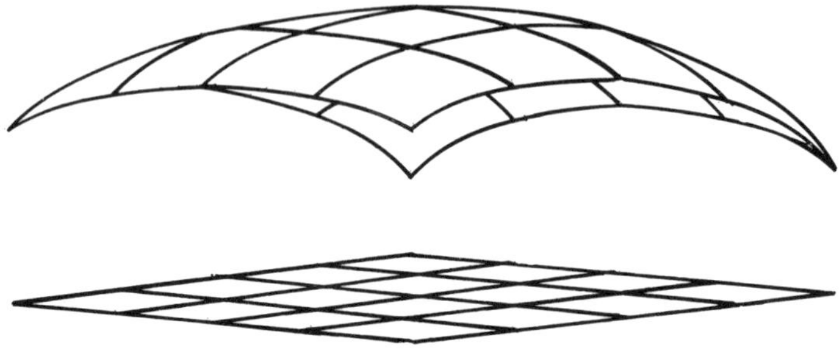

Figure 14.19. Axonometric View of the Optimal Plastic Metal Shell [Reitman (1985)].

of symmetry (the circle and the square), the optimal shape of surface has the same number of axes of symmetry, whereas for the rectangle and ellipse the lifting of the symmetry restriction allows us to save part of the material (up to 34%). The shape in Fig. 14.17 could be lighter by 9.5% with two axes of symmetry even though the shape has four axes of symmetry: horizontal, vertical, and both diagonal axes. It makes this nontrivial fact possible: sometimes a mid-surface with two axes of symmetry is efficient for a shell with a plan view having four axes of symmetry. However, other shapes of a nonsymmetrical surface have given no saving of material. Certainly, these conclusions may be true only conditionally, since some corrections may occur because of multiextremality of the problem. However, the low dimension of the optimization problem (only three to four parameters!) permits us to consider the solution as likely to be true.

The symmetry of the exact optimal design has not been generally proved even for axisymmetric plans, and we must look for other surprising phenomena in this field. For instance, there is no full explanation of the discrepancy between degrees of symmetry for the functions z and F. The last two columns of Table 14.7 show the relative costs of covering a unit area by the shells; they estimate the efficiency of each kind of shell. The result is poorest for the square shape of the plan view of the shell: it requires 40% more material than the circle, whereas for other noncircular plans the ratio is rather uniform. Figure 14.18 illustrates the distribution of internal forces in the optimal metal shell; the distribution is almost uniform but differs slightly from being symmetrical with four axes of symmetry, whereas the shape of the mid-surface is symmetrical with four axes of symmetry.

14.5.2. Plastic Concrete Shells

The statement of the problem for plain concrete differs very little from the problem for a metal shell: the von Mises plasticity condition is to be replaced by an adequate condition that takes into consideration the properties of the material (the BGK condition for plain concrete, for instance, or the Karpenko–Morley condition for reinforced concrete). The results of the calculations are presented in Table 14.7. We only discuss the analysis. The first observation shows that the highest shells are the most efficient, and they are more efficient than shallow shells. Moreover, the assumption of shallowness at the limiting slope of 35° strongly distorts the results of the optimization, but its influence is not pronounced at the limit slope of 18°. The error of this assumption is connected with both the distinction between the horizontal projections of the membrane forces and (to a greater degree) the oblique forces; the difference between them has been discussed in Section 14.2. In metal shells this effect is less noticeable, and replacement of the forces by their horizontal projections is possible at a greater value of the slope. Some other observations for concrete shells coincide with the conclusions for the optimization of the metal shells; for instance, the rectangular shell is much more efficient than the square one and could compete with the circular shell. But the quantitative results are different.

It is instructive to trace the changes in the results that are defined by this type of material. We underline the differences between the relative consumption of concrete and metal in the shell: the difference is as much as 32% for the square and 20% for the rectangle. The explanation is simple: plain concrete is 35–40% stronger when it is equally compressed in two perpendicular directions, as compared with the strength under a one-dimensional compression in the plane stress state. (At least, this is true in relation to the BGK plasticity condition.) For metals, this effect is not so pronounced, if there is any at all. Certainly, this effect also depends on the degree of asymmetry of the shells considered: it is more noticeable for axisymmetric shells, but it vanishes for arches.

The reported results are related to the assumption of perfect plasticity and perfect behavior of the supporting edge of the shell. In fact, the strength of any support is restricted, no matter if the support is a kind of soil or a

Table 14.7. The Amount of a BGK Material in the Optimal Shell[a]

		Volume, V		Volume/area	
Shape	Area	Shallow	Nonshallow	Shallow	Nonshallow
Square (2 × 2)	4	6.14	3.01	1.54	0.75
Circle ($r = 1$)	3.14	3.45	1.80	1.09	0.57
Rectangle (2 × 1)	2	2.17	1.10	1.08	0.55

[a] From Reitman (1983).

supporting structure. It generates doubts about the costs of the supporting structures. To resolve these doubts, we must take into account the cost of the supporting arch and its ties. The forces and bending moments of the arch could be related to shear forces in the shell and the locking force in the supporting arch, which is an additional variable of the problem. The appropriate calculations have been made in the thesis of Gudzenchuk [see Gudzenchuk and Reitman (1977)]. The stress state in rectangular shells has been approximated by a double sine series, and the results obtained were essentially different. But they were close to the results described earlier when costs of the arch and tie were made to vanish. Figure 14.19 illustrates an optimal design of the shell made by a plotter.

14.5.3. Elastic Shell with Uniform Thickness

We consider shallow elastic shells of uniform thickness and given surface area. We want to find the mid-surface shape function $z(x, y)$ that maximizes the buckling load q uniformly distributed over the plan view. Following Plaut (1986), who referred to Young (1985), we may formulate the elastic optimization problem for a shallow shell by Rayleigh's method. The boundary of the shell is given in the form of the rectangle $x = \pm A$ and $y = \pm B$ in the horizontal plane; on this boundary $z = 0$. The volume of a shell with uniform thickness ($H =$ const) may be derived from the general formula (14.86) by the relation

$$C = 4H \int_{x=0}^{A} \int_{y=0}^{B} \sqrt{1 + z_{,x}^2 + z_{,y}^2}\, dx\, dy. \tag{14.100}$$

If a shell is shallow, then $z_{,x}$ and $z_{,y}$ are each much smaller than 1 in magnitude; and, in the same way as in the axisymmetric problem, the volume will be approximately given by

$$C = V = 4ABH + 2H \int_{x=0}^{A} \int_{y=0}^{B} (z_{,x}^2 + z_{,y}^2)\, dx\, dy. \tag{14.101}$$

The problem is to find the shape $z(x, y)$ of the mid-surface with the volume of material in the whole shell specified so that the uniformly distributed lateral load q will be greatest without violating the behavioral constraints. The edge of the shell is simply supported or clamped on the rectangular boundary described. After introducing the new nondimensional variables, we easily obtain the equilibrium equation (compare with Section 14.2)

$$\Delta\Delta W - (H/D)\mathrm{PU}(W - z, F) - q = 0 \tag{14.102}$$

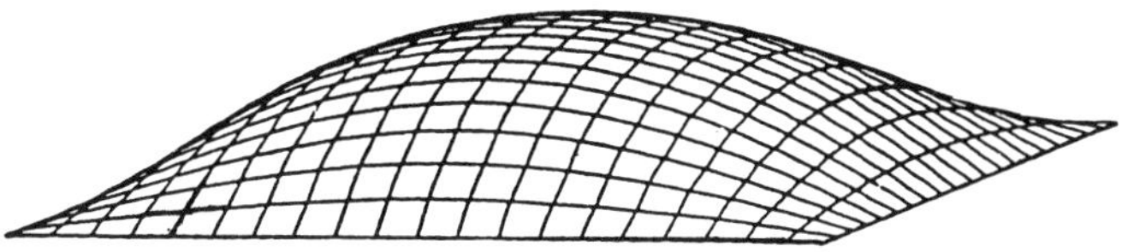

Figure 14.20. Axonometric View of the Optimal Elastic Shell with Simply Supported Edge [Plant, Johnson, and Parbery (1984)].

and the compatibility equation

$$(H/D)\Delta\Delta F - \mathrm{PU}(W, z) + \tfrac{1}{2}\mathrm{PU}(W, W) = 0, \qquad (14.103)$$

where Laplace's and Pucher's operators in the Cartesian frames are used, as defined in Section 14.2. Both equations contain the unknown functions nonlinearly. To obtain the solution, we used a finite difference discretization and a Runge–Kutta method with variable steps (the Runge–Kutta–Fehlberg method) in the form of the standard code.

We used the piecewise cubic Hermite polynomial for interpolating the design function; details of the procedure are given for the axisymmetric problem by Plaut, Johnson, and Parbery (1984). The forms of the optimal shells are presented in Figs. 14.20 and 14.21, where nondimensional values do not permit us to judge such important characteristics as the size of the boundary layer and the slope of the shells. However, in spite of the previous remark of the authors about the axisymmetric problem (see Section 14.3), at least one design effect is apparently true: the involvement of bending forces in the accumulation of energy in the boundary zone. However thin the shell is, at least part of the material in the vicinity of the edge has to be used in bending; for shells of practical importance this boundary zone could be of a width commensurate with the shell thickness.

The optimal designs obtained have been compared to a reference shell with uniform thickness and shape defined by the equation

$$z(x, y) = A_1 \sin(\pi x/2A) \sin(\pi y/2B), \qquad (14.104)$$

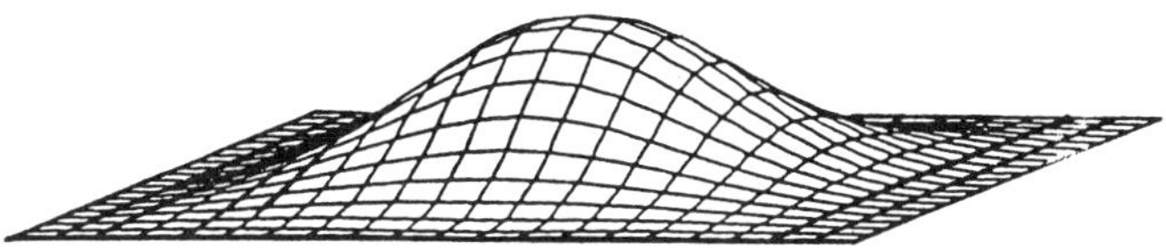

Figure 14.21. Axonometric View of the Optimal Elastic Shell with Clamped Edge [Plaut, Johnson, and Parbery (1984)].

where $2A$ and $2B$ are the sides of the rectangular plan view and A_1 is a constant. The optimal critical load is 4 to 50% higher than for shells with simply supported edges, and 25 to 68% higher than for shells with clamped edges. (The maximal buckling load q has been uniformly distributed over the plan view of the shell in both cases.) These ranges of differences are now easily explained: the sinusoidal shape of the reference shell is similar to the optimal shape in shells with simply supported edges, but is different in shells with clamped edges. This difference is especially noticeable for the lower bound of the possible ranges: 4% versus 25%.

14.5.4. Cylindrical Shells with Variable Cross Section

The optimal section of a long cylindrical shell has been studied by many researchers. They usually assumed a constant thickness for the shell and applied Bernoulli's hypotheses. We now consider the simplest case of a shell having the greatest bending strength for a given width s_1 of the initially plane sheet of material. Moreover, due to the constant thickness H and specified width of the section c, the problem reduces to the optimal inextensible bending of a given rectangular sheet as defined by the function $z(x)$. But this plane sheet of material has to be laterally bent to provide the section with the greatest plastic strength modulus W_p while the width of the shell c is specified in advance and $s_1 > c$ (Fig. 14.22). The last inequality enables the designer to have an admissible shell cross section. The optimal shell will have the maximal bending plastic strength modulus W_p [the solution is due to Rzhanitsyn (1960)]:

$$W_p = 2H \int_0^{s_1} z\, ds \to \max, \tag{14.105}$$

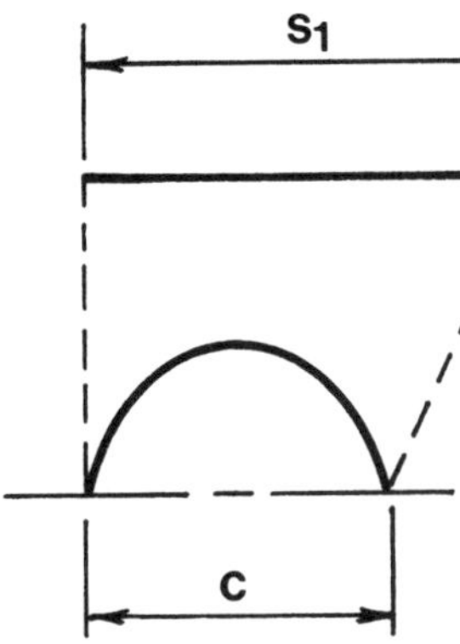

Figure 14.22. Optimal Section of a Long Cylindrical Shell: c Is the Width of the Shell and s_1 Is the Width of the Rectified Wave [Rzhanitsyn (1960)].

where $z(x)$ is the unknown vertical coordinate of the middle surface of the section. In Cartesian coordinates Eq. (14.105) takes the form

$$W_p = \int_0^c y\sqrt{1 + z'^2}\, dx \to \max \tag{14.106}$$

(the constant factor $2H$ is omitted). The isoperimetric condition must also hold:

$$\int_0^c \sqrt{1 + z'^2}\, dx = s_1, \tag{14.107}$$

where s_1 is the length of the rectified wave.

Equations (14.106) and (14.107) define an isoperimetric problem of the calculus of variations. To resolve this problem, we construct the functional

$$\Phi = \int_0^c [\sqrt{1 + z'^2} + \lambda\sqrt{1 + z'^2}]\, dx, \tag{14.108}$$

where λ is a Lagrange multiplier. The Euler–Lagrange equation for Φ is

$$\sqrt{1 + z'^2} - \frac{d}{dx}\frac{(\lambda + z)z'}{\sqrt{1 + z'^2}} = 0. \tag{14.109}$$

Hence

$$\sqrt{1 + z'^2} - \frac{z''^2 + z'^4 + zz'' + \lambda z'}{(1 + z'^2)^{3/2}} = 0. \tag{14.110}$$

Let $p = z'$. Then (14.110) can be written as the first-order equation

$$1 + p = (z + \lambda)p\frac{dp}{dz}, \tag{14.111}$$

where it is easy to separate the variables:

$$\frac{p\, dp}{1 + p^2} = \frac{dz}{z + \lambda}. \tag{14.112}$$

Integration gives

$$p = \frac{dz}{dx} = \sqrt{C(z+\lambda)^2 - 1}, \tag{14.113}$$

where C is an arbitrary constant. Again separating variables in (14.113) and integrating give

$$x = A\cosh^{-1}\left(\frac{z+\lambda}{A}\right) - B, \tag{14.114}$$

where the constant $A = C^{-1/2}$ and B is a new constant. Solving (14.114) for z leads to

$$z = A\cosh\left(\frac{x+B}{A}\right) - \lambda. \tag{14.115}$$

At $x = c/2$, $z' = 0$; hence $\sinh[(c/2 + B)/A] = 0$ and $B = -c/2$. At $x = c$,

$$A\cosh(c/2A) - \lambda = 0$$

or

$$\lambda = A\cosh(c/2A). \tag{14.116}$$

Finally,

$$z = A\left[\cosh\left(\frac{x - c/2}{A}\right) - \cosh\left(\frac{c}{2A}\right)\right]. \tag{14.117}$$

This hyperbolic cosine line is a chain line, or catenoid, where A may be found from the isoperimetric condition (14.107), which leads to the transcendental equation

$$\sinh(c/2A) = s_1/2A. \tag{14.118}$$

This result underlines the similarity of this problem to the problem of a chain under a vertical distributed load. A similar equation can be derived for the elastic cylindrical shell if z in the integrand of (14.105) is replaced by z^2. The possible discontinuity of the function $z'(x)$ is prevented by the Weierstrass–Erdmann condition, since the optimal middle line must be normal to the symmetry axis.

Zyczkowski and Kruzelecki (1975) studied the same problem from the point of view of buckling instead of strength. But direct comparison of results is impossible, since the statement of the problem includes no constraint on the width of the shell.

14.6. Special Problems of Shell Optimization

14.6.1. Plastic Shells under Impulse Load

Martin's bounds help to simplify the design of shells under dynamic loading in a similar way to Gvozdev's theorems, which help to do this for statical loads. We show the procedure for an annular plate. The distributed impulse hits the annular plate with clamped edge $r = r_0$ and free edge $r = r_1$ (Fig. 14.23). At the initial moment this impulse gives velocity v_0 to all points of the plate. According to Martin's second bound [Martin (1964)], optimization of this plate under dynamic loading is equivalent to the optimization of the same plate under a static load:

$$Q = -\pi \int_{r_1}^{r_0} \frac{rmv_0^2\,dr}{2\pi u_0} = \frac{mv_0^2(r_0^2 - r_1^2)}{4u_0}, \tag{14.119}$$

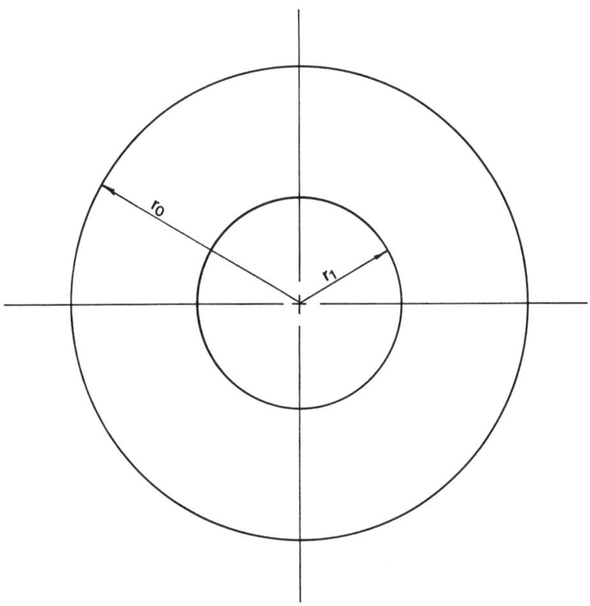

Figure 14.23. Annular Plate under Dynamic Loading.

where u_0 is a specified limit displacement of the inner edge and m is the mass of the plate per unit area. The optimality condition, similar to Section 14.3, is

$$C = \int_{r_1}^{r_0} (M_\theta + M_\phi) r \, dr \to \min. \tag{14.120}$$

The Hamiltonian of the resulting system after maximization according to Pontryagin's maximum principle (see Section 14.3) leads to the solution:

$$\begin{aligned} &\text{for } r_1 \leqslant r \leqslant r_0/2, \qquad M_\phi = 0, \quad M_\theta = -Qr_1, \\ &\text{for } r_0/2 \leqslant r \leqslant r_0 \qquad M_\theta = 0, \quad M_\phi - Qr_1(1 - r_0/2r). \end{aligned} \tag{14.121}$$

Notice the interesting result which was indicated by Rozvany (1976) for static loading (see also Volume 1, p. 217): the switchpoint is at the mid-radius, and its position does not depend on the load. If the switchpoint $r = r_0/2$ is outside the plate, as for a relatively thin annulus, the whole domain is defined by a single analytical expression without the switchpoint.

Similarly, we can consider rigid-plastic axisymmetrical shells, and mathematical programming provides the approximate parameters for arbitrary shells [see Reitman and Shapiro (1976)]. Although this approach is similar to limit design, it is different: contrary to Gvozdev's theorem, it cannot guarantee the exact optimal solution with given constraints, but it gives reasonable bounds for this solution.

14.6.2. Ill-Posed Problems

Ill-posed problems are typical in tricky fields such as the optimal design of shells and even simpler structures, including annular plates (see Volume 1, pp. 37–39), and even the bending of beams. Consider the following optimization problem for a beam of rectangular section with width b and height h. The problem is to find b and h such that, for a given moment $M = Rbh^2/6 = \text{const}$, the section area $A = bh$ will be a minimum. This problem is subject to degeneration: as $h \to \infty$ and $b \to 0$, the area $A \to 0$ for each finite value of moment M and yield stress σ_0. The reason is that the strength modulus $W = bh^2/6$ for an elastic material (or $W_p = bh^2/4$ for a perfectly plastic material) is proportional to h^2, but A is proportional to h. This consideration has been extended to any physical law relating the strains and stresses [see, for instance, Reitman and Shapiro (1976)]. The same is true for the beam stiffness (proportional to h^3) or the moment of inertia of any rectangular beam section. But this fact of degeneration, well established for the bending of beams, has been easily forgotten in the study of the same beams when used as stiffeners of elastic or plastic plates.

The classical problem of a solid annular plate has been studied for degeneration in Volume 1. It was shown that, even in the axisymmetric case, annular stiffeners could formally support any load with no expenditure of material at any finite values of circumferential plastic moments M_ϕ. To restore physical sense to the problem, we must remember that such degeneration is caused by ignoring radial shear forces: in fact, the weightless and infinitely high Megaref rings will slide along each other and bear no load. Olhoff (1973) studied this effect for elastic plates, and Lurie *et al.* (1984) considered a more sophisticated degeneration in any flexural solid structure (Section 14.6.3).

The procedure is standard: for any elastic flexural structure with constraints on buckling or natural frequency, each form of buckling will be blocked by a Megaref–Olhoff stiffener of no weight. The stiffeners themself may buckle too, but each form of buckling will also be blocked by a similar stiffener. The plate will degenerate into a porous medium of no weight capable of bearing any loads without buckling.

Now the Megaref–Olhoff effect is well understood, and some optional measures can exclude it by using different constraints. Following are some of these constraints:

1. Constraints on the lateral shear force; for instance, the constraint for an annular plate

$$V_r = dM_r/dx \leqslant Q_0; \tag{14.122}$$

2. Constraints on the thickness:

$$H_1 \leqslant H \leqslant H_2; \tag{14.123}$$

and

3. Constraints on the slope of the mid-surface in any direction. The constraints do not work in some exotic cases, but, in general, use of such constraints leads to a sound structure.

An example of the first constraint is given in Nemirovsky (1975). A particular case of the second constraint has been used in Plaut, Johnson, and Parbery (1984), where prescribing a constant thickness H prevents degeneration. But first we must accept the fact that all cases of ill-posed statements of the problems existing in plates also exist in shells. For instance, in Fig. 14.24 we can see the degeneration of ill-posed models for axisymmetric shells. That is why we used only membrane models in the optimization

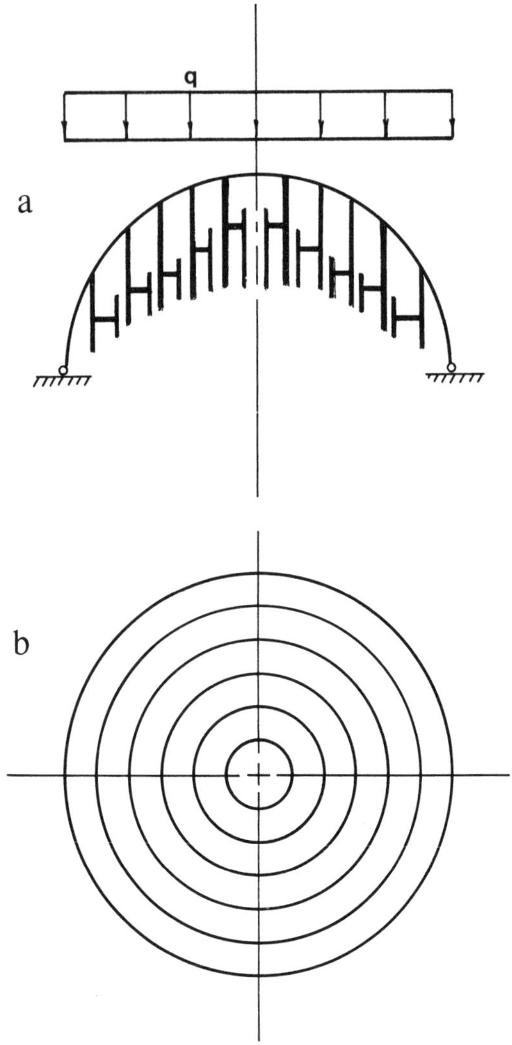

Figure 14.24. Degeneration of a Bending-Resistant Double-Curvature Shell under Uniform Transverse Load: (a) Vertical Section. (b) Horizontal View of Concentric Stiffeners.

of shells and specified the given moment arm of the internal couple (Section 14.3). But there are several classes of ill-posed problems specifically for shells, and we shall discuss them.

The third constraint helps prevent the formation of M–O asymmetric stiffeners, since the stiffeners lead to infinite values not only for the function

z but also for its derivatives. The intermediate values of $H(H_1 < H < H_2)$ that have been obtained in many published solutions are not correct if they are not between stiffeners with height H_2 [see the detailed discussion by Olhoff (1986); Fig. 14.25 is taken from this source]. In fact, if $H < H_2$ in an optimal solid plate, we can remove some material and make the stiffeners of height H_2 without decreasing the strength or stiffness of the structure. So, such a plate (or moment-resistant shell) will not be optimal. The stiffeners may be in one or two directions (in the latter case they form a waffle structure), but this phenomenon absolutely rules out any pillowlike or similar optimal designs of plates and shells.

14.6.2.1. Shallow Membrane Shells. Generally, the optimization of membrane shells is not an ill-posed problem, although sometimes imposing a constraint on the slope or the rise is reasonable, since otherwise the optimal shells are too high. Under the ordinary definition of shell shallowness (the shell mid-surface is identified with the horizontal plane, and the actual forces in shells are identified with their horizontal projections), the shape-optimization problem is ill-posed: the shell rise will grow to infinity, whereas its surface area will be limited and equal to the projection area. Hence, the shallowness must be provided by constraints, since otherwise the optimal solution will degenerate. But if the shallowness is stated only as an overall average for the whole shell in the integral form

$$I = \tfrac{1}{2}\int_{S_p} (z_{,x}^2 + z_{,y}^2)\, dS \leqslant \int_{S_p} dS = A_p, \qquad (14.124)$$

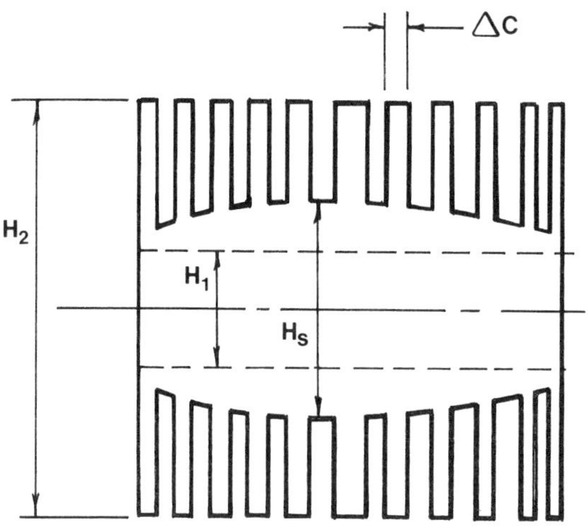

Figure 14.25. Formation of Stiffeners for the Plate with Limited Height.

it can lead to a function of unbounded variation. In particular, even Dirac's delta may be selected as a function $z(x, y)$ at arbitrary locations. In the process of such optimization the premises of the theory of shallow shells will be violated long before the optimal solution is achieved. So, optimization of shallow membrane shells with variable thickness and variable mid-surface inevitably brings us to the domain where basic postulates of the theory of shallow shells will be violated. As we noted, one way to overcome this difficulty was used by Plaut, Johnson, and Parbery (1984), who optimized a uniform thickness shell. (In our solution described in Section 14.5, both kinds of constraints were used.)

14.6.2.2. Nonshallow Membrane Shell. We consider the optimization problem for a membrane solid shell, allowing for a variable thickness H. If no restrictions on shear forces and thickness are stated, unbounded artificial bending moments may also occur. The thickness decreases, and this leads to the same form of degeneration as in the previous plate problem. The bending resistance is growing, if waves of the shells are growing in their heights and transforming into M–O stiffeners (see Fig. 14.26). If a proper theory of shells is adopted, this may prevent degeneration, since the lateral shear forces are now replaced by membrane forces.

14.6.2.3. Variation of the Plan View. When a shell with given plan view area is optimized, thus minimizing the material volume, a new form of

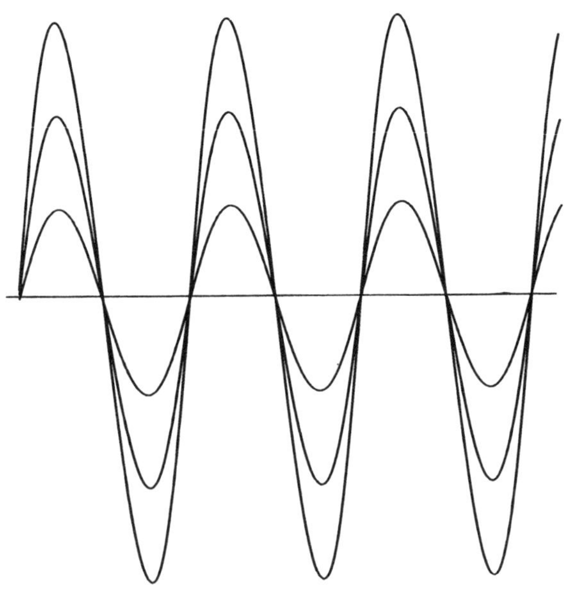

Figure 14.26. Transformation of Waves to Stiffeners for a Membrane Shell.

degeneration starts to appear: an increase of the length of one dimension of the plan view. Since the volumes of a beam (for given bending load) and an arch (for given pressure) are proportional to the square of the span, increasing the length while decreasing the shell's plan view width causes the plan to degenerate into a straight line of vanishing width. Actually, this line need not be straight: it is enough for it to have only a finite number of corners or very curved segments. (We assume such a segment of plan view to have large curvature if $R \ll L$, where R is a radius of the plan view and L is the plan view span.) If this condition is met, each form of the plan view with vanishing span (for instance, the spiral of Fig. 14.27) will give a vanishing total volume of material.

To the collection of ill-posed continuous structures already mentioned in Volume 1, we may add a disk with a round hole under homogeneously distributed pressure. If we try to find the minimal distribution of material that supports the pressure, we tend immediately toward a growing flange on the boundary of the hole. But this degeneration is not the sign of an ill-posed statement of the problem: the volume of material will remain finite.

14.6.3. Theoretical Problems of Shell Optimization

14.6.3.1. Plastic Shells. Only physical and numerical considerations have been discussed. However, some *a priori* theoretical studies of an

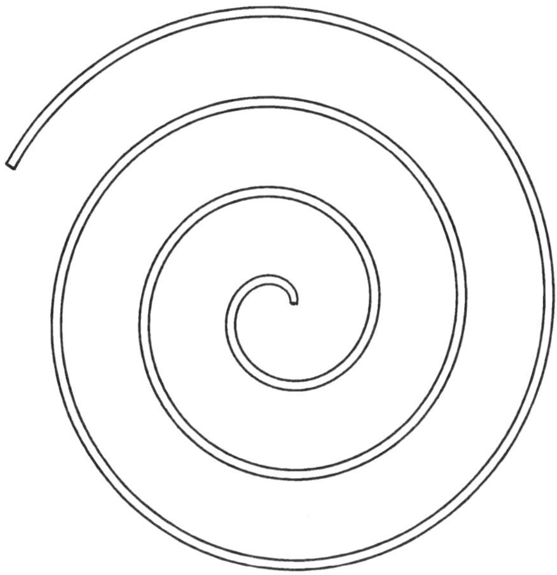

Figure 14.27. A Possible Form of the Optimal Plane View of a Shell: Irregular Spiral.

optimization problem help to avoid some time-consuming exercises, which serve as no more than a bare introduction to the actual design process. Some research has been devoted to the theoretical studies of the shell optimization problems. For instance, Reitman (1967) studied the classification of some plasticity problems, including disk optimization. The type of the partial differential system (elliptic, hyperbolic) may help to anticipate like difficulties in solution that the researcher could meet. If $F(\sigma_{\alpha\beta}) = 0$ is the plasticity criterion in the plane stress state, where α and β are characteristic directions, then we can classify the problem by studying the characteristic equation

$$\det\begin{bmatrix} H & 0 & 0 & \sigma_\alpha \\ 0 & 0 & H & \sigma_{\alpha\beta} \\ F_{,\sigma_\alpha} & F_{,\sigma_\beta} & F_{,\sigma_{\alpha\beta}} & 0 \\ F_{,,\sigma_\alpha\sigma_\beta} & F_{,,\sigma_\beta\sigma_\beta} & F_{,,\sigma_\beta\sigma_{\alpha\beta}} & 0 \end{bmatrix} = 0. \qquad (14.125)$$

For Tresca's condition (inclined segment) $\sigma_{\alpha\beta} = 0$ and the problem is hyperbolic everywhere. The characteristic directions are then collinear with the principal stress directions. Von Mises' criterion gives the equation $\sigma_\alpha^2 + 4\sigma_{\alpha\beta}^2 = 0$, or $\sigma_\alpha = 0$ and $\sigma_{\alpha\beta} = 0$; hence almost everywhere (except at the intersection points of the coordinate axes and the Mises ellipse) the problem is elliptic. A similar study has been made for all other plasticity criteria for disks, plates, and shells, but we do not discuss them except to note that, for the linearized plasticity condition in plain concrete, there are nondegenerate elliptic and hyperbolic zones as well as parabolic isolated points.

14.6.3.2. Elastic Shells. The regularization of ill-posed problems has been traced in physical examples, but it permits a theoretical consideration, as was shown by Lurie, Fedorov, and Cherkaev (1984). A layered composite plate is assembled from initially isotropic compounds or orthotropic compounds, which differ only in the orientation of the principal axes of elasticity. For a wide class of problems with weakly continuous cost functions, the problem of optimal orientation of the principal axes is analyzed. In general cases, such problems may have no solutions because of "gliding regimes" and similar reasons, although the necessary conditions for an optimum are satisfied. Hence, the problems need a regularization that consists basically of constructing a procedure for approximately solving a slightly different problem. Technically, the problem is reduced to the construction of an elasticity tensor, and this problem is much simpler when this tensor is partly defined.

As particular problems, control of the dilatation and shear moduli has been discussed, including problems with isoperimetric restrictions. For all cases the existence of a solution is proved, and the corresponding necessary extremal conditions are constructed. Although the approach, essentially based on functional analysis, is common, the derivation has been restricted to thin plates.

Much more sophisticated studies have been made for optimal elastic shells. For instance, the complex general study of the height for uniform shells was developed by Chenais (1987). She studied the existence and properties of solutions of elastic shell optimization problems. The duality of the problems of minimal cost and maximal efficiency of the structures have also been studied.

Exercise 14.1

Find the angle 2ϕ for the optimal spherical dome at internal pressure and compare the result with the optimal arch. Which of the two structures is more efficient? Which structure has a greater optimal height and why?

Exercise 14.2

Formulate the problem of optimal design for an anisotropic double curvature shell under different assumptions for the objective function.

Exercise 14.3

Construct an objective function of the optimization problem for a homogeneous shell with variable standardization level, allowing for the scale economy. Assume that the unit cost of a standardized finite element of the shell is the function

$$C_1 = c + a\,e^{-b(n-1)},$$

where n is the number of identical elements, c is the unit cost at complete standardization ($n \to \infty$), $c + a$ is the cost of a single element structure (at $n = 1$), and b measures the standardization efficiency.

Exercise 14.4

State one of the alternative formulations of the plate optimization problem, allowing for the lateral shear forces Q_x and Q_y. Is this problem subject to degeneration without constraints on thickness?

Exercise 14.5

Optimize the reinforcing stiffener of a rectangular plate under a lateral uniformly distributed force accounting for the strength and stability. The thickness of the plate is given. The cross section of the stiffener is a rectangle with constant sides that are to be found. Discuss the problem of degeneration.

Exercise 14.6

State the dynamic programming problem for optimization of a shell, allowing for its own weight.

Exercise 14.7

Write Bellman's equation for the dynamic programming problem of axisymmetric shell optimization. The material of the shell is anisotropic with variable anisotropy factor. What is the difference between the two versions of the problem, with and without allowing for the shell's own weight?

Exercise 14.8

Resolve the optimization problem for an axisymmetric elastic shell with given lowest frequency, using Pontryagin's maximum principle. Is the solution unique?

14.7. The Maximum Principle

14.7.1. The Optimal Control Problem

Structural engineers are not as likely as system engineers to have encountered the maximum principle as a tool for optimization problems, so we present a brief statement of the simplest optimal control problem and the formalization that the maximum principle provides for its solution. The statement of the optimal control problem comes naturally for systems described by a state or phase vector function $\mathbf{x}(t)$ of time having n components x_i, $i = 1, 2, \ldots, n$. The change in these functions with time is governed by ordinary differential equations

$$\frac{dx_i}{dt} = f_i(x_1, x_2, \ldots, x_n; u_1, u_2, \ldots, u_k), \tag{14.126}$$

where the u_h, $h = 1, 2, \ldots, k$, are the components of a control or design parameter vector $\mathbf{u}(t)$ that is generally constrained to belong to a prescribed region $\mathbf{\Omega}$ of k-dimensional control space. Note that derivatives of $\mathbf{u}$ do not appear. No constraints are applied to $\mathbf{x}$, except possibly boundary conditions at the ends (t_0, t_1) of the interval of concern.

Let some cost functional C of the state and control be given:

$$C = \int_{t_0}^{t_1} f_0(\mathbf{x}(t), \mathbf{u}(t))\, dt. \tag{14.127}$$

The optimal control problem is to find the optimal control function $\mathbf{u}(t)$ and the corresponding optimal state trajectory $\mathbf{x}(t)$ that minimize the cost C while satisfying the constraint equations (14.126), with $\mathbf{u}$ contained in $\boldsymbol{\Omega}$ and $\mathbf{x}$ satisfying any end conditions.

This "dynamical" or "evolutionary" statement of the problem can be adapted to problems of frame structures by replacing time by the length along a beam or arch and by using displacements, bending moments, etc., as state variables and stiffness and mass distribution as control or design functions.

The absence of t as an explicit variable on the right sides of (14.126) is not restrictive. If any f_i depends on t, then replace it in all such functions by a new state variable x_{n+1} and add the additional constraint equation $dx_{n+1}/dt = 1$. Similarly, the integral cost can be replaced by introducing a state variable $x_0(t)$ satisfying the differential equation $dx_0/dt = f_0(\mathbf{x}, \mathbf{u})$ and initial condition $x_0(t_0) = 0$; the optimization problem is then that of maximizing $x_0(t_1)$ subject to the constraints, including the differential equation for x_0 in (14.126).

14.7.2. The Maximum Principle

Introduce auxiliary variables similar to Lagrange multipliers called the *dual state variables* $\psi_0(t), \psi_1(t), \psi_2(t), \ldots, \psi_n(t)$, which satisfy the conjugate differential equations

$$\frac{d\psi_i}{dt} = -\sum_{j=0}^{n} \frac{\partial f_i(\mathbf{x}, \mathbf{u})}{\partial x_i}, \qquad i = 0, 1, \ldots, n. \tag{14.128}$$

Also introduce the Hamiltonian $H(t, \mathbf{x}, \mathbf{u}, \boldsymbol{\psi})$by

$$H = \sum_{i=0}^{n} \psi_i f_i(\mathbf{x}, \mathbf{u}); \tag{14.129}$$

the original and conjugate equations (14.126) and (14.128) have the form of the canonical system formed from H for $i = 0$ to n:

$$\frac{dx_i}{dt} = \frac{dH}{\partial \psi_i}, \qquad \frac{d\psi_i}{dt} = -\frac{\partial H}{\partial x_i}. \tag{14.130}$$

The maximum principle states that a solution to the optimal control problem must be such that the optimal control vector $\mathbf{u}(t)$ maximizes the Hamiltonian H at each time t while the state and dual vectors [satisfying (14.130) as t changes] are considered constant at each instant. If the maximum is smooth, occurring when the control lies inside its admissible domain $\mathbf{\Omega}$, then the problem reduces to one in the classical calculus of variations. (Indeed, if the region $\mathbf{\Omega}$ is open, then the maximum principle coincides with Weierstrass's necessary condition for an extremum.) When the maximum occurs for the control on the boundary of $\mathbf{\Omega}$, it is generally not smooth.

The maximum principle is often used as a constructive principle to find candidates for the optimal control $\mathbf{u}(t)$ and the associated state trajectories $\mathbf{x}(t)$. If this defines a unique control and associated state trajectory, we may be able to verify that sufficient as well as necessary conditions have been met. But, generally, we can only expect to satisfy necessary conditions.

For different subintervals of (t_0, t_1), the analytical expressions for $\mathbf{u}$ and $\mathbf{x}$ may be different; $\mathbf{x}(t)$ must be continuous, but $\mathbf{u}(t)$ need only be piecewise continuous and bounded. The values t^*, where $\mathbf{u}(t)$ may be discontinuous [and $\mathbf{x}(t)$ changes analytical form] are called *switchpoints*: their number is finite. For linear control problems where all f_i are linear, the number of such points is $n - 1$.

14.7.3. Bibliographic Comments

The maximum principle is due to several scientists, most notably Hestenes in the 1940s [see Hestenes (1966)] and Pontryagin and his students [Pontryagin *et al.* (1962)]; certainly the work of the latter school is responsible for bringing it to common notice among the engineering community.

As noted, the control $\mathbf{u}$ may be discontinuous but must be bounded. This tacit assumption of classical control theory prevents the use of the maximum principle for problems where any u behaves like a Dirac delta function. Methods for dealing with this case are given in Bell and Jacobson (1975). Attempts have been made to extend the methods of control theory to systems governed by partial differential equations; see Carmichael (1981) and Lurie (1985) for surveys.

15

Optimal Design of Elastic Frame Structures

15.1. Introduction

The earliest theoretical studies of optimal structural design are by Maxwell and Michell. In 1869, Maxwell (1890) determined that, for a given set of forces applied at specified points to a fully stressed braced framework, the difference in volume between tension and compression members is a constant regardless of the configuration.

On the basis of this theorem, Michell (1904) showed that unique geometries could be obtained for pin-jointed frames of absolute minimum weight. Although Michell's theory has little practical application, since it is not a direct design method, it was the first example in which the minimum-weight design problem was expressed in terms of form and behavior parameters.

In general, three groups of design variables can be identified (compare Section 12.3): *size* variables, which define the sizes (excluding lengths) of the structural members; *configuration* variables, which are typically the spatial coordinates of the intersections of the structural members (nodes); and *topological* variables, which define which members are included in the structure. We now briefly review these types.

The overwhelming majority of the work in optimal structural design has involved only sizing types of variables [see, e.g., Schmit and Mallet (1963), Dwyer (1974), Schmit and Miura (1978)]. There are many methods for sizing optimization. Schmit has been a major advocate of MP methods in structural design. He recommended a steepest descent type of procedure in nonlinear programming. Cornell, Reinschmidt, and Brotchie (1965) have devised a linear program to select the best set of design variables that minimize the cost without violating constraints on stresses and deflections. Gallagher and Gellatly (1966) used the direct stiffness method for analysis. In the redesign process the steepest descent technique directs the desired changes in design parameters.

Brown and Ang (1966) employed the gradient projection method of nonlinear programming. They obtained a minimum-weight design of a two-story, single-bay frame subject to elastic design constraints. Majid and Elliott (1971) also used a gradient method to find minimum-weight designs for a family of fixed-base pitched-roof frames. Majid (1974) and Majid and Anderson (1968, 1972) developed an iterative procedure that involves repeated elastic–plastic analysis of the frame. Starting with a set of lower-bound sections, the frame is analyzed for maximum moments and axial loads. The members are then redesigned so that the frame is elastic under working loads and no hinges form in the columns. Column instability and torsional buckling are also prevented. This procedure is repeated until the final design has the required strength.

Cammaert (1971) and Nakamura (1976) developed the minimum-weight elastic design of multistory unbraced steel frames by using nonlinear programming. But the size of frames that can be tackled by this approach is limited. The major part of Cammaert's work deals with the use of dynamic programming (DP) in the optimal plastic design of multistory frames; this approach will be considered in Chapter 17. Lee and Kapton (1974), Leimbach and Liekweg (1977), and Atrek (1984) developed cost (weight) optimization of steel frames and trusses. Grierson and Chiu (1982) have considered planar frameworks composed of beam and column elements under combined axial and bending stresses. They have approximated the stiffness and strength properties of the elements as linear functions of the sizing variables.

Configuration variables have received less attention, although several techniques have been put forth that include these variables [see, e.g., Vanderplaats and Moses (1972) or Kunar and Chan (1976)]; however, they generally increase the complexity of the problem. Many examples of shape (configuration) optimization of beams, arches, grid shells, and cable nets can be found in Brandt (1984, 1989). Pedersen (1973, 1981) and Pedersen and Jorgensen (1984) have considered shape optimization of bar structures subjected to multiple loading conditions. They developed a unified optimization approach that combines finite element and linear programming.

The salient feature that differentiates these first two sets of variables from topological variables is that topological variables are usually considered to be discrete and sizing and configuration variables are usually considered to be continuous. Therefore, gradient techniques may be used for sizing and configuration variables, but may not, in general, be used for topological variables. Despite its difficulty, the topological problem has remained of great interest because it is the basic problem that must be solved before either the configuration or sizing problem can be attacked. Basically, two approaches can be used in handling the topological problem: members can be deleted from an existing design, or they may be added to a minimal design. The latter method is philosophically very satisfying since the structural layout is constructed by starting from a minimal amount of information (e.g., the load points and support points). Unfortunately, this approach is more difficult and, as a result, has had limited success. One approach for this was suggested by Porter Goff (1966), who used dynamic programming (see Chapter 17) to grow a structure from a load point. This approach requires the development of a criterion that will identify at some node of the structure what member should be added to produce the minimum total weight. Since only part of the structure is known at that time, this criterion must be independent of the entire structure and so must rely, to some extent, on *a priori* knowledge of what the structure will look like. The approach suggested by Porter Goff for a two-dimensional frame structure was to start from the single load point and move to the single spatial point with the lowest moment generated by the external forces. This line of attack does not seem to have been followed in subsequent papers.

A second approach was developed by Spillers (1966, 1971, 1975) and Spillers and Friedland (1972). They chose to expand their structure by bisecting each of a predefined set of exterior members with a new member. Optimizations are carried out on each new configuration to determine which member should be added to the base structure. This approach entails numerous sizing-type optimizations, which may make it impractical. In common with virtually all topological methods, the members to be added in their study essentially serve the purpose of shear members. Since, in general, the results tend toward adding many members, we might speculate that these approaches tend to imply that shear webs are more efficient than discrete members.

The removal method has been adopted by researchers who are trying to develop more general-purpose codes. Two basic approaches have been used. In one approach, during the configuration phase, if nodes move within a prescribed distance of each other they are coalesced into one node and a member is removed [see, e.g., Thomas and Brown (1977)]. A second approach has been used by several authors in various forms [see, e.g., Dorn, Gomory, and Greenberg (1964) or Lipson and Gwinn (1977)]. In one form the areas of the members are allowed to go to zero; therefore, these members automatically remove themselves. A second form examines the forces in the members, and the first member to reach a prescribed minimum value is the member chosen for deletion. The technique of allowing the member to shrink to zero allows the problem to be treated in a continuous fashion, but poses a serious limitation for some practical problems. In many problems minimum sizes are imposed by manufacturing constraints. If the design stops between this value and zero for any member, the design is of little use. In general, we can expect the design to stop in this condition.

Because of the very high cost of the basic structural analysis, we are forced to discard exhaustive search as an approach in either deletion or addition. Therefore, we must develop and rely on heuristic methods to identify which member should be added or deleted to improve the final design. For truss problems, Sheu and Schmit (1972), for example, took advantage of the LP approach to plastic collapse design to identify likely candidates for removal from the original elastic problem. We cannot, in general, expect to obtain proofs that heuristic methods will find the global (or even local) optimum. In fact, those methods that have been presented in the literature usually do not indicate how well they have performed on simple problems for which the optimum is known. In addition, heuristic methods will introduce some possibility of choosing the wrong members to delete or add. For example, it has been pointed out that most candidates for deletion or addition are basically stiffening members. When using a deletion method based on stress criteria, we can take false steps, and this possibility must be considered in the heuristic. Such algorithms for both deletion and addition methods for frames and stiffened structures have been developed by Bennett (1980). For sizing optimization he used the feasible-directions algorithm with reciprocal design variables and a Taylor series expansion of the active constraints. Haftka and Grandhi (1986) presented a survey of structural shape optimization with an emphasis on techniques dealing with shape optimization of the boundaries of two- and three-dimensional bodies.

In the sequel the sizing optimization problem for skeletal structures is formulated and solved by iterative NLP methods.

15.2. Skeletal Structures

15.2.1. Problem Formulation

Optimal design of skeletal structures can be formulated in the following way. Find the cross-sectional areas of elements (e.g., frame bars) A_i, $i = 1, 2, \ldots, n$, that minimize the weight of the structure

$$F(A_i) = \sum_{i=1}^{n} A_i l_i \gamma_i \tag{15.1}$$

and satisfy the following side and behavioral constraints:

$$
\begin{aligned}
g_1 &= 1 - F_{\text{new}}/F_{\text{old}} \geqslant 0, \\
g_{i1} &= A_i - A_i^{\text{l}} \geqslant 0, \qquad i = 1, \ldots, n, \\
g_{i2} &= A_i^{\text{u}} - A_i \geqslant 0, \\
g_{j3} &= \delta_j^{\text{u}} - \delta_j \geqslant 0, \qquad j = 1, \ldots, m_1, \\
g_{i4} &= 1 - (\sigma_{\text{act}}/\sigma_{\text{adm}})_i \geqslant 0.
\end{aligned}
\tag{15.2}
$$

In these expressions, l_i and γ_i are the length and bulk density of element i, respectively; A_i^{l} and A_i^{u} are the lower and upper bounds for the cross-sectional area of element i, respectively. The number of displacement checkpoints is denoted by m_1, δ_j and δ_j^{u} are the actual and upper-bound displacements, respectively, σ_{act} is the actual active stress, and σ_{adm} is the corresponding allowable stress. For each element, we identify up to 18 stress components, which we calculate for each loading condition. Displacements and stresses are analyzed by the matrix displacement method and second-order theory for each loading condition.

The total number of constraints is $m = 3n + m_1 + 1$. The moment of inertia J, the section modulus W, and the effective cross-sectional area A_s with respect to shear stresses are related to A as follows:

$$J = a_1 A^{b_1}, \quad W = a_2 A^{b_2}, \quad A = a_3 A^{b_3}, \tag{15.3}$$

where a_i and b_i are specific coefficients and exponents characterizing a group of elements [see Fleury (1984), Lawo (1981), or Chiu (1982)].

15.2.2. Optimization Algorithms

Using the penalty function method (see Section 10.6), we first convert the constrained optimization problem into an unconstrained one. To solve an unconstrained optimization problem, we apply four optimization algorithms: Rechenberg's evolution strategy (EVOL) [see Rechenberg (1973) or Schwefel (1977)], Rosenbrock's algorithm of rotating coordinates with internal (barrier) and external penalty functions (ROSE1, ROSE2) [see Machura and Maluwa (1973) or Schwefel (1977)], and the complex strategy by Box (BOX) [see Richardson and Kuester (1973) or Schwefel (1977)]. The four algorithms are direct search algorithms without gradients.

15.2.3. Numerical Examples

Based on Kos (1983) and Jendo, Lawo, and Thierauf (1984), five examples of framed structure optimization are presented. The topologies of the frame structures and the loading conditions are described in Figs. 15.1a–15.5a, respectively. The optimization results are illustrated by the bending moment distributions for initial (Figs. 15.1b–15.5b) and optimal (Fig. 15.1c–15.5c) structures, respectively.

15.2.4. Comparison of Optimization Algorithms

Tables 15.1–15.5 show some characteristic results of the optimization process. Computations were carried out on a Siemens 2000 computer. Independent of starting point and problem size, EVOL proved to be the

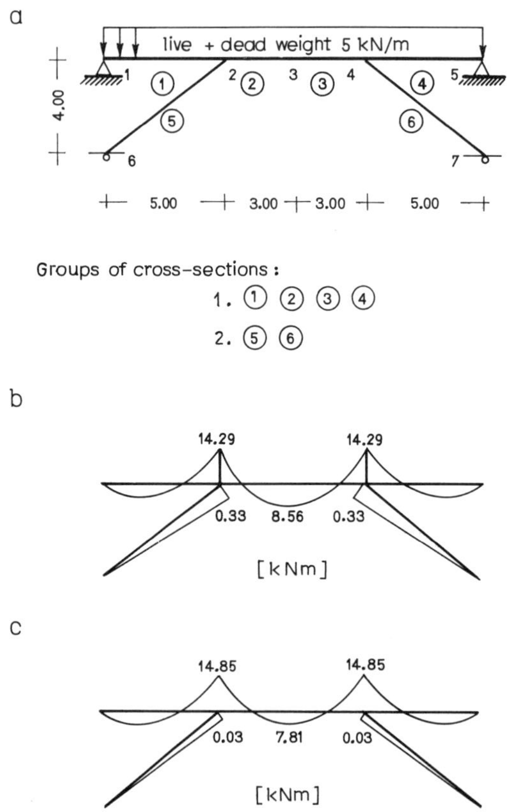

Figure 15.1. (a) Sketch of Frame (Example 1). (b) Distribution of Bending Moments in Initial Structure. (c) Distribution of Bending Moments in Optimal Structure.

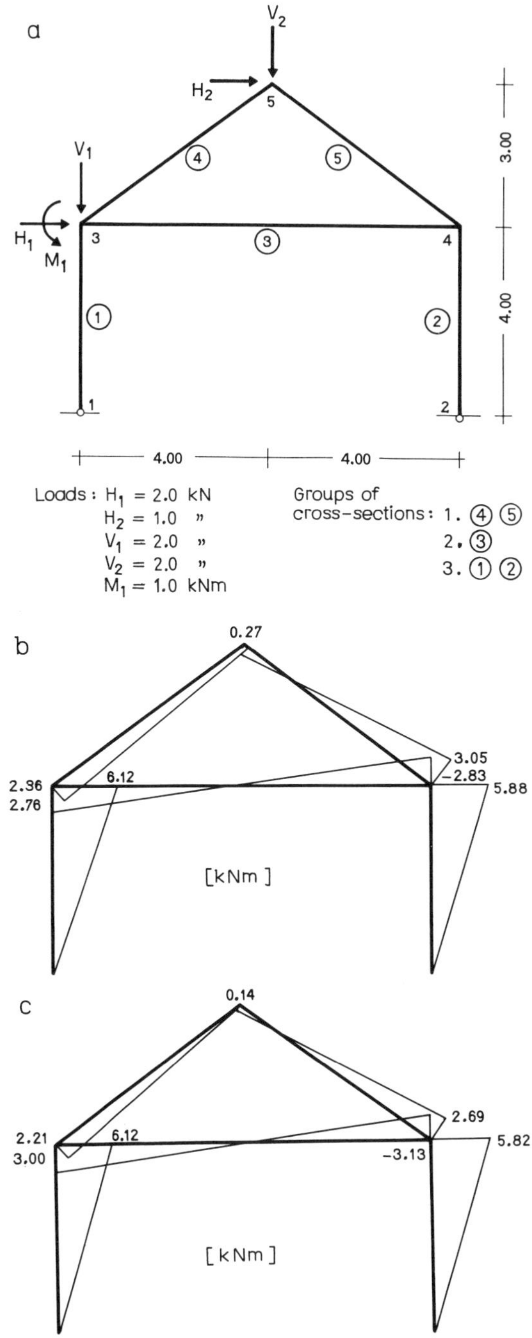

Figure 15.2. (a) Sketch of Frame (Example 2). (b) Distribution of Bending Moments in Initial Structure. (c) Distribution of Bending Moments in Optimal Structure.

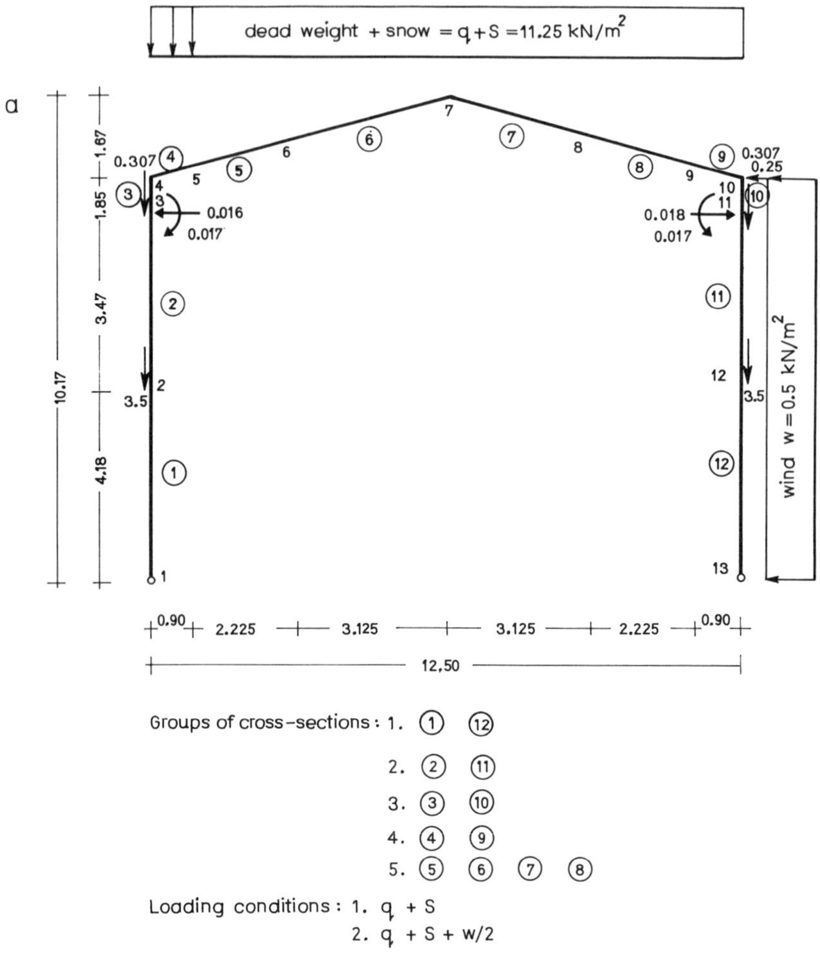

Figure 15.3. (a) Sketch of Frame (Example 3). (b) Distribution of Bending Moments in Initial Structure. (c) Distribution of Bending Moments in Optimal Structure.

most reliable method. BOX was satisfactory only for problems with up to five design variables. ROSE1 and ROSE2 showed good improvement of the objectives in less computing time than the others did, especially when the starting points were not close to the local minimum.

The use of an external penalty function (ROSE2) has the advantage that it can be started with an infeasible point and that, in many cases, not all constraints must be calculated at every point. This is especially important for the very "costly" stress and displacement constraints calculations. But,

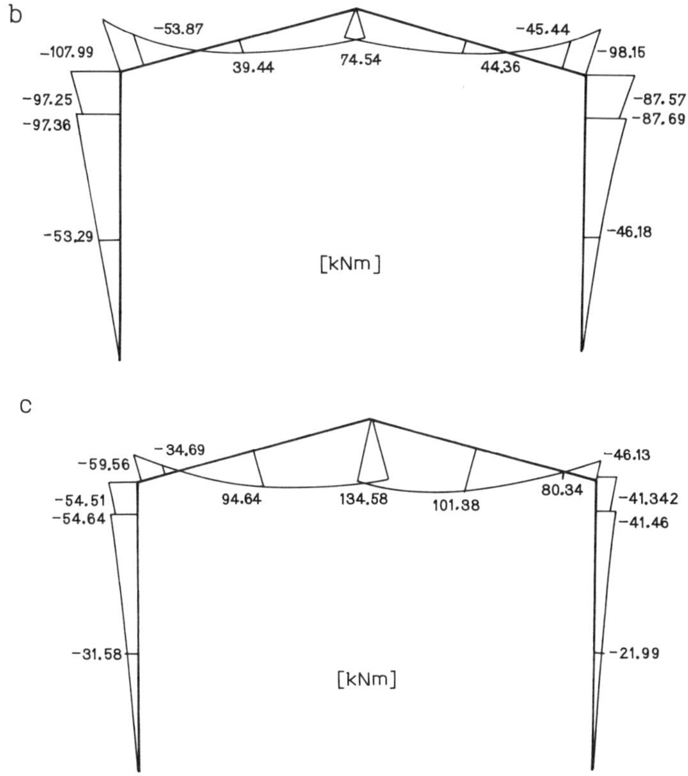

Figure 15.3. Continued.

in general, we can conclude that none of the four nongradient methods is superior to the others for the general problem of optimal structural design.

15.3. Optimal Design of Tanker Transverse Web Frames

15.3.1. Review of Papers

The first steps in optimal design of ship structures with transverse web frames were made by Evans (1958) in the late 1950s and by Evans and Khoushy (1963) at the beginning of the 1960s. They analyzed the influence of parameters such as ship width, draft, and spacing of web frames on the unit weight of the tanker. Emtell and Ovrebo (1965) also considered the optimal design of web frames in an oil tanker, using systematic variation of design variables. Murphy, Sabat, and Taylor (1965) and Mandel and

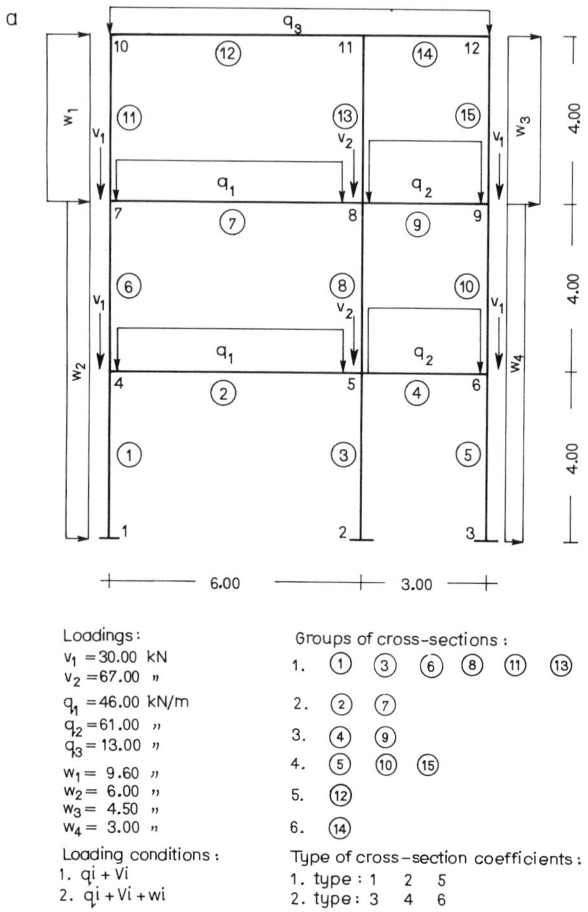

Figure 15.4. (a) Sketch of Frame (Example 4). (b) Distribution of Bending Moments in Initial Structure. (c) Distribution of Bending Moments in Optimal Structure.

Leopold (1966) have applied computer techniques and optimization methods in ship design.

Kavlie *et al.* (1966) and Kavlie, Kowalik, and Moe (1966) have applied a penalty function method and satisfied Norwegian code requirements in ship optimization. Moe and Lund (1968), Moe (1969a, b), Lund (1971), and Pettersen (1973) have used finite element methods and the sequential unconstrained minimization technique (SUMT) in optimal ship design. Kavlie and Moe (1969) and Miura, Kavlie, and Moe (1975) have also discussed NLP techniques in optimal grillage design, using the force method in static analysis. A similar problem has been discussed by Moses and

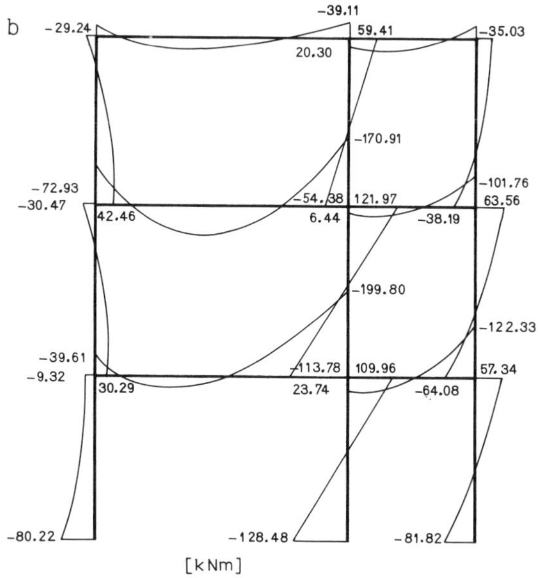

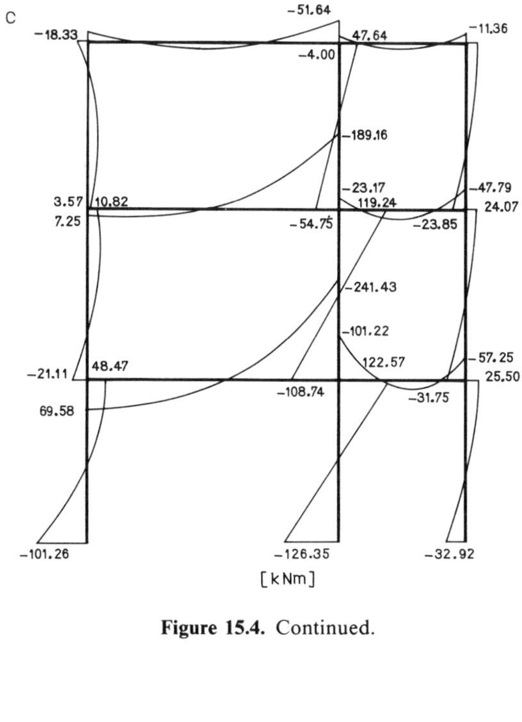

Figure 15.4. Continued.

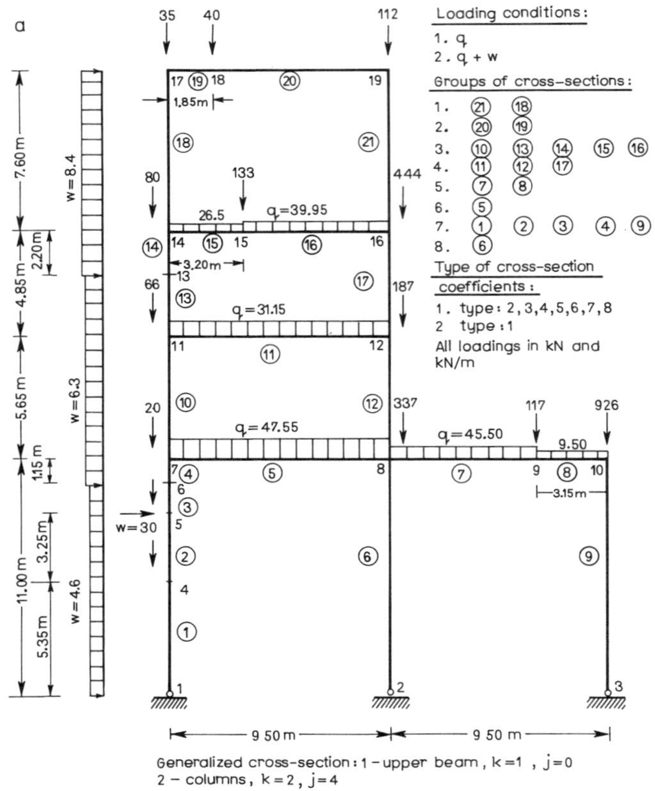

Figure 15.5. (a) Sketch of Frame (Example 5). (b) Distribution of Bending Moments in Initial Structure. (c) Distribution of Bending Moments in Optimal Structure.

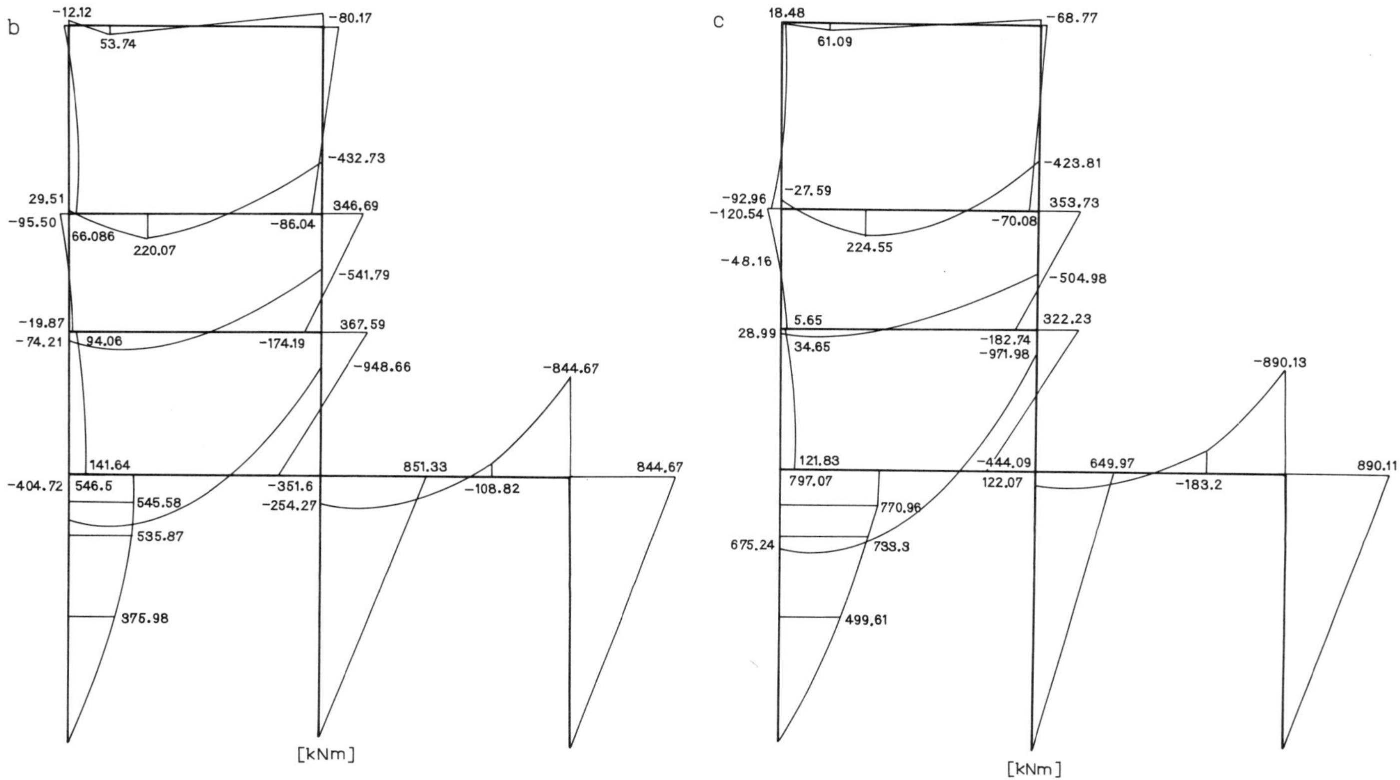

Figure 15.5. Continued.

Table 15.1. Computer Time (CPU in Seconds)

	Algorithm			
Example	EVOL	ROSEN1	ROSEN2	BOX
1	16.15	17.90	18.20	22.32
2	8.70	9.90	11.40	2.23
3	310.80	61.90	37.10	446.86
4	308.30	35.90	34.50	175.36
5	330.90	121.60	122.10	1161.60

Table 15.2. Final Value of Weight (in Tons)

	Algorithm			
Example	EVOL	ROSEN1	ROSEN2	BOX
1	0.34109	0.31394	0.31394	0.35153
2	0.55517	0.55343	0.55324	0.55515
3	1.14670	1.59910	1.20910	1.27770
4	4.43890	4.24130	4.24130	7.47340
5	25.09300	20.90400	20.90400	26.49700

Onoda (1969), using the displacement method in static analysis. Mandel and Chryssostomidis (1972) have applied a system theory and decomposition methods for ships and other complex systems.

Akita and Kitamura (1970) and Kitamura (1971) have also studied optimization of ship structures, using NLP techniques. They considered the optimal design of box and I section girders as well as elastoplastic grillages. Fox and Miura (1971) have proposed an approximate analysis technique to decrease the number of reanalysis steps. Finifter (1972) and Finifter and

Table 15.3. Number of Weight Function Evaluations

	Algorithm			
Example	EVOL	ROSEN1	ROSEN2	BOX
1	180	82	82	147
2	272	84	95	17
3	205	241	27	177
4	257	370	30	83
5	334	20	20	102

Table 15.4. Number of Constraint Evaluations

	Algorithm			
Example	EVOL	ROSEN1	ROSEN2	BOX
1	161	103	103	108
2	241	153	158	8
3	205	300	256	127
4	223	436	316	66
5	321	320	320	85

Mansour (1974) have discussed the optimization of web frames of a tanker with isolated ballast systems. Nowacki, Brusis, and Swift (1970), Folkers (1973), and Evans (1974) have considered a tanker preliminary design as a constrained optimization problem. Jendo, Nowacki, and Yi-Suang Tein (1974) and Jendo (1979) have considered the optimal design of tanker transverse web frames, using the displacement method and nonlinear programming. Moussouros (1976) has applied the finite element method in optimal design of transverse web frames. Woodhead (1972) and Smith and Woodhead (1973) have used sequential linear programming in optimal design of tanker transverse web frames.

We now look at the formulation and solution of the optimal design of tanker transverse web frames.

15.3.2. Problem Formulation

15.3.2.1. Design Variables. We assume that the topology (layout) of the web frame is given (see Fig. 15.6). Each member of the web frame consists of a web plate and two flanges, called the *free flange* and the *plate flange*. The free flange faces the interior of the hole or web structure and can be freely dimensioned within given limits. The plate flange is the effective part of the shell, bulkhead, or deck plating to which the web frame is

Table 15.5. Number of Effective Iterations

	Algorithm			
Example	EVOL	ROSEN1	ROSEN2	BOX
1	17	27	27	31
2	29	31	35	6
3	27	6	18	11
4	32	6	21	13
5	11	13	13	9

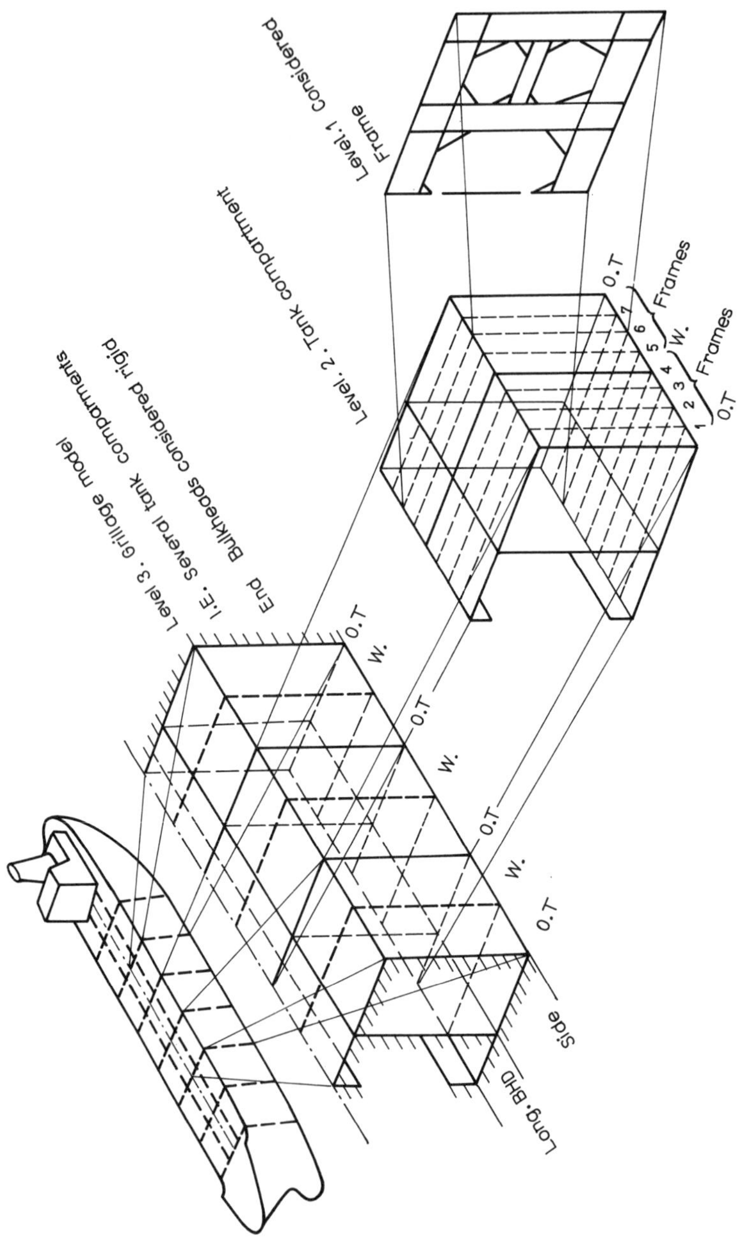

Figure 15.6. Perspective Sketch of Structural Layout of Tanker Ship.

attached. We assume that the plate flange is a fixed user input, since it is usually determined from considerations other than frame action. This leaves three design variables for the cross section of each web frame member: web height, web thickness, and flange area (Fig. 15.7).

We further assume that, if any brackets or end fairings are present in a member, the scantlings of these end reinforcements depend on the scantlings of the cross section of the mid-portion of the member in a manner specified by the user (Fig. 15.7). That is, web height and web thickness of the reinforced end portions, where present, are defined as multiples f_1 and f_2, respectively, of the corresponding mid-portion scantlings. The factors f_1 and f_2 are user inputs, presently applied uniformly to all reinforced member end sections in the structure. This entire scheme allows approximate handling of brackets, etc., without introducing more than three independent variables per member.

15.3.2.2. Objective Function. For ships, one can argue in favor of either weight or cost, depending on the shipyard's or owner's immediate goals. We have chosen a weight minimization for greater simplicity. The weight

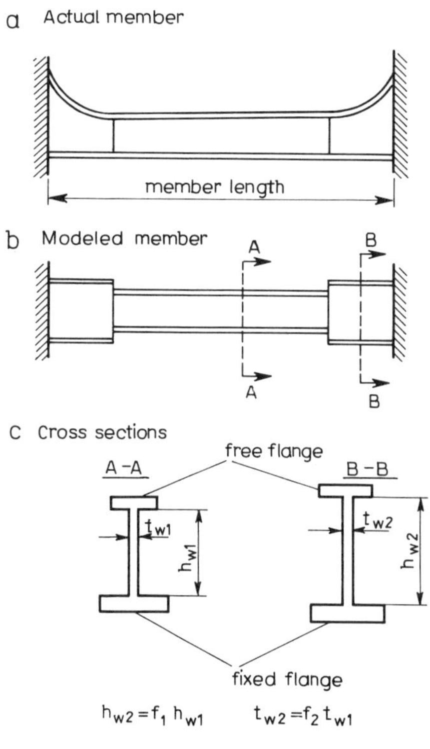

Figure 15.7. Modeling of Member with Rigid Ends: (a) Actual Member. (b) Modeled Member. (c) Cross Sections.

of the web frame, by our definition, does not include the weight of the plate flange (shell, etc.), since it is invariant during the optimization. This must be kept in mind when we study the output. In this way, the weight of the web frame can be written as

$$V = \sum_{i=1}^{n} A_i l_i = \sum_{i=1}^{n} (A_f + h_w t_w + h_p w_p)_i l_i, \tag{15.4}$$

where A_f is the free flange area (design variable), h_w is the depth of the web in centimeters or inches (design variable), t_w is the thickness of the web in centimeters or inches (design variable), w_p is the width of the plate flange in centimeters or inches (fixed user input), and h_p is the thickness of the plate flange in centimeters or inches (fixed user input).

15.3.2.3. Optimization Constraints. The following side constraints are imposed on each member of the frame:

$$\begin{aligned} t_{\min} &\leq t_w \leq t_{\max}, \\ h_{\min} &\leq h_w \leq h_{\max}, \\ A_{\min} &\leq A_f \leq A_{\max}. \end{aligned} \tag{15.5}$$

All lower and upper range limits for these quantities are user inputs.

The only behavioral constraints are on the maximum combined stress in each frame member; this must be bounded by the allowable combined stress. Stresses are calculated in each member in all positions where maximum internal loads occur. Maximum bending moments may occur at the supports or in the field. Maximum shear forces are expected at the supports because the loads are distributed along the elements. At those positions where concentrated forces or moments occur, a new breakpoint (joint) will be introduced, leaving only distributed loads between joints in the model. The influence of axial forces upon the stresses is included. The maximum bending stresses are calculated where maximum moments may occur (at the supports and in the field). Shear stresses are calculated at the supports. These critical points of the member are shown in Fig. 15.8 and

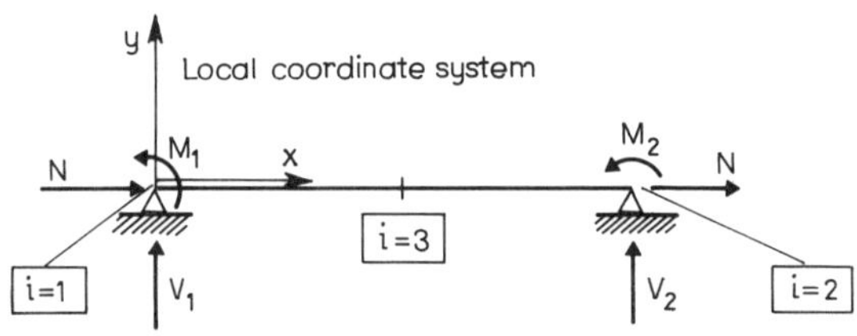

Figure 15.8. Typical Frame Member.

are denoted by $i = 1, 2, 3$. At the ends of members we also note the positive directions of the member reactions in the local coordinate system. We determine the properties of a typical web frame cross section by assuming an unbalanced section (see Fig. 15.9). In a typical member cross section the critical stresses can be expected at points $j = 1, 2, 3$, shown in Fig. 15.9. At each of these nine points, we compute σ_{ij} and τ_{ij}, $i, j = 1, 2, 3$, where σ_{ij} is the combined direct axial and bending stress at section i and level j in that section, and τ_{ij} is the shear stress there. We then calculate the corresponding stress intensity (psi or MPa)

$$f_{ij} = (\sigma_{ij}^2 + 3\tau_{ij}^2)^{1/2}. \tag{15.6}$$

We then compare the highest value among these nine stress intensities compared with the allowable stress

$$f_{\text{adm}} = \alpha f_{\text{ym}}/Q, \tag{15.7}$$

where f_{ym} is the specified lower yield point of mild steel, Q is the conversion factor for higher-strength material (for mild steel unity may be taken), and α is a coefficient taken as 0.85 for transverse, vertical, and horizontal supporting members. The normalized stress constraints can be written in the form

$$g_{ij} = 1 - f_{ij}/f_{\text{adm}} \geqslant 0 \qquad \text{from } f_{ij} \leqslant f_{\text{adm}}. \tag{15.8}$$

Let us establish a list of stresses for the nine points corresponding to the nine (i, j) choices as before (see Figs. 15.8 and 15.9). First, we consider the left end of the member ($i = 1$) and write formulas for combined axial and bending stress and for shear stress. We denote compression by a minus

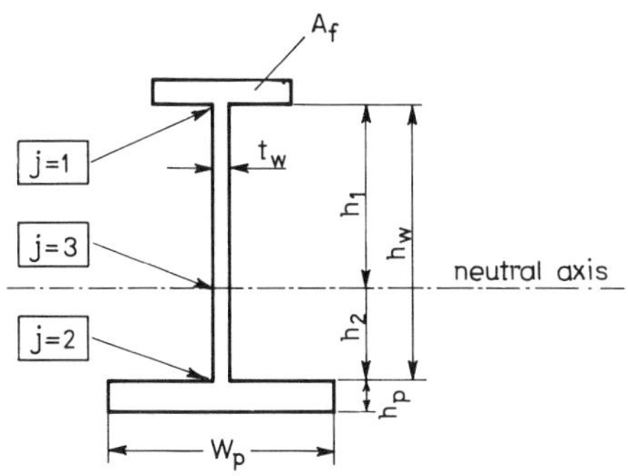

Figure 15.9. Unbalanced Section.

sign (−) and tension by a plus sign (+). Using this convention and remembering that member reactions are given in local coordinates as shown in Fig. 15.8, we can write the stresses σ_{ij} and τ_{ij}:

$$\begin{aligned} \sigma_{11} &= -N/A + M_1/W_1, \\ \sigma_{12} &= -N/A - M_1/W_2, \\ \sigma_{13} &= -N/A, \end{aligned} \tag{15.9}$$

where W_1 and W_2 are the section moduli of the upper and lower parts of the cross-sectional areas (see Fig. 15.9), and

$$\begin{aligned} \tau_{11} &= T_1 S_1/t_w J_1, && \text{where } S_1 = A_f h_1, \\ \tau_{12} &= T_1 S_2/t_w J_1, && \text{where } S_2 = A_p h_2, \\ \tau_{13} &= T_1 S_3/t_w J_1, && \text{where } S_3 = A_f h_1 + t_w h_1^2/2, \end{aligned} \tag{15.10}$$

where h_1 and h_2 are distances from the neutral axis to the upper and lower plates, respectively (see Fig. 15.9). Now consider the right end of the member ($i = 2$) and write stress formulas for this case. Using the same sign convention, we have

$$\begin{aligned} \sigma_{21} &= N/A - M_2/W_1, \\ \sigma_{22} &= N/A + M_2/W_2, \\ \sigma_{23} &= N/A. \end{aligned} \tag{15.11}$$

The shear stresses can be calculated as before.

Now we consider the section $i = 3$ at which a maximum bending moment may occur. To calculate bending stresses at this section, we can use the same formulas as at section $i = 1$. We take the axial forces (N) from section $i = 1$ and the bending moment (M) as $M_{\max}$, determined by a separate procedure.

At each point we can now calculate the combined stress intensities (f_{ij}), compare them, and substitute the largest one into the constraint inequalities

$$g_{ij} = 1 - f_{ij}/f_{\text{adm}} \geqslant 0. \tag{15.12}$$

We use only one active constraint per member, because only the dominant stress at any cross section is stored in computer memory, although all are calculated. Which stress constraint controls a cross section may vary as the design evolves.

15.3.3. Idealization of the Structure

A perspective view of the structural layout of a tanker ship with a few levels of decomposition is shown in Fig. 15.6. A typical midship section of a large tanker is shown in Fig. 15.10a. Figure 15.10b shows a structural model that is assumed to describe the essential characteristics of the structure in Fig. 15.10a. The structural model comprises slender members, that is, members that can be represented by their centroidal axes. Joints or nodes are assumed to be located at the intersections of the centroidal lines of the individual members.

A symmetry condition exists at the centerline of the deck and bottom girders. These nodes are fixed with respect to rotation and horizontal movement, but are subject to elastic restraint in the vertical direction. The nodes of the side shell and longitudinal bulkhead at the bottom are free

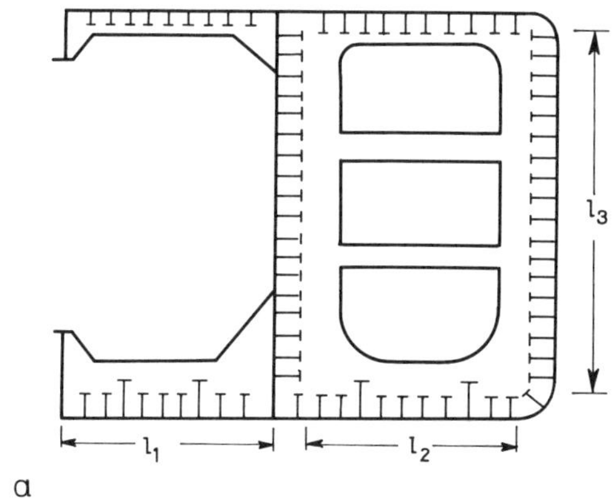

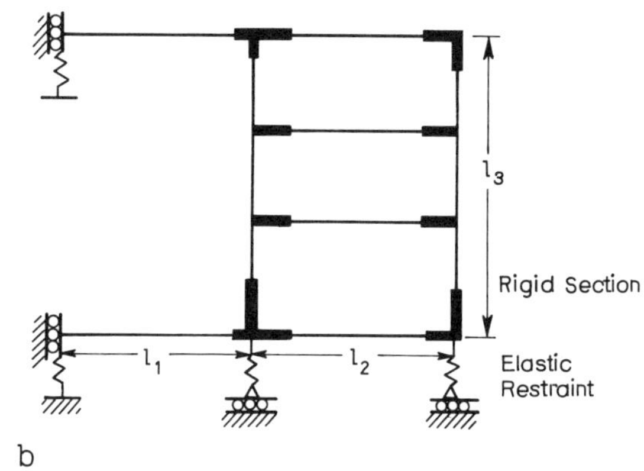

Figure 15.10. (a) Typical Midship Section. (b) Its Structural Model.

with respect to rotation and horizontal movement, but are subject to elastic restraint in the vertical direction. We introduce elastic restraints or springs at all of these nodes to account for the supporting effects of the side shell, longitudinal bulkhead, and longitudinal girders. We neglect the elastic support of girders of depth less than the transverse webs.

Rigid sections are introduced in the structural model, as shown in Figs. 15.10b and 15.11. Portions of the frame where no beam bending can occur (e.g., at locations of intersecting webs and heavy brackets) are considered to be rigid sections. Introducing rigid sections reduces the elastic length of a beam member. A sketch similar to Fig. 15.10b, showing the nodes, rigid sections, elastic restraints, and support points of the structure to be analyzed, should be the first step in the web frame analysis.

15.3.4. Properties of Elastic Structural Members

In the context of design optimization, the user is mainly concerned with the scantling definition of his initial design and the optimal design

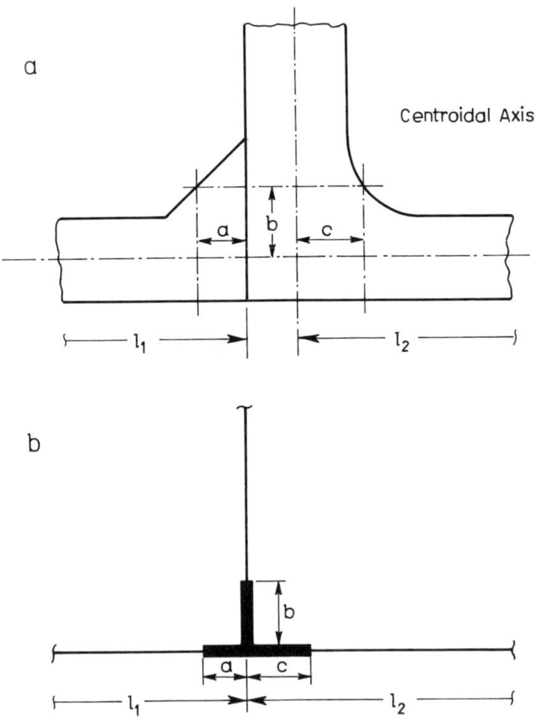

Figure 15.11. Rigid Section Lengths: (a) Actual System. (b) Substitution System.

obtained as a result. The calculation of effective cross-sectional area and moment of inertia for each member is performed automatically from the given scantlings and is updated every time the design is changed. The user does not get involved in this process, but he or she should be aware of how the elastic properties of the structural members are calculated by the program.

15.3.4.1. Calculation of Span Properties (Mid-Portion). The properties of the member spans are determined by assuming an unbalanced section (see Fig. 15.9):

$$A = t_w h_w + A_f + w_p h_p,$$

$$J = \frac{t_w h_w^3}{12} + t_w h_w \left(\frac{h_w}{2} + h_p - h_2\right)^2 + \frac{w_p h_p^3}{12} + w_p h_p \left(h_2 + \frac{h_p}{2}\right)^2 + A_f h_1^2, \tag{15.13}$$

where h_1, h_2 are the distances from the neutral axis to the upper and lower flanges, respectively (see Fig. 15.9).

15.3.4.2. Properties of Rigid Sections (End Portions). It was mentioned earlier that the scantlings of the rigid end portions of the members are treated as being dependent on the scantlings of the mid-portion. The user who must choose a fixed ratio of these scantlings initially can be guided by some general modeling rules for rigid sections.

The lengths of the rigid sections must not be less than the unsupported span depth of the intersecting members. Where large brackets support the member, the lengths of the rigid sections may be measured to the centers of the brackets, as shown in Fig. 15.11.

The design program user must specify two factors to define the degree of end reinforcement: one for web height and one for web thickness. The program derives the properties of the rigid end cross sections from the neighboring free variable segments as follows (Fig. 15.7): $h_{wr} = f_1 h_{wf}$, $t_{wr} = f_2 t_{wf}$, where h_{wr}, t_{wr} are the web height and web thickness of the rigid section, respectively; h_{wf}, t_{wf} are the web height and web thickness of the free section, respectively; f_1 and f_2 are the bracket height and thickness factors, which will be user inputs (in our examples both are assumed to be 1.25).

15.3.5. Properties of Elastic Supports

Since the entire structural model of the web frame is assumed to be supported on a system of springs, we need only evaluate the stiffnesses of the springs relative to one another. We can do this by evaluating the stiffnesses of the side shell, longitudinal bulkhead, and center girders under the same arbitrary loading system. The elastic springs shown in Fig. 15.10b are represented by equivalent structural members having axial stiffnesses equal to the elastic restraints. The evaluation of the relative stiffnesses of the equivalent structural members is described next.

15.3.5.1. Side Shell and Longitudinal Bulkhead. The axial stiffnesses of the equivalent structural members corresponding to the relative in-plane stiffnesses of the side shell and longitudinal bulkhead are determined from shear deformation considerations. The idealization of the structural system under a unit load is shown in Fig. 15.12a. The deflection at the mid-span due to shear is

$$\delta_{\text{shear}} = Ps/4GA. \tag{15.14}$$

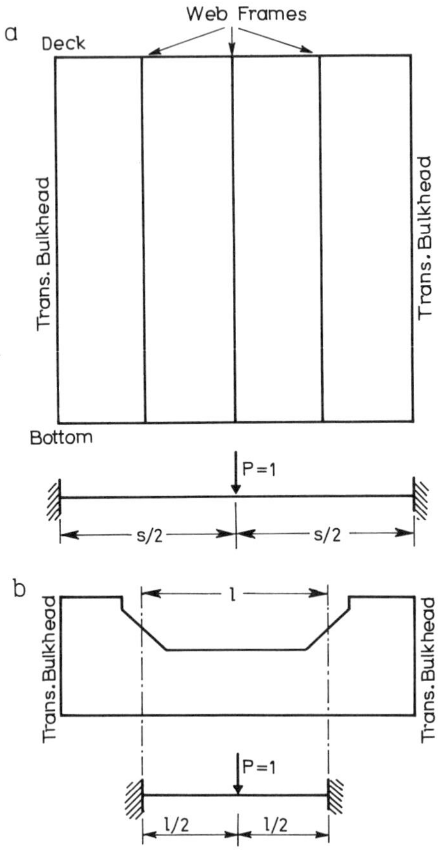

Figure 15.12. Loading System for Elastic Supports Evaluation: (a) Longitudinal Bulkhead or Side Shell. (b) Center Girder.

The deflection of an equivalent structural member under an axially applied load is

$$\delta_{\text{axial}} = PL/AE. \tag{15.15}$$

Setting $\delta_{\text{shear}} = \delta_{\text{axial}}$, we have

$$Ps/4GA = PL/AE. \tag{15.16}$$

Assuming $E/G = 2.6$, we get

$$A = 1.54 A_w L/s. \tag{15.17}$$

The length L of the equivalent structural member is arbitrarily assigned. We choose $L = s/10$ to simplify Eq. (15.17). The resultant formulas for calculating the properties of the equivalent structural members are $A = 0.154 A_w$, $L = 0.1s$, $J = 1$, where A_w is the effective shear web area (effective depth × average thickness of side shell or longitudinal bulkhead) in square centimeters or square inches, s is the transverse bulkhead spacing in centimeters or inches, and L is the length of the equivalent structural member in centimeters or inches. The elastic springs do not transmit moments, so the moment of inertia J of the equivalent structural members should be zero. But due to certain limitations in the computer program, J is arbitrarily assigned a value of 1, which, when compared to the inertias of other members, is negligible.

15.3.5.2. Deck and Bottom Center Girders. We consider both shear and bending deformations when evaluating the relative stiffnesses of the center girders. Since the deck and bottom center girders lie in the longitudinal centerline plane of symmetry, only one half of the girder properties are used to evaluate the stiffnesses. The structural idealization of a center girder under the same unit load system as before is shown in Fig. 15.12b. The deflection at the midspan due to bending is

$$\delta_{\text{bending}} = 2Pl^3/192EJ_g. \tag{15.18}$$

Hence the total deflection at the mid-span due to shear and bending is

$$\delta_{\text{total}} = 2(Pl/4GA_w + Pl^3/192EJ_g). \tag{15.19}$$

Equating the axial deflection of the equivalent structural member to the total deflection of the girder gives

$$2(Pl/4GA_w + Pl^3/192EJ_g) = PL/AE. \tag{15.20}$$

Thus

$$A = \frac{1}{2(El/4GA_wL + l^3/192J_gL)}. \tag{15.21}$$

Since L is arbitrary, we select $L = l/10$ to simplify Eq. (15.21). The resulting values for the properties of the equivalent structural members are

$$A = \frac{1}{(13/A_w + l^2/9.6J_g)}, \qquad L = 0.1l, \quad J = 1, \tag{15.22}$$

where A_w is the effective shear web area of the girder in square centimeters or square inches, l is the elastic length of the center girder in centimeters or inches, usually measured between the centers of the brackets supporting the girder, J_g is the moment of inertia of the center girder taken as a balanced section in (centimeters)^4 or (inches)^4, and L is the length of the equivalent structural member in centimeters or inches.

15.3.6. Determination of Loadings

We will investigate three conditions of loading at light draft. In condition 1, the center tank is full and pressed to 8 ft above the deck at the side; the wing tanks are empty. Light draft is taken as one third of the full-load operating draft.

In condition 2, the center tank is empty; the wing tanks are full and pressed to 8 ft above the deck at the side. Light draft is again one third of the full-load operating draft.

In condition 3, both the center tank and wing tanks are full and pressed to 8 ft above the deck at the side. Again, light draft is one third of the full-load operating draft.

The three loading conditions are shown in Fig. 15.13. Referring to Figs. 15.10 and 15.11, note that the lengths l_2 and l_3 are not the actual breadth and depth, respectively, of the wing tanks, since these lengths are measured between centroidal axes. To account for the loss of loadings on these lengths, we may apply concentrated point loads at the end points of l_2 and l_3.

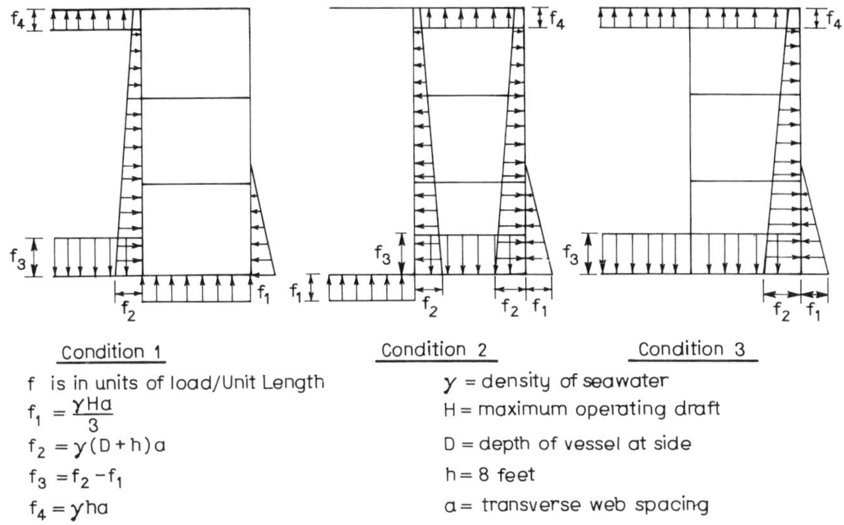

Figure 15.13. Static Loading Conditions.

Usually the distance from the centroidal axis to the flange of a member is quite small, and the application of point loads may be neglected.

15.3.7. Definition of Structural Topology for the Computer Program

For solving the optimization problem of tanker transverse web frames, a computer program has been written [see Jendo, Nowacki, and Yi-Suang Tein (1974) and Jendo (1979)] to analyze and optimize structures as plane frames in which the forces acting on the frame and its resultant displacements are in the plane of the structure. In frame analysis, rigid joints are assumed; i.e., the angle between members before deformation remains unchanged even after deformation takes place (see Fig. 15.14). The program can handle structures of up to 30 members with pinned or rigid supports, with prismatic or nonprismatic members, and subjected to concentrated loads.

The program requires that support joints be completely free or fixed against rotation and deflection. Since these extreme conditions are seldom realized in practice, a judicious selection of the type of supports is imperative. It is also possible to create a more realistic model of the structure by substituting equivalent structural members for restraints, resulting in elastic supports that are neither pinned nor rigid.

15.3.7.1. Global Coordinate System. To describe the structure to the computer, we use a global coordinate system, chosen such that the positive

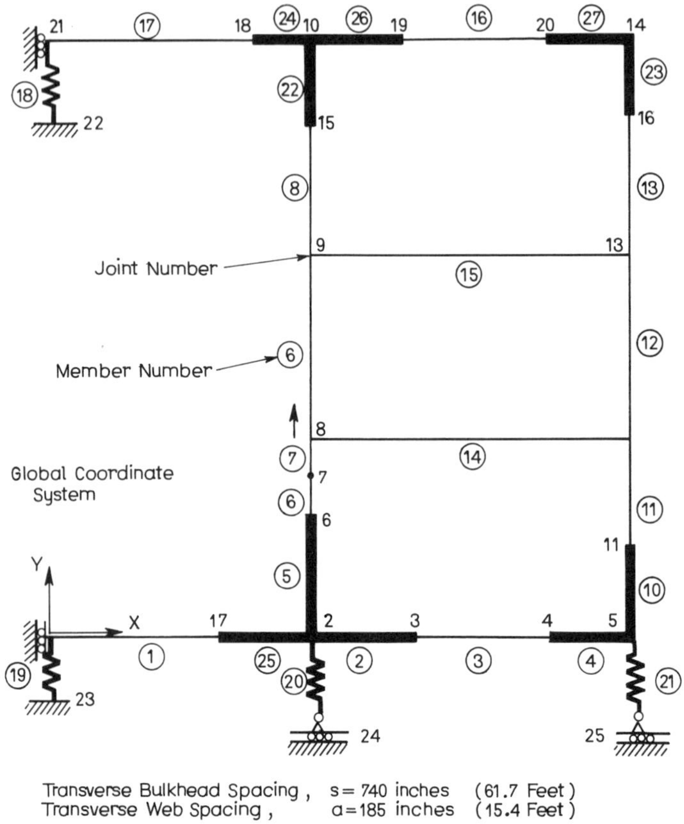

Figure 15.14. Numbering of Members and Nodes.

X-coordinate is directed to the right and the positive Y-coordinate is directed up (Fig. 15.14). All joint data, such as joint coordinates and joint loads, are specified in relation to the global system.

15.3.7.2. Local Coordinate System. The local coordinate system is associated with each member, and all member data, such as member loads, are specified in terms of this system. The local X-axis coincides with the axis of the member, and its positive direction is from the starting joint toward the end joint. The local positive Y-axis is directed at 90° from the positive X-direction, measured counterclockwise (Fig. 15.8).

15.3.7.3. Identification of Joints and Members. Joints and members are referred to by identification numbers assigned by the user. The numerical order of data for joints and members must obey the following restrictions in connection with the present version of the program:

1. The numbering should start with 1 and proceed consecutively, so the highest number does not exceed the specified total number of joints or members, respectively.
2. The consecutive numbering of the members must be arranged in such a sequence that in the direction of increasing member count the fixed plate flange will always lie on the right side of the neutral axis. This will avoid ambiguities with respect to the section modulus calculation.

In addition to the complete member count, which includes free members, rigid members, and elastic supports, we need a free member count. This count identifies members subject to optimization, and should parallel the sequence of the complete member count.

Joints must be placed at each support and connecting point of members. Additional joints may be introduced in the following circumstances:

1. If part of the member is nonprismatic, that is, if sectional properties are not constant. This part of the member should be represented by several members varying stepwise in sectional properties, since the program assumes that each member is prismatic.
2. If output is desired at certain points, for example, at locations of concentrated loads.
3. To avoid overlong, slender members in combination with short, stiff members.
4. If structure and loads are symmetrical, a joint may be introduced at the center of symmetry to reduce the amount of input data.

15.3.7.4. Joint Releases. A fixed support is a support joint that is restrained rotationally as well as in the X- and Y-directions. The introduction of pins and rollers releases some of the foregoing restraints. The term *joint release* is used to describe the nature of the effective restraints at a support. Joint releases are only applicable to support joints.

A pinned support is restrained in the X- and Y-directions but is free to rotate. There is only one release.

A roller support is restrained in the X-direction or Y-direction but is free to rotate. Joints 24 and 25 in Fig. 15.14 are restrained in the Y-direction, released in the X-direction, and are free to rotate.

At a center of symmetry, the joint is taken as restrained against rotation, but is free to move either in the X- or Y-direction. There is only one release.

15.3.8. Computer Program

The flowchart of the optimization package is shown in Fig. 15.15. The web frame design program has two options for the optimization technique:

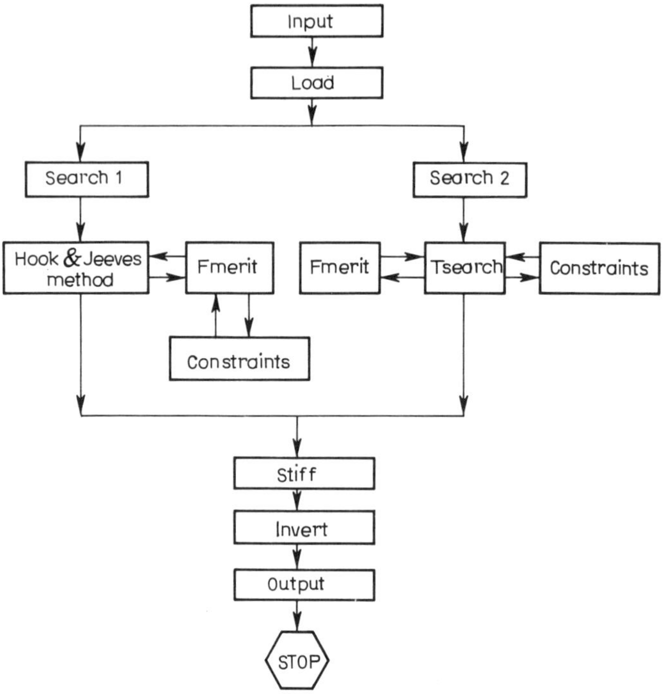

Figure 15.15. Flowchart of Optimization Package.

tangent search method (TSEARCH) [see Hilleary (1966)] and external penalty function with Hook and Jeeves direct search (EPT) [see Hook and Jeeves (1962), Zangwill (1967b), and Wangdahl (1972)]. Both methods should lead to the same optimum, if a unique optimum exists (unimodal case), although they might differ in efficiency. In our experience, TSEARCH is slightly more efficient.

An optimization run will always improve the design, but the result must be carefully interpreted as to whether it is truly an "optimum" in the mathematical sense. A run can fail to reach the optimum for basically two reasons. First, the search has not yet converged because it was terminated before the specified tolerance was reached. This may happen when the computing time or the number of function evaluations is limited. Second, the function has more than one optimum (multimodal case), and the search has arrived at a local, rather than global, optimum.

The user can test the optimum by using a "unimodality test" option provided with the program. When this option is invoked, the program simply performs a systematic variation in the design variables around the current optimum, $\hat{x}_i = x_i \pm \Delta x_i$, in all i design variable directions and prints out

Table 15.6. Characteristic Values of Three-Member Frame Optimization (Example 1)

Optimization method	Value of objective function (in^3)		Number of base points	Evaluation number		Value of constraint	Design variables[a]					
	Initial	Minimal		Objective function	Constraint		Depth		Thickness		Flange area	
TSEARCH	40.000	13.466	5	23	36	0.0301	6.00	5.48	0.50	0.23	0.50	0.04
EPT	40.000	13.187	29	171		0.0043	6.00	6.09	0.50	0.20	0.50	0.05

[a] The first value listed under each variable is the initial value. The second is the optimal value.

Table 15.7. Characteristic Values of 10-Member Frame Optimization (Example 2)

Optimization method	Value of objective function (cm^3)		Number of base points	Evaluation number		No. of design variables	No. of behavioral constraints
	Initial	Minimal		Objective function	Constraints		
TSEARCH	18710544	5296919	19	505	830	18	6

Table 15.8. Behavioral Constraints for 10-Member Frame (Example 2)

Bar number	Values of behavioral constraints	
	Initial	Final
1	0.60143	0.00015
2	0.73679	0.10057
3	0.68165	0.00166
4	0.74549	0.01486
5	0.86471	0.64667
6	0.80547	0.04325

the measure of merit and constraint function values in the vicinity of the optimum. The user can then decide whether sufficient convergence was achieved, and he will see which constraints are preventing further improvements.

To test for unimodality, one must start the search repeatedly from different starting points (initial designs), each time applying the foregoing

Table 15.9. Design Variables for 10-Member Frame (Example 2)

Design variable number	Values of design variables			
	Upper bound	Initial	Lower bound	Optimal
1	381.0	323.9	63.5	268.3
2	12.7	9.2	0.8	3.2
3	315.0	294.1	39.1	233.4
4	381.0	281.9	63.5	118.4
5	12.7	2.1	0.8	0.8
6	315.0	39.1	39.1	39.1
7	381.0	371.3	63.5	300.4
8	12.7	11.6	0.8	4.3
9	315.0	39.1	39.1	42.3
10	381.0	226.1	63.5	183.2
11	12.7	3.6	0.8	0.8
12	315.0	39.1	39.1	39.1
13	381.0	113.0	63.5	63.5
14	12.7	4.4	0.8	0.8
15	315.0	39.1	39.1	39.1
16	381.0	225.0	63.5	154.0
17	12.7	6.0	0.8	0.8
18	315.0	39.1	39.1	39.1

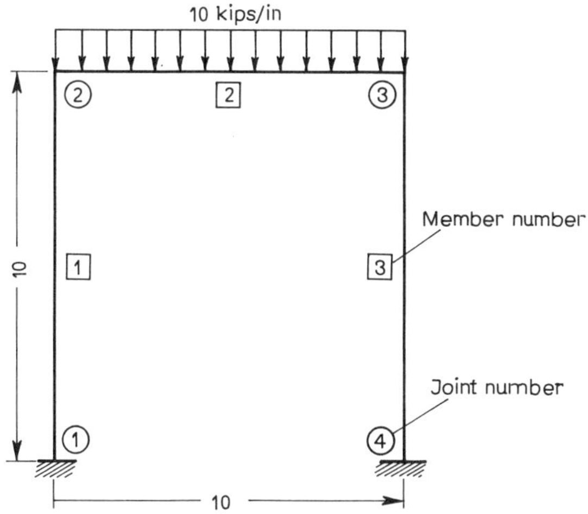

Figure 15.16. Sketch of Three-Member Frame.

test to the final result. Comparison will show whether identical solutions are obtained.

The success of the present design program depends on the computing time one is willing to invest. The most time-consuming part is the calculation of the stresses for each new design by a matrix inversion step. This step

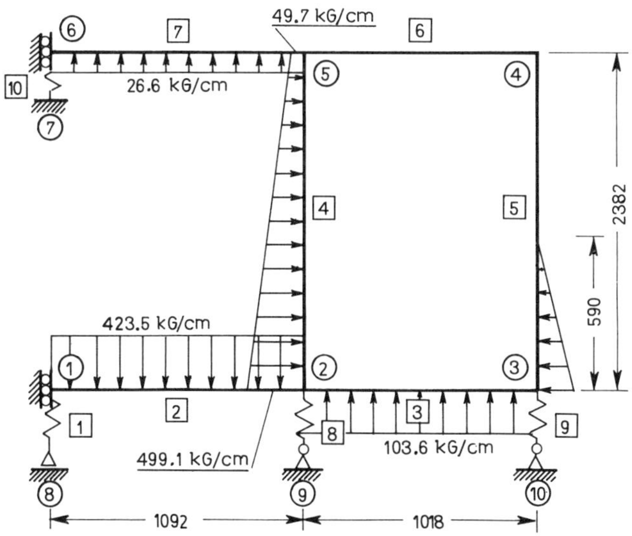

Figure 15.17. Sketch of 10-Member Web Frame.

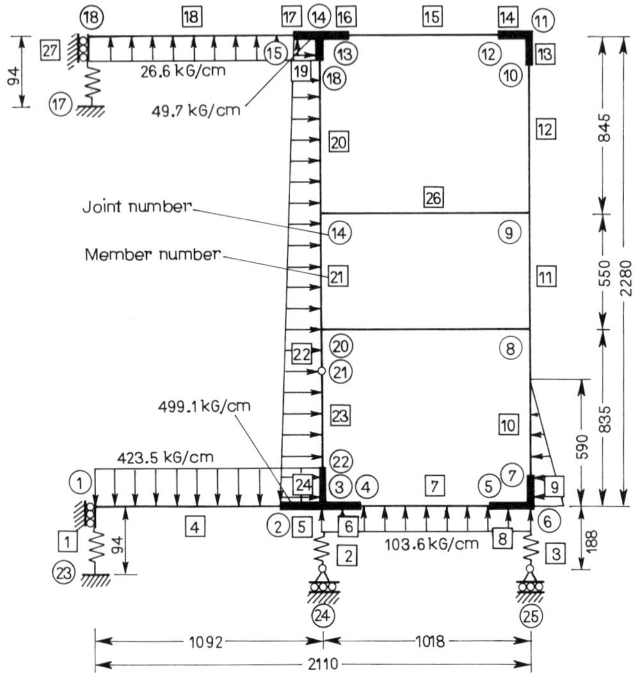

Figure 15.18. Sketch of 27-Member Web Frame.

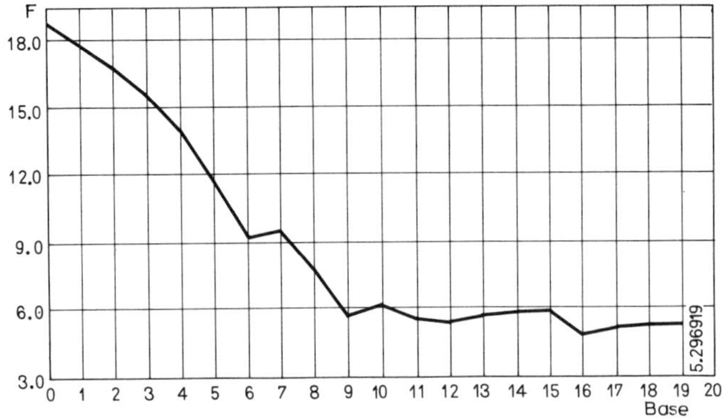

Figure 15.19. Iteration History for 10-Member Web Frame.

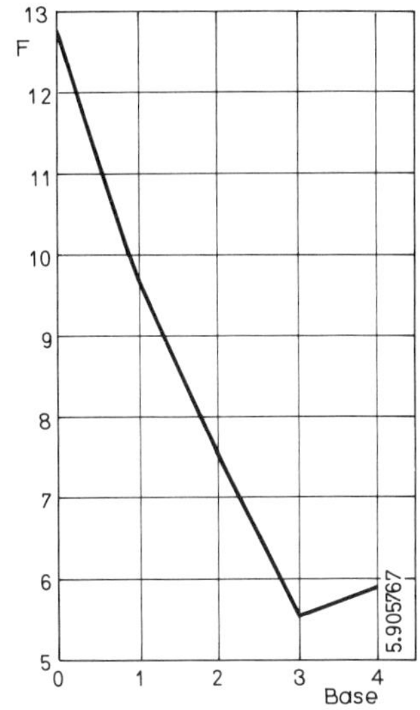

Figure 15.20. Iteration History for 27-Member Web Frame.

tends to become prohibitively expensive for the larger frame systems, especially in connection with the great number of associated design variables. This stage can be improved by using the Choleski decomposition method instead of the matrix inversion method.

To control this problem as much as possible in the short run, the user must use good judgment in choosing the initial design, step widths, and tolerances. The stop/restart option can help accelerate the learning process and prevent wasteful computing.

15.3.9. Numerical Examples

Let us consider three examples of optimization of frame structures: frames of 3, 10, and 27 members. A complete optimization was performed for the three-member frames (Table 15.6). and 10-member frames (Tables 15.7–15.9), and a few initial cycles of optimization were done for the 27-member system. The difference between the 10- and 27-member systems is simply that the first one does not include rigid members. The layouts of the three frames are shown in Figs. 15.16–15.18. The iteration history for 10 (full) and 27 (four iterations) is shown in Figs. 15.19 and 15.20, respectively. The complete listing of runs for these examples can be found in Jendo (1979).

16

Geometric Programming

16.1. Introduction

Essentially, geometric programming (GP) consists of a sequence of operations upon a set of terms that are expressed as generalized positive polynomials or, briefly, *posynomials*. An example of a posynomial containing three terms and three variables is

$$F(\mathbf{x}) = ax_1x_2x_3^{-2} + bx_1^{-1}x_2^{0.3}x_3 + cx_3. \tag{16.1}$$

A posynomial may thus be defined as a function consisting of a sum of terms, each of which consists of a positive coefficient multiplied by products of variables with each variable raised to an arbitrary power. The requirement that the coefficients be positive ensures the convexity of the problem and, hence, the identity of a global minimum with a local minimum, if one exists.

A feature that distinguishes geometric programming from other programming techniques is that it first yields the optimal value of the objective function and only then is the optimal design variable vector found, if needed. Another advantage of the method is that it often reduces a complicated optimization problem to a set of simultaneous linear algebraic equations.

The geometric programming method first came to light when Zener (1961) observed that a sum of component costs can sometimes be minimized almost by inspection when each cost depends on products of the design variables, with each variable raised to a known power. An account of this early work and its subsequent development is given by Duffin, Peterson, and Zener (1967) and Zener (1971). Some extensions to the original theory are discussed by Duffin and Peterson (1973) and by Wilde and Beightler (1967). A modern treatment is given by Peterson (1970, 1976), Beightler and Phillips (1976), Rao (1978), Ecker (1980), Kirsch (1981), and Morris (1982). A state-of-the-art review was conducted by Ecker (1980), and a review and comparative testing of available algorithms was compiled by Sin and Recklaitis (1981). Computational aspects of geometric programming are discussed by Bradley (1975, 1978), Ecker, Gochet, and Smeers (1978), and Dembo and Rijckaert (1978).

The first change in the prototype formulation was the lifting of the sign restriction on the posynomial coefficients [Passy and Wilde (1967)]. The generalized polynomial defined as the difference of two posynomials was called a *signomial.* Peterson (1976) defined generalized geometric programming in n-dimensional Euclidean space and showed that several mathematical programming problems (e.g., linear, quadratic, signomial) may be transformed into a geometric programming problem. A generalized geometric programming formulation in infinite-dimensional space was constructed by Scott and Jefferson (1977). Diaz, Papalambros, and Taylor (1984) combined the approach of Peterson (1976) and Scott and Jefferson (1977) with some theorems and concepts of convex functionals.

Burns (1987) presented a monomial method for solving mixed equality–inequality constrained generalized geometric programs. He shows that the method is equivalent to a Newton–Raphson procedure carried out in log space on the equality constraints at the same time that a cutting-plane procedure is performed on a convex region contained within the feasible region of the inequality constraints.

16.2. Unconstrained Geometric Programming

In the absence of constraints, a geometric programming problem can be stated as follows:

Find a vector $\mathbf{x}^T = (x_1, x_2, \ldots, x_n)$ that minimizes the function

$$F(\mathbf{x}) = \sum_{j=1}^{N} U_j(\mathbf{x}) = \sum_{j=1}^{N} \left(c_j \prod_{i=1}^{n} x_i^{a_{ij}} \right)$$

$$= \sum_{j=1}^{N} (c_j x_1^{a_{1j}} \cdots x_n^{a_{nj}}), \tag{16.2}$$

where $c_j > 0$, $x_i > 0$, and a_{ij} are real constants.

Several procedures can be used to solve this problem; we outline two: one based on the differential calculus, and one involving arithmetic-geometric inequalities.

16.2.1. Solution by Differential Calculus

The necessary conditions for a minimum of the function F are

$$\begin{aligned}\frac{\partial F}{\partial x_k} &= \sum_{j=1}^{N} \frac{\partial U_j}{\partial x_k} \\ &= \sum_{j=1}^{N} \left(c_j x_1^{a_{1j}} \cdots x_{k-1}^{a_{k-1,j}} a_{kj} x_k^{a_{kj}-1} x_{k+1}^{a_{k+1,j}} \cdots x_n^{a_{nj}}\right) \\ &= 0, \qquad k = 1, 2, \ldots, n. \end{aligned} \tag{16.3}$$

Multiplying the kth equation by x_k, we get

$$\begin{aligned} x_k \frac{\partial F}{\partial x_k} &= \sum_{j=1}^{N} a_{kj}\left(c_j x_1^{a_{1j}} x_2^{a_{2j}} \cdots x_{k-1}^{a_{k-1,j}} x_k^{a_{kj}} x_{k+1}^{a_{k+1,j}} \cdots x_n^{a_{nj}}\right. \\ &= \sum_{j=1}^{N} a_{kj} U_j(\mathbf{x}) = 0, \qquad k = 1, 2, \ldots, n. \end{aligned} \tag{16.4}$$

It can be shown that the necessary conditions (16.3) or (16.4) are also sufficient for a minimum of F. To this end, it is enough to prove that the Hessian matrix of F is positive definite. We leave this task to the reader.

Suppose that $\hat{\mathbf{x}}$ is the optimal vector, i.e., the one that yields the minimum of F, $\hat{F} = F(\hat{\mathbf{x}})$. By (16.4) we have

$$\sum_{j=1}^{N} a_{kj} U_j(\hat{\mathbf{x}}) = 0, \qquad k = 1, 2, \ldots, n. \tag{16.5}$$

Dividing this equation by $\hat{F}$ gives

$$\sum_{j=1}^{N} \hat{\Delta}_j a_{kj} = 0, \qquad k = 1, 2, \ldots, n, \tag{16.6}$$

where the quantities $\hat{\Delta}_j$ are defined as

$$\hat{\Delta}_j = U_j(\hat{\mathbf{x}})/\hat{F} = \hat{U}_j/\hat{F} \tag{16.7}$$

so that

$$\sum_{j=1}^{N} \hat{\Delta}_j = \frac{1}{\hat{F}} \sum_{j=1}^{N} \hat{U}_j = 1. \tag{16.8}$$

Equations (16.6) are called the *orthogonality conditions*, and Eq. (16.8) is called the *normality condition.*

The minimum value $\hat{F}$ can be determined by the following procedure. By the normality condition, we can write $\hat{F}$ as

$$\hat{F} = (\hat{F})^1 = \hat{F}^{\sum_{j=1}^{N} \hat{\Delta}_j} = (\hat{F})^{\hat{\Delta}_1}(\hat{F})^{\hat{\Delta}_2} \cdots (\hat{F})^{\hat{\Delta}_N}. \tag{16.9}$$

Since, from Eq. (16.7),

$$\hat{F} = \hat{U}_1/\hat{\Delta}_1 = \hat{U}_2/\hat{\Delta}_2 = \cdots = \hat{U}_N/\hat{\Delta}_N, \tag{16.10}$$

we can write Eq. (16.9) as

$$\hat{F} = (\hat{U}_1/\hat{\Delta}_1)^{\hat{\Delta}_1}(\hat{U}_2/\hat{\Delta}_2)^{\hat{\Delta}_2} \cdots (\hat{U}_N/\hat{\Delta}_N)^{\hat{\Delta}_N}. \tag{16.11}$$

Now if we substitute

$$\hat{U}_j = c_j \prod_{i=1}^{n} (\hat{x}_i)^{a_{ij}}, \qquad j = 1, 2, \ldots, N, \tag{16.12}$$

Eq. (16.11) becomes

$$\begin{aligned}
\hat{F} &= \left\{\left(\frac{c_1}{\hat{\Delta}_1}\right)^{\hat{\Delta}_1}\left\{\prod_{i=1}^{n} (\hat{x}_i)^{a_{i1}}\right\}^{\hat{\Delta}_1}\right\} \cdots \left\{\left(\frac{c_N}{\hat{\Delta}_N}\right)^{\hat{\Delta}_N}\left\{\prod_{i=1}^{n} (\hat{x}_i)^{a_{iN}}\right\}^{\hat{\Delta}_N}\right\} \\
&= \left\{\prod_{j=1}^{N}\left(\frac{c_j}{\hat{\Delta}_j}\right)^{\hat{\Delta}_j}\right\}\left\{\prod_{j=1}^{N}\left\{\prod_{i=1}^{n} (\hat{x}_i)^{a_{ij}}\right\}^{\hat{\Delta}_j}\right\} \\
&= \left\{\prod_{j=1}^{N}\left(\frac{c_j}{\hat{\Delta}_j}\right)^{\hat{\Delta}_j}\right\}\left\{\prod_{i=1}^{n} (\hat{x}_i)^{\sum_{j=1}^{N} a_{ij}\hat{\Delta}_j}\right\} \\
&= \prod_{j=1}^{N}\left(\frac{c_j}{\hat{\Delta}_j}\right)^{\hat{\Delta}_j},
\end{aligned} \tag{16.13}$$

because $\sum_{j=1}^{N} a_{ij}\hat{\Delta}_j = 0$ for each i.

Thus, once $\hat{\Delta}_j$ are known, we can find $\hat{F}$ from Eq. (16.13). To determine $\hat{\Delta}_j$, we have $n + 1$ equations in N unknowns, namely, Eqs. (16.6) and (16.8). The number $N - n - 1$ is called the *degree of difficulty* of the program (in the constrained geometric programming, N denotes the total number of terms in all the posynomials, and n is the number of design variables). If $N - n - 1 = 0$, the problem is said to be of zero degree of difficulty. Indeed, the orthogonality and normality conditions then provide exactly as many equations as there are unknowns, so $\hat{\Delta}_j$ can be determined uniquely. When $N > n + 1$, the number of unknowns exceeds the number of equations; in this case, the program should be solved by the Cauchy inequality method (discussed later). Finally, when the degree of difficulty is negative, geometric programming is inapplicable.

Having found $\hat{\Delta}_j$ and $\hat{F}$, we can determine the optimal design vector $\hat{\mathbf{x}}$ by solving the system of nonlinear equations

$$\hat{U}_j = \hat{\Delta}_j \hat{F} = c_j \prod_{i=1}^{n} (\hat{x}_i)^{a_{ij}}, \qquad j = 1, 2, \ldots, N. \tag{16.14}$$

To this end, we divide Eq. (16.14) by c_j and take logarithms of both sides. The result is

$$\ln(\hat{\Delta}_j \hat{F}/c_j) = a_{1j} \ln \hat{x}_1 + a_{2j} \ln \hat{x}_2 + \cdots + a_{nj} \ln \hat{x}_n. \tag{16.15}$$

Putting $w_i = \ln \hat{x}_i$, $i = 1, 2, \ldots, n$, we obtain N linear equations in n unknowns w_i:

$$\begin{aligned} a_{11}w_1 + a_{21}w_2 + \cdots + a_{n1}w_n &= \ln(\hat{\Delta}_1 \hat{F}/c_1), \\ a_{12}w_1 + a_{22}w_2 + \cdots + a_{n2}w_n &= \ln(\hat{\Delta}_2 \hat{F}/c_2), \\ &\vdots \\ a_{1N}w_1 + a_{2N}w_2 + \cdots + a_{nN}w_n &= \ln(\hat{\Delta}_N \hat{F}/c_N). \end{aligned} \tag{16.16}$$

If there are more equations than unknowns, any subsystem of n linearly independent equations can be chosen to determine w_i; hence

$$\hat{x}_i = e^{w_i}, \qquad i = 1, 2, \ldots, n. \tag{16.17}$$

16.2.2. Solution Based on Cauchy's Inequality

Another important method for solving geometric programming problems is a procedure involving Cauchy's inequality between the arithmetic and geometric means of n nonnegative numbers:

$$(x_1 + x_2 + \cdots + x_n)/n \geqslant (x_1 x_2 \cdots x_n)^{1/n}. \tag{16.18}$$

Equality between the two means holds only when $x_1 = x_2 = \cdots = x_n$. An equivalent form of inequality (16.18), more useful in geometric programming, is

$$\Delta_1 u_1 + \Delta_2 u_2 + \cdots + \Delta_N u_N \geqslant (u_1)^{\Delta_1}(u_2)^{\Delta_2}\cdots(u_N)^{\Delta_N} \tag{16.19}$$

with

$$\Delta_1 + \Delta_2 + \cdots + \Delta_N = 1. \tag{16.20}$$

If we apply this inequality to the objective function (16.2), setting $U_i(\mathbf{x}) = u_i \Delta_i$, $i = 1, 2, \ldots, N$, with $\Delta_1 + \Delta_2 + \cdots + \Delta_N = 1$ (the normality condition), we obtain

$$U_1 + U_2 + \cdots + U_N \geqslant (U_1/\Delta_1)^{\Delta_1}\cdots(U_N/\Delta_N)^{\Delta_N}. \tag{16.21}$$

We call the left side of inequality (16.21), i.e., the original function $F(\mathbf{x})$, the *primal function*, and the right side the *predual function.*

Using the relations

$$U_j = c_j \prod_{i=1}^{n} x_i^{a_{ij}}, \qquad j = 1, 2, \ldots, N, \tag{16.22}$$

we can represent the predual function as

$$\begin{aligned}
\left(\frac{U_1}{\Delta_1}\right)^{\Delta_1}\cdots\left(\frac{U_N}{\Delta_N}\right)^{\Delta_N} &= \left(c_1 \prod_{i=1}^{n} \frac{x_i^{a_{i1}}}{\Delta_1}\right)^{\Delta_1}\cdots\left(c_N \prod_{i=1}^{n} \frac{x_i^{a_{iN}}}{\Delta_N}\right)^{\Delta_N} \\
&= \left(\frac{c_1}{\Delta_1}\right)^{\Delta_1}\cdots\left(\frac{c_N}{\Delta_N}\right)^{\Delta_N}\left\{\left(\prod_{i=1}^{n} x_i^{a_{i1}}\right)^{\Delta_1}\cdots\left(\prod_{i=1}^{n} x_i^{a_{iN}}\right)^{\Delta_N}\right\} \\
&= \left(\frac{c_1}{\Delta_1}\right)^{\Delta_1}\cdots\left(\frac{c_N}{\Delta_N}\right)^{\Delta_N}\left\{x_1^{\sum_{j=1}^{N} a_{1j}\Delta_j}\cdots x_n^{\sum_{j=1}^{N} a_{nj}\Delta_j}\right\} \\
&= \left(\frac{c_1}{\Delta_1}\right)^{\Delta_1}\cdots\left(\frac{c_N}{\Delta_N}\right)^{\Delta_N},
\end{aligned} \tag{16.23}$$

provided that the weights Δ_j are selected so as to satisfy the orthogonality condition

$$\sum_{j=1}^{N} a_{ij}\Delta_j = 0, \qquad i = 1, 2, \ldots, n. \tag{16.24}$$

Thus, inequality (16.21) becomes

$$U_1 + U_2 + \cdots + U_N \geqslant (c_1/\Delta_1)^{\Delta_1}(c_2/\Delta_2)^{\Delta_2} \cdots (c_N/\Delta_N)^{\Delta_N}. \tag{16.25}$$

The expression on the right side is called the *dual function.* We denote it by $\nu(\Delta_1, \ldots, \Delta_N)$ and write inequality (16.25) as

$$F(\mathbf{x}) \geqslant \nu(\boldsymbol{\Delta}).$$

It follows that the maximum of the dual function is less than or equal to the minimum of the primal function. Now the central result of the theory, which we state here without proof, is that these two values are equal.

In this way the unconstrained minimization of the primal function can be accomplished by maximizing the dual function

$$\nu(\boldsymbol{\Delta}) = \prod_{j=1}^{N} \left(\frac{c_j}{\Delta_j}\right)^{\Delta_j} \tag{16.26}$$

or its logarithm, ln $\nu(\boldsymbol{\Delta})$, whichever is convenient, subject to the constraints given by the normality condition

$$\sum_{j=1}^{N} \Delta_j = 1 \tag{16.27}$$

and the orthogonality conditions

$$\sum_{j=1}^{N} a_{ij}\Delta_j = 0, \qquad i = 1, 2, \ldots, n. \tag{16.28}$$

If the degree of difficulty of the problem is zero, there will be a unique solution for Δ_j. When $N > n + 1$, the constraints can sometimes be used to eliminate $n + 1$ variables from the dual function, which can subsequently be maximized with respect to the remaining $N - n - 1$ independent Δ_j's.

Example 16.1. We illustrate the unconstrained minimization problem with zero degree of difficulty by the well-known "gravel box" example given in Duffin, Peterson, and Zener (1967). The optimization problem can be formulated as follows. Find x_1, x_2, and x_3 that minimize

$$f(\mathbf{x}) = 80x_1x_2 + 40x_2x_3 + 20x_1x_3 + 80(x_1x_2x_3)^{-1}. \tag{16.29}$$

Comparing $f(\mathbf{x})$ with the general posynomial (16.2), we see that $c_1 = 80$, $c_2 = 40$, $c_3 = 20$, $c_4 = 80$, and

$$\begin{bmatrix} a_{11} & a_{12} & a_{13} & a_{14} \\ a_{21} & a_{22} & a_{23} & a_{24} \\ a_{31} & a_{32} & a_{33} & a_{34} \end{bmatrix} = \begin{bmatrix} 1 & 0 & 1 & -1 \\ 1 & 1 & 0 & -1 \\ 0 & 1 & 1 & -1 \end{bmatrix}.$$

The orthogonality and normality conditions (16.27) and (16.28) take the following form:

$$\Delta_1 + \Delta_3 - \Delta_4 = 0, \tag{16.30}$$

$$\Delta_1 + \Delta_2 - \Delta_4 = 0, \tag{16.31}$$

$$\Delta_2 + \Delta_3 - \Delta_4 = 0, \tag{16.32}$$

$$\Delta_1 + \Delta_2 + \Delta_3 + \Delta_4 = 1. \tag{16.33}$$

From (16.30) and (16.31), we obtain

$$\Delta_4 = \Delta_1 + \Delta_3 = \Delta_1 + \Delta_2, \qquad \Delta_2 = \Delta_3. \tag{16.34}$$

Thus we find that $\Delta_1 = \Delta_2 = \Delta_3$, and Eq. (16.34) gives $\Delta_4 = \Delta_1 + \Delta_3 = 2\Delta_1$. Finally, Eq. (16.33) leads to the unique solution

$$\hat{\Delta}_1 = \hat{\Delta}_2 = \hat{\Delta}_3 = \tfrac{1}{5} \qquad \text{and} \qquad \hat{\Delta}_4 = \tfrac{2}{5}.$$

Thus we obtain the optimal value of the objective function from Eq. (16.13) as

$$\begin{aligned} \hat{f} &= \left(\frac{80}{1/5}\right)^{1/5}\left(\frac{40}{1/5}\right)^{1/5}\left(\frac{20}{1/5}\right)^{1/5}\left(\frac{80}{2/5}\right)^{2/5} \\ &= (4 \times 10^2)^{1/5}(2 \times 10^2)^{1/5}(1 \times 10^2)^{1/5}(4 \times 10^4)^{1/5} \\ &= (32 \times 10^{10})^{1/5} \\ &= 200. \end{aligned}$$

We found the optimal value of the objective function before finding the optimal vector $\hat{\mathbf{x}}$. To find $\hat{\mathbf{x}}$, we write Eq. (16.14) in the following form:

$$\hat{U}_1 = 80\hat{x}_1\hat{x}_2 = \hat{\Delta}_1\hat{f} = (1/5)200 = 40, \tag{16.35}$$

$$\hat{U}_2 = 40\hat{x}_2\hat{x}_3 = \hat{\Delta}_2\hat{f} = (1/5)200 = 40, \tag{16.36}$$

$$\hat{U}_3 = 20\hat{x}_1\hat{x}_3 = \hat{\Delta}_3\hat{f} = (1/5)200 = 40, \tag{16.37}$$

$$\hat{U}_4 = 80/(\hat{x}_1\hat{x}_2\hat{x}_3) = \hat{\Delta}_4\hat{f} = (2/5)200 = 80. \tag{16.38}$$

We thus have

$$\hat{x}_2 = 1/2\hat{x}_1, \qquad \hat{x}_1 = 2/\hat{x}_3, \qquad \hat{x}_2 = 1/\hat{x}_3,$$
$$1/\hat{x}_1\hat{x}_2\hat{x}_3 = 1 = 2/\hat{x}_3, \qquad \hat{x}_3 = 2, \tag{16.39}$$
$$\hat{x}_1 = 1, \qquad \hat{x}_2 = 1/2, \qquad \hat{x}_3 = 2.$$

Notice that there is one redundant equation among Eqs. (16.35)-(16.38), which is not needed for the solution of $\hat{x}_i$ $(i = 1, 2, 3)$. We could have used Eqs. (16.16) and (16.17) to get the solution in Eq. (16.39).

Exercise 16.1

Find the solution of the optimization problem described in Example 16.1 by using Eqs. (16.16) and (16.17) in the final stage.

16.3. Constrained Geometric Programming

16.3.1. Posynomial Programs

A constrained minimization problem with the objective function and the constraints all in the form of posynomials can be stated as follows:

Find a vector $\mathbf{x}^{\mathrm{T}} = (x_1, x_2, \ldots, x_n)$ that minimizes the objective function

$$F(\mathbf{x}) = \sum_{j=1}^{N_0} c_{0j} \prod_{i=1}^{n} x_i^{a_{0ij}} \tag{16.40}$$

and satisfies the constraints

$$g_k(\mathbf{x}) = \sum_{j=1}^{N_k} c_{kj} \prod_{i=1}^{n} x_i^{a_{kij}} \leqslant 1, \qquad k = 1, 2, \ldots, m, \tag{16.41}$$

or

$$g_k(\mathbf{x}) = \sum_{j=1}^{N_k} c_{kj} \prod_{i=1}^{n} x_i^{a_{kij}} \geqslant 1, \qquad k = 1, 2, \ldots, m, \tag{16.42}$$

where the coefficients c_{0j} $(j = 1, 2, \ldots, N_0)$ and c_{kj} $(k = 1, 2, \ldots, m; j = 1, 2, \ldots, N_k)$ are positive numbers, the exponents a_{kij} $(k = 0, 1, \ldots, m; i = 1, 2, \ldots, n; j = 1, 2, \ldots, N_k)$ are any real numbers, m is the number of constraints, N_0 is the number of terms in the objective function, and N_k $(k > 0)$ is the number of terms in the kth constraint. The design variables $x_1, \ldots, x_n$ are assumed to be positive.

For a unified notation, it will be convenient to denote the objective function as

$$x_0 = g_0(\mathbf{x}) = F(\mathbf{x}) \tag{16.43}$$

and to rewrite the constraints as

$$f_k = \sigma_k\{1 - g_k(\mathbf{x})\} \geqslant 0, \qquad k = 1, 2, \ldots, m, \tag{16.44}$$

where $\sigma_k = \pm 1$, according as $g_k(\mathbf{x}) \leqslant 1$ or $g_k(\mathbf{x}) \geqslant 1$.

We call the problem just formulated the *primal* problem. To solve it, we construct an equivalent problem, called its *dual*, which will have linear constraints and therefore will often be easier to solve. For this purpose, just as in the unconstrained case, we introduce new variables w_i by the relations

$$x_i = e^{w_i}, \qquad i = 0, 1, \ldots, n; \tag{16.45}$$

w_i can take any real values. Next, we define the variables Δ_{kj} as

$$\begin{aligned} \Delta_{0j} &= c_{0j} \prod_{i=1}^{n} \frac{x_i^{a_{0ij}}}{x_0} > 0, \qquad j = 1, 2, \ldots, N_0, \\ \Delta_{kj} &= c_{kj} \prod_{i=1}^{n} x_i^{a_{kij}} > 0, \qquad k = 1, 2, \ldots, m, \quad j = 1, 2, \ldots, N_k, \end{aligned} \tag{16.46}$$

so that, by definition of x_0,

$$\sum_{j=1}^{N_0} \Delta_{0j} = 1 \tag{16.47}$$

and, if the kth constraint is active,

$$\sum_{j=1}^{N_k} \Delta_{kj} = 1. \tag{16.48}$$

After applying logarithms to Eq. (16.45) we can restate the primal problem as

Minimize $w_0 = \ln x_0$, subject to the equality constraints

$$\sum_{j=1}^{N_0} \Delta_{0j} = 1, \tag{16.49}$$

$$\ln\left(\frac{\Delta_{0j}}{c_{0j}}\right) = -w_0 + \sum_{i=1}^{n} a_{0ij} w_i, \qquad j = 1, 2, \ldots, N_0, \tag{16.50}$$

$$\ln\left(\frac{\Delta_{kj}}{c_{kj}}\right) = \sum_{i=1}^{n} a_{kij} w_i, \qquad j = 1, 2, \ldots, N_k, \quad k = 1, 2, \ldots, m, \tag{16.51}$$

and the inequality constraints

$$f_k = \sigma_k\left(1 - \sum_{j=1}^{N_k} \Delta_{kj}\right) \geqslant 0, \qquad k = 1, 2, \ldots, m; \tag{16.52}$$

the variables Δ_{kj} are positive, whereas $w_0, w_1, \ldots, w_n$ can take any real values.

We can use the Lagrange multiplier method to find the minimum of the function w_0. We thus form the Lagrange function and apply the Kuhn–Tucker conditions, and we obtain a system of equations for a saddle point at which the primal function F attains a minimum and another function ν, called the dual, attains a maximum.

If all the constraints are of the form $g_k(\mathbf{x}) \leqslant 1$, which ensures strict convexity of the objective function g_0 with respect to the transformed variables $w_1, w_2, \ldots, w_n$, the primal-dual relation is

$$F(\mathbf{x}) \geqslant \min F = \max \nu \geqslant \nu(\boldsymbol{\lambda}). \tag{16.53}$$

we omit the details of the derivation and just state the dual problem obtained.

Find the vector of dual variables

$$\boldsymbol{\lambda}^{\mathrm{T}} = (\lambda_{01}, \ldots, \lambda_{0N_0}, \lambda_{11}, \ldots, \lambda_{1N_1}, \ldots, \lambda_{m1}, \ldots, \lambda_{mN_m})$$

that maximizes the dual function

$$\nu(\boldsymbol{\lambda}) = \prod_{k=0}^{m} \prod_{j=1}^{N_k} \left\{\frac{c_{kj}}{\lambda_{kj}} \sum_{l=1}^{N_k} \lambda_{kl}\right\}^{\lambda_{kj}} \tag{16.54}$$

subject to the normality constraint

$$\sum_{j=1}^{N_0} \lambda_{0j} = 1, \tag{16.55}$$

the orthogonality constraints

$$\sum_{k=0}^{m} \sum_{j=1}^{N_k} a_{kij}\lambda_{kj} = 0, \qquad i = 1, 2, \ldots, n, \tag{16.56}$$

and the nonnegativity conditions

$$\lambda_{kj} \geqslant 0, \qquad j = 1, 2, \ldots, N_k, \quad k = 0, 1, \ldots, m. \tag{16.57}$$

Comparing the primal and the dual programs, we note the following characteristics:

1. The coefficients c_{kj} occurring in the dual function $\nu(\boldsymbol{\lambda})$ are the coefficients of the posynomials $g_k(\mathbf{x})$, $k = 0, 1, \ldots, m$.
2. The vector $\boldsymbol{\lambda}$ has as many components as there are terms in all the posynomials $g_0, g_1, \ldots, g_m$. Associated with each term in g_k is a corresponding Δ_{kj}.
3. Every factor $\{\sum_{i=1}^{N_k} \lambda_{ki}\}^{\lambda_{kj}}$ of $\nu(\boldsymbol{\lambda})$ originates from an inequality constraint $g_k(\mathbf{x}) \leqslant 1$; none comes from the primal function $g_0(\mathbf{x})$, since, by the normality condition, $\sum_{j=1}^{N_0} \lambda_{0j} = 1$.
4. The coefficient matrix $\{a_{kij}\}$ of the orthogonality conditions is the exponent matrix for the primal program.

The theory of geometric programming is applicable to programs with positive terms only. If we tried to construct a dual function to a primal in which some of the terms are negative, the result would involve fractional powers of negative numbers, and hence imaginary numbers. Passy and Wilde (1967) developed a method by which an inequality-constrained polynomial problem with coefficients of an arbitrary sign can be transformed into a problem involving posynomials only, and hence a geometric programming problem susceptible to the primal-dual techniques described here. The transformation introduces some new variables, but an equal number of terms is also added to keep the degree of difficulty of the problem unchanged.

Avriel and Williams (1970) extended the method of geometric programming to include an even larger class of problems. They allowed the objective function and constraints to be rational functions of posynomial terms. The method, called *complementary geometric programming*, consists of successively approximating the rational functions involved by posynomials and solving the corresponding sequence of ordinary geometric programs. The successive solutions converge to a local minimum of the original problem. However, unlike ordinary geometric programs, complementary geometric programs are not in general convex, and therefore a local minimum need not be global.

Exercise 16.2

Minimize

$$f = x_1 x_2^2 x_3^{-1} + 2x_1^{-1} x_2^{-3} x_4 + \tfrac{5}{2} x_1 x_3$$

subject to

$$3x_1^{-1}x_3x_4^{-2} + 4x_3x_4 \leqslant 1,$$
$$5x_1x_2 \leqslant 1.$$

Exercise 16.3

Based on Duffin, Peterson, and Zener (1967, problem 8, page 15), find the dimensions of the stiffest rectangular beam that can be cut from a circular cylindrical log of radius R (i.e., max bh^3 or min $1/bh^3$ subject to $(b^2 + h^2)/R^2 \leqslant 1$).

16.3.2. Signomial Programs

So far we have considered the case where each posynomial term $U_i(\mathbf{x}) = c_i x_1^{a_{i1}} x_2^{a_{i2}} \cdots x_n^{a_{in}}$ has a positive coefficient c_i. Now, following Ecker (1980), we relax that condition and consider some of the consequences. Many programs arising from important applications fail to be posynomial programs only because some coefficients c_i are negative. Such programs are called *signomial* and were first studied by Passy and Wilde (1967) and Blau and Wilde (1969). Duffin and Peterson (1973) show that any signomial program can be transformed into an equivalent reversed posynomial program of the form

$$\min g_0(\mathbf{x}) \tag{16.58}$$

subject to

$$g_k(\mathbf{x}) \leqslant 1, \qquad k = 1, 2, \ldots, m \tag{16.59}$$

and

$$g_k(\mathbf{x}) \geqslant 1, \qquad k = m + 1, \ldots, q, \tag{16.60}$$

where $g_k(\mathbf{x})$ is a posynomial for $k = 1, 2, \ldots, q$. To see how such a reformulation is possible, consider, for example, a constraint of the form

$$h_1(\mathbf{x}) - h_2(\mathbf{x}) \leqslant 1, \tag{16.61}$$

where each $h_i(\mathbf{x})$ is a posynomial. Notice that $\mathbf{x}$ satisfies (16.61) if and only if there exist a single variable $s > 0$ such that $(\mathbf{x}, s)$ satisfies

$$h_1(\mathbf{x}) \leqslant s \leqslant h_2(\mathbf{x}) + 1. \tag{16.62}$$

Now note that (16.62) is equivalent to the two constraints

$$s^{-1}h_1(\mathbf{x}) \leqslant 1 \qquad \text{and} \qquad s^{-1}h_2(\mathbf{x}) + s^{-1} \geqslant 1,$$

which are in a form consistent with (16.59) and (16.60). For complete reformulation details for all possible cases, see Duffin and Peterson (1973).

Although posynomial programs have an equivalent reformulation as convex programs, it is clear from (16.59) and (16.60) that no such reformulation exists in the signomial case. The reversed constraints $g_k(\mathbf{x}) \geqslant 1$ give rise to a nonconvex feasible region. In general, local minima for signomial problems are not global minima.

Signomial programs have been studied extensively. Abrams and Bunting (1974) developed the notion of a reduced signomial program, and show that, under certain conditions, equivalent programs can be obtained by deleting certain terms and constraints. Avriel and Williams (1970), Passy (1971), and Pascual and Ben-Israel (1970) simultaneously developed methods for obtaining local minimizing points for signomial programs. These methods involve the concept of condensed posynomials. Condensed posynomials were also used by Duffin (1970) to obtain another proof for the duality theory of posynomial programming.

We will consider these methods and condensed posynomials in more detail later; but, essentially by using condensed posynomials, these authors showed how a signomial program can be approximated by a posynomial program. The solution to the posynomial program is then used to generate new condensed posynomials, and a better approximation to the signomial program is obtained. Avriel and Williams (1970) provide a convergence proof for their algorithm. Duffin and Peterson (1973) extend this work by showing that the study of algebraic programs can be reduced to the study of a family of approximating linear programs. Algebraic programs involve real-valued functions that are generated solely by addition, subtraction, multiplication, division, and extraction of roots.

The dual of the reversed posynomial programs has linear constraints similar to those in the posynomial case; however, the log of its objective function is not concave. In fact, its objective function is convex in some of its variables while concave in others. The rather complete duality theory that exists for posynomial programs does not carry over to signomial programs. Most algorithms to date for signomial programs have not considered solving its dual directly [although this is essentially the approach of Blau and Wilde (1971) and Rijckaert and Martens (1976)]. As indicated, most signomial algorithms consider instead a sequence of approximating posynomial programs.

16.4. Computational Methods for Posynomial Programs

16.4.1. Dual Methods

In this section, we review some of the computational algorithms for solving posynomial programs. We will not consider the specifics of all existing methods, but instead we will try to give the reader a basic understanding of primal and dual methods by focusing on some selected methods in detail. We begin by considering so-called dual methods that solve the *dual program* (D); we then use the duality relations to obtain the solution to the *primal posynomial program* (P).

The first published algorithm for posynomial programming was the dual method of Frank (1966). Since that time, several different types of dual algorithms have been investigated. Blau and Wilde (1971) developed a method that solves the Kuhn–Tucker (1951) necessary and sufficient optimality conditions for the dual posynomial program (D). A similar approach was taken by Rijckaert and Martens (1976).

Another class of methods uses the fact that the constraints for the dual program (D) are linear; therefore the dual can be reformulated so that the number of variables is equal to the dimension of the set of all solutions to the linear constraints. This reformulation, called the *reduced dual*, was first studied by Duffin, Peterson, and Zener (1967). Templeman, Wilson, and Winterbottom (1972) and Bradley (1975) have investigated this approach. Jefferson (1974) has also developed a Newton algorithm for solving the reduced dual. Dinkel, Kochenberger, and McCarl (1974) report on a modified Newton–Raphson algorithm for the reduced dual. Beck and Ecker (1975) proposed certain modifications to overcome the numerical difficulties associated with a direct application of the concave simplex method of Zangwill (1967a). Ecker, Gochet, and Smeers (1976) have adapted the reduced gradient method of Wolfe (1967) to solve the logarithmic dual problem. Dembo (1978b) has investigated modifications to overcome the nondifferentiability problem in adapting a modified Newton algorithm to solve the dual program.

16.4.2. Primal Methods

Several methods have been developed for solving program P directly, although initially more attention was given to dual methods. Recklaitis and Wilde (1974) used a general-purpose nonlinear programming algorithm of Wilde and Beightler (1967) to solve the separable convex form of program (P). Gochet and Smeers (1974) developed a cutting-plane algorithm that solves the convex reformulation via a sequence of linear programs. They present a proof of convergence and show that the cuts they develop are deeper than the cuts generated by the method of Kelley (1960) for solving general convex programs. Dinkel, Elliott, and Kochenberger (1977) study four different cutting plane approaches for solving (P) directly. Ecker and Zoracki (1976) use the fact, established by Duffin and Peterson (1972a), that any posynomial program can be reformulated so that each posynomial contains at most two terms, and they develop a method that approximates (P) to any desired degree of accuracy by a single linear program.

The notion of a condensed posynomial program, introduced by Duffin (1970), plays a crucial role in many primal methods. As mentioned earlier, the initial methods for signomial programs developed by Avriel and Williams (1970), Passy (1971), and Pascual and Ben-Israel (1970) all use the concept of a condensed posynomial. Also, Avriel, Dembo, and Passy (1975) present algorithms for posynomial and signomial programs using condensed posynomials. We will consider some details of condensed programs, and we illustrate the general approach taken by primal methods that use condensed posynomials.

The arithmetic-geometric mean inequality that played such a crucial role in developing the duality theory for posynomial programs is also used in obtaining a condensed posynomial. Recall that this inequality states that, for any vector $\mathbf{w} > 0$ and any nonnegative vector of weights $\boldsymbol{\varepsilon}$ whose components sum to unity, we have [see (16.21)]

$$\sum_i w_i \geqslant \prod_i \left(\frac{w_i}{\varepsilon_i}\right)^{\varepsilon_i}, \tag{16.63}$$

provided $(w_i/\varepsilon_i)^{\varepsilon_i}$ is defined to be 1 when $\varepsilon_i = 0$. Given a posynomial $g_k(\mathbf{x}) = \sum_i U_i(\mathbf{x})$ and $\varepsilon_i \geqslant 0$ with $\sum_i \varepsilon_i = 1$, we define a condensed posynomial $\bar{g}_k$ by

$$\bar{g}_k(\mathbf{x}) = \prod_i \left\{\frac{U_i(\mathbf{x})}{\varepsilon_i}\right\}^{\varepsilon_i}. \tag{16.64}$$

If each term $U_i(\mathbf{x})$ in $g_k(\mathbf{x})$ has the form [see (16.22)]

$$U_i(\mathbf{x}) = c_i x_1^{a_{i1}} x_2^{a_{i2}}, \ldots, x_n^{a_{in}}, \tag{16.65}$$

then [see (16.23)]

$$\bar{g}_k(\mathbf{x}) = \bar{c}_k x_1^{\bar{a}_{k1}}, \ldots, x_n^{\bar{a}_{kn}}, \tag{16.66}$$

where

$$c_k = \prod_i \left(\frac{c_i}{\varepsilon_i}\right)^{\varepsilon_i} \quad \text{and} \quad \bar{a}_{kj} = \sum_i a_{ij}\varepsilon_i. \tag{16.67}$$

Thus, the condensed posynomial $\bar{g}_k$ is also a posynomial, but it has a single posynomial term. The general approach of methods that use condensed posynomials is to approximate the original program (P) in (16.43) and (16.44) by a posynomial program where the objective function and each constraint consist of a single-term posynomial. After taking logarithms, we see that the approximating problem is equivalent to a linear program. Essentially, this method fits into a framework of the cutting-plane methods of Kelley (1960), for which convergence can be shown. For complete details of this approach, see Avriel, Dembo, and Passy (1975) and Dembo (1972a, b).

16.5. Computational Methods for Signomial Programs

Any signomial program can be reformulated as an equivalent reversed posynomial program, as defined in (16.58)–(16.60). One approach to solve signomial problems is to condense the posynomial in each reversed constraint and obtain a posynomial program of the form (P) that approximates the original signomial program. After solving the posynomial program by one of the methods in the previous section, we use the solution to generate a better approximation.

Following Duffin (1970) and Ecker (1980), we illustrate one such approach. Suppose a program S of the form (16.58)–(16.60) contains a single reversed constraint

$$g_l(\mathbf{x}) \geqslant 1. \tag{16.68}$$

Let $\bar{g}_l(\mathbf{x})$ be a single-term posynomial obtained by condensing g_l with an arbitrary set of weights $\boldsymbol{\varepsilon}$, using the arithmetic-geometric mean inequality. Let $\bar{S}$ denote the program obtained from S, where (16.68) is replaced by

$$\bar{g}_l(\mathbf{x}) \geqslant 1 \tag{16.69}$$

or, equivalently,

$$\bar{g}_l(\mathbf{x})^{-1} \leqslant 1. \tag{16.70}$$

Since $\bar{g}_l(\mathbf{x})$ is a single-term posynomial, (16.70) is a standard posynomial constraint and $\bar{S}$ is a posynomial program that approximates the signomial program S. Moreover, the arithmetic-geometric inequality implies that $\bar{g}_l(\mathbf{x}) \leq g_l(\mathbf{x})$. Thus, if $\mathbf{x}$ is feasible for $\bar{S}$, then $\mathbf{x}$ is feasible for S. Therefore, an optimal solution to $\bar{S}$ will be feasible but not necessarily optimal for S. The minimum value for $\bar{S}$, $M(\bar{S})$, is an upper bound on the minimum value for S, $M(S)$.

Suppose that $\hat{\mathbf{x}}$ is optimal for $\bar{S}$. Define a new set of weights by

$$\varepsilon_i = U_i(\hat{\mathbf{x}})/g_l(\hat{\mathbf{x}}), \tag{16.71}$$

using the terms $U_i(\mathbf{x})$ in $g_l(\mathbf{x})$. Use these weights to define a new condensed posynomial $\bar{\bar{g}}_l(\mathbf{x})$ and form the program $\bar{\bar{S}}$, where $\bar{\bar{g}}_l(\mathbf{x}) \geq 1$ replaces $g_l(\mathbf{x}) \geq 1$ in S. Since $\bar{\bar{g}}_l(\hat{\mathbf{x}}) = g_l(\hat{\mathbf{x}})$ and since $\hat{\mathbf{x}}$ is feasible for S, it follows that $\hat{\mathbf{x}}$ is feasible for $\bar{\bar{S}}$. The minimum value for $\bar{\bar{S}}$, $M(\bar{\bar{S}})$, therefore satisfies

$$M(S) \leq M(\bar{\bar{S}}) \leq M(\bar{S}).$$

This defines an iterative process for generating a sequence of posynomial programs whose minimum values are monotone decreasing upper bounds on the desired minimum value for S. Recall, however, that, in general, program S may have local minima that are not global minima, and the foregoing process may converge to such a point. Avriel and Williams (1970) develop a similar approach and show that under certain assumptions the process converges to a local minimum. Avriel, Dembo, and Passy (1975) use the Avriel–Williams approach to develop a highly successful algorithm. Duffin and Peterson (1972b) show how to use the arithmetic-harmonic mean inequality to develop a similar iterative process for solving signomial programs. Dinkel, Kochenberger, and McCarl (1974) compare the Avriel–Williams method, which uses the arithmetic-geometric mean, with the method suggested by Duffin and Peterson, which uses the arithmetic-harmonic mean inequality. They report that the computational effort for the former appears to be less than the latter.

For a detailed discussion of current computer software available for signomial and posynomial programming, see Dembo (1978a), where a computational comparison of many existing methods is given. Also, Sarma *et al.* (1978) compare various computational strategies for posynomial programs. Rijckaert and Martens (1978a) present some additional computational comparisons for posynomial and signomial codes.

General-purpose nonlinear programming methods have been used to solve posynomial and signomial programs, and some computational comparisons have been made. Dembo (1978a) and Dembo and Rijckaert (1978) compare several general-purpose algorithms with several special geometric programming codes. They found, for example, that the best times for the general-purpose codes were consistently between 1.4 and 5 times slower than the best times for the specialized geometric programming codes. Also, Ratner, Lasdon, and Jain (1978) used their generalized reduced gradient (GRG) method to solve test problems given by Dembo (1976) and Rijckaert and Martens (1978a). They found that the GRG code competes well with the special-purpose codes, although GRG failed to solve two of Dembo's problems. Fiacco and Ghaemi (1979) report successful solution of a special class of posynomial problems, using the SUMT algorithm of Fiacco and McCormick (1968).

Gochet, Loute, and Solow (1974) report on some computational experience in using the fixed-point algorithm of Eaves and Saigal (1972) and a GRG algorithm applied to the primal problem. There have been some investigations of methods for finding global minima for signomial problems. Falk (1973) proposes a branch and bound algorithm that converges to global minima for (nonconvex) signomial programs. Gochet and Smeers (1975) also investigate a branch and bound method for reversed posynomial programs, which involves approximations of the reversed constraints. They also provide a proof of convergence. Passy (1974) considered a similar approach. Some early work on this problem was done by Smeers (1972).

16.6. Applications of Geometric Programming to Structural Design

16.6.1. Optimization of Trusses

We consider the simple two-bar pinned truss with tubular steel members under a given vertical load $2P$ (see Fig. 16.1). This problem was solved by

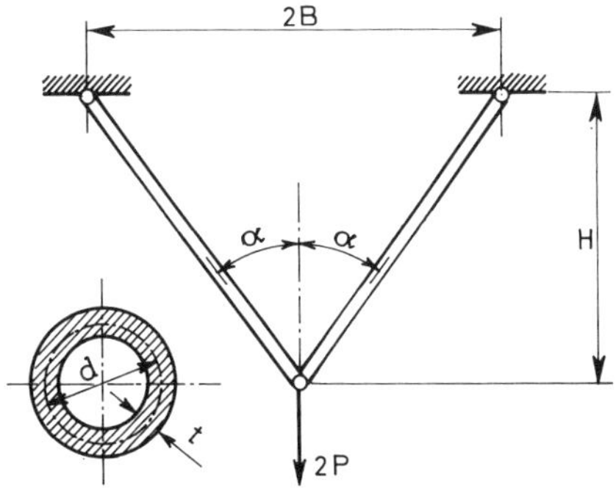

Figure 16.1. Example of Truss.

Morris (1972) in a similar illustrative role. The problem is to find the minimum weight of the truss by selecting appropriate values for the mean diameter for the tubes d and the height H subject to the constraint that the stresses in the tubes should not exceed the yield stress. The following values are used: $P = 1.5 \times 10^5$ N, $t = 2.54 \times 10^{-3}$ m (0.1 in.), $B = 0.762$ m (30 in.), $\sigma_0 = 4 \times 10^8$ N/m^2, $\gamma = 8.14 \times 10^4$ N/m^3.

Thus the primal problem consists of minimizing the weight

$$F = 2\pi\, dt\, \gamma (B^2 + H^2)^{1/2} = 1.299 \times 10^3 d(0.5806 + H^2)^{1/2}[\mathrm{N}], \tag{16.72}$$

subject to

$$P/\pi\, dt \cos\alpha = P(B^2 + H^2)^{1/2}/\pi t\, dH \leq \sigma_0,$$

which may be written as

$$0.047(0.5806 + H^2)^{1/2}/dH \leq 1.$$

The objective function and constraint can be transformed into posynomials by introducing a related function together with an additional independent variable and constraint. The related function for (16.72) is then

$$F = 1.299 \times 10^3 zd,$$

where $z = (0.5806 + H^2)^{1/2}$ is the new independent variable, and the new constraint is

$$0.5806/z^2 + H^2/z^2 \leq 1.$$

Introducing the notation $x_1 = z$, $x_2 = H$, $x_3 = d$, we state the primal problem as

$$\min F = 1.299 \times 10^3 x_1 x_3 \tag{16.73}$$

subject to

$$0.047 x_1 x_2^{-1} x_3^{-1} \leq 1, \tag{16.74}$$

$$0.5806 x_1^{-2} + x_1^{-2} x_2^2 \leq 1. \tag{16.75}$$

Using (16.40), we write the objective function as

$$g_0(\mathbf{x}) = F(\mathbf{x}) = \sum_{j=1}^{1} U_j(\mathbf{x}) = c_{01} x_1^{a_{011}} x_2^{a_{021}} x_3^{a_{031}}, \tag{16.76}$$

because $N_0 = 1$ is the number of terms in the primal function. Comparing (16.76) and (16.73), we see that

$$c_{01} = 1.299 \times 10^3, \quad a_{011} = 1, \quad a_{021} = 0, \quad a_{031} = 1.$$

There are two constraints in this example ($k = 1, 2$); therefore, $m = 2$. The first constraint ($k = 1$) has only one term and can be written as

$$g_1(\mathbf{x}) = \sum_{j=1}^{1} U_j(\mathbf{x}) = c_{11} x_1^{a_{111}} x_2^{a_{121}} x_3^{a_{131}},$$

where

$$c_{11} = 0.047, \quad a_{111} = 1, \quad a_{121} = -1, \quad a_{131} = -1.$$

The second constraint ($k = 2$) has two terms and can be written as

$$g_2(\mathbf{x}) = \sum_{j=1}^{2} U_j(\mathbf{x}) = U_1 + U_2$$
$$= c_{21} x_1^{a_{211}} x_2^{a_{221}} x_3^{a_{231}} + c_{22} x_1^{a_{212}} x_2^{a_{222}} x_3^{a_{232}},$$

where

$$c_{21} = 0.5806, \quad a_{211} = -2, \quad a_{221} = 0, \quad a_{231} = 0,$$
$$c_{22} = 1, \qquad a_{212} = -2, \quad a_{222} = 2, \quad a_{232} = 0.$$

Three variables (x_1, x_2, x_3) occur in this example; therefore, $n = 3$. The total number of terms in the primal problem is $N = 4$. The degree of difficulty of the problem is $D = N - n - 1 = 4 - 3 - 1 = 0$.

The associated dual problem can be stated according to (16.54)-(16.57) as follows. Find the vector of dual variables

$$\boldsymbol{\lambda}^{\mathrm{T}} = (\lambda_{01}, \lambda_{11}, \lambda_{21}, \lambda_{22})$$

that maximizes the dual function (16.54)

$$\nu(\boldsymbol{\lambda}) = \prod_{j=1}^{N_0=1} \left(\frac{c_{0j}}{\lambda_{0j}} \sum_{l=1}^{N_0=1} \lambda_{0l} \right)^{\lambda_{0j}} \prod_{j=1}^{N_1=1} \left(\frac{c_{1j}}{\lambda_{1j}} \sum_{l=1}^{N_1=1} \lambda_{1l} \right)^{\lambda_{1j}} \prod_{j=1}^{N_2=2} \left(\frac{c_{2j}}{\lambda_{2j}} \sum_{l=1}^{N_2=2} \lambda_{2l} \right)^{\lambda_{2j}}$$
$$= \left(\frac{c_{01}}{\lambda_{01}} \lambda_{01} \right)^{\lambda_{01}} \left(\frac{c_{11}}{\lambda_{11}} \lambda_{11} \right)^{\lambda_{11}} \left[\frac{c_{21}}{\lambda_{21}} (\lambda_{21} + \lambda_{22}) \right]^{\lambda_{21}} \left[\frac{c_{22}}{\lambda_{22}} (\lambda_{21} + \lambda_{22}) \right]^{\lambda_{22}},$$

subject to the dual constraints of:

nonnegativity,

$$\lambda_{01} \geqslant 0, \quad \lambda_{11} \geqslant 0, \quad \lambda_{2j} \geqslant 0, \qquad j = 1, 2;$$

normality,

$$\sum_{j=1}^{N_0=1} \lambda_{0j} = \lambda_{01} = 1; \tag{16.77}$$

orthogonality,

$$\sum_{k=0}^{m} \sum_{j=1}^{N_k} a_{kij}\lambda_{kj} = 0, \qquad i = 1, 2, \ldots, n,$$

or, in extended form,

$$\begin{aligned} a_{011}\lambda_{01} + a_{111}\lambda_{11} + a_{211}\lambda_{21} + a_{212}\lambda_{22} &= 0, \\ a_{021}\lambda_{01} + a_{121}\lambda_{11} + a_{221}\lambda_{21} + a_{222}\lambda_{22} &= 0, \\ a_{031}\lambda_{01} + a_{131}\lambda_{11} + a_{231}\lambda_{21} + a_{232}\lambda_{22} &= 0. \end{aligned} \tag{16.78}$$

Substituting appropriate values a_{kij} into Eqs. (16.78) gives

$$\begin{aligned} \lambda_{01} + \lambda_{11} - 2\lambda_{21} - 2\lambda_{22} &= 0, \\ -\lambda_{11} + 2\lambda_{22} &= 0, \\ \lambda_{01} - \lambda_{11} &= 0. \end{aligned}$$

The solution to these equations is straightforward, since the problem is one with zero degrees of difficulty. Hence the maximizing vector for the dual problem is

$$\hat{\lambda}_{01} = 1, \quad \hat{\lambda}_{11} = 1, \quad \hat{\lambda}_{21} = \tfrac{1}{2}, \quad \hat{\lambda}_{22} = \tfrac{1}{2}.$$

and the maximum value of the dual function [and thus the minimum for (16.73)] is

$$\begin{aligned} \nu(\boldsymbol{\lambda}) = (1299)^1(0.047)^1\{(0.5806/0.5)(0.5 + 0.5)\}^{1/2} \\ \cdot \{(1/0.5)(0.5 + 0.5)\}^{1/2} = 93 \text{ N}. \end{aligned} \tag{16.79}$$

The values of the design variables can be obtained by solving simultaneously Eq. (16.46); i.e.,

$$\hat{\Delta}_{0j} = \hat{\lambda}_{0j} \equiv \frac{c_{0j} \prod_{i=1}^{n} (\hat{x}_i)^{a_{0ij}}}{\hat{x}_0}, \qquad j = 1, 2, \ldots, N_0,$$

and

$$\hat{\Delta}_{kj} = \frac{\hat{\lambda}_{kj}}{\sum_{j=1}^{N_k} \lambda_{kj}} = c_{kj} \prod_{i=1}^{n} (\hat{x}_i)^{a_{kij}}, \qquad j = 1, 2, \ldots, N_k, \quad k = 1, 2, \ldots, m,$$

or, in extended form,

$$\begin{aligned} 1 &\equiv \frac{1299(\hat{x}_1)^1(\hat{x}_3)^2}{93}, \\ 1 &\equiv 0.047(\hat{x}_1)^1(\hat{x}_2)^{-1}(\hat{x}_3)^{-1}, \\ \tfrac{1}{2} &\equiv 0.5806(\hat{x}_1)^{-2}, \\ \tfrac{1}{2} &= 1(\hat{x}_1)^{-2}(\hat{x}_2)^2. \end{aligned}$$

Hence, the solution to the primal problem is $\hat{x}_1 = 1.07759$ m, $\hat{x}_2 = H = 0.762$ m, $\hat{x}_3 = d = 0.0665$ m. The minimum value of the weight of the structure is

$$F(\mathbf{x}) = (1299)(1.07759)(0.0665) = 93 \text{ N}.$$

Achieving a solution vector $\hat{\boldsymbol{\lambda}}$ to the dual problem by using only Eqs. (16.77)-(16.78) focuses attention on an important invariance property of geometric programs of zero degree of difficulty. This states that solution vectors to the dual program are independent of the coefficients c_{kj}. Thus, if one solution has already been found, we can obtain a variation in the maximum value of the dual function for variations in the primal coefficients c_{kj} simply by substituting the modified c_{kj} values into (16.79) and calculating $\nu(\hat{\boldsymbol{\lambda}})$ with the aid of the known $\hat{\boldsymbol{\lambda}}$'s, since these remain the same. In this way, one complete solution giving the maximum value to a given problem is the key to a whole family of related maxima (or minima).

16.6.2. Optimization of Plate Girders

Burns and Ramamurthy (1983) and Abuyounes and Adeli (1986) applied generalized geometric programming to the minimum-cost (weight) design of welded plate girders. Based on the latter paper, we present a problem formulation and some results of optimization of plate girders.

The design variables for a simply supported plate girder (Fig. 16.2) made of a given type of steel are the web height h, web thickness t_w, flange width b_f, flange thickness t_f, stiffener thickness t_s, stiffener width b_s, stiffener spacing s, and number of stiffeners N_{st}.

The design parameters are those known values that do not change during the optimization process. For a simply supported welded plate girder, design parameters are the bending moment M, shearing force V, laterally unsupported length L_u, and yield strength F_y.

For a stiffened plate girder with uniform cross section the objective function may be written as

$$W = \rho\{ht_wL + 2b_ft_fL + hb_st_sN_{st}\},$$

in which ρ is the density of the material and L is the span length of the plate girder. For an unstiffened plate girder, the objective function may be simplified to

$$W = \rho L\{ht_w + 2b_ft_f\}.$$

The design constraints based on American Institute of Steel Construction (AISC) specifications are expressed in a suitable form for application of the generalized geometric programming (GGP) technique. They are concerned with allowable bending stress, allowable shear stress, aspect ratios of the web and flange plates, and dimensions of intermediate transverse stiffeners when provided.

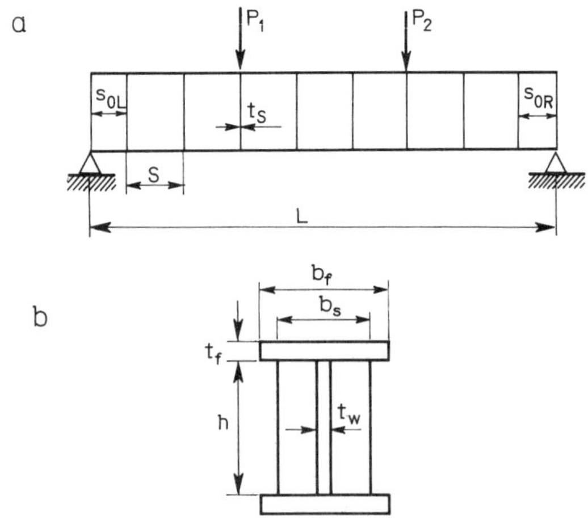

Figure 16.2. Design Variables for a Simply Supported Plate Girder.

16.6.2.1. Bending Stress Constraints. The maximum bending stress f_b should not exceed the allowable bending stress F_b:

$$f_b \leqslant F_b.$$

For girders with full lateral support against lateral torsional buckling, if no reduction in the allowable stress is required (i.e., $h/t_w < 760/(F_b)^{1/2}$), then

$$0.5M_{\max}h/0.6F_yI + M_{\max}t_f/0.6F_yI \leqslant 1,$$

where $M_{\max}$ is the maximum bending moment in the girder (k-in.) and I is the moment of inertia of the cross section about the major axis (in.^{4}).

If reduction in the allowable bending stress is required, i.e.,

$$h/t_w \geqslant 760/(F_b)^{1/2},$$

then

$$\frac{0.5M_{\max}h}{0.6F_yI} + \frac{M_{\max}t_f}{0.6F_yI} + \frac{0.0005h^2}{b_ft_f} - \frac{0.38t_wh}{(0.6F_y)^{1/2}b_ft_f} \leqslant 1.$$

16.6.2.2. Shear Stress and Stiffener Spacing Constraints. The largest average web shear f_ϑ should not exceed the allowable shear stress F_ϑ:

$$f_\vartheta \leqslant F_\vartheta.$$

For unstiffened plate girders,

$$V_{\max}h/83{,}150t_w^3 \leqslant 1 \qquad \text{when } h/t_w > 546/(F_y)^{1/2}$$

and

$$V_{\max}/152t_w^2(F_y)^{1/2} \leqslant 1 \qquad \text{when } h/t_w < 548/(F_y)^{1/2}.$$

For stiffened plate girders, at end panels,

$$2.89V_m/F_yt_whC_\vartheta \leqslant 1$$

and

$$C_\vartheta/1.1560 \leqslant 1.$$

Here V_m is the maximum shear in the panel or segment under consideration (kips), and C_ϑ is the ratio of the shear stress at buckling to shear yield stress:

$$C_\vartheta = \frac{45{,}000k}{F_y(h/t_w)^2}, \qquad \text{for } C_\vartheta \leqslant 0.8,$$

$$C_\vartheta = \frac{190}{(h/t_w)}\left\{\frac{k}{F_y}\right\}^{1/2} \qquad \text{for } C_\vartheta > 0.8,$$

with

$$k = 4.0 + \frac{5.34}{(s/h)^2} \qquad \text{for } \frac{s}{h} < 1,$$

$$k = \frac{4.0}{(s/h)^2} + 5.34 \qquad \text{for } \frac{s}{h} > 1,$$

where s is the spacing between intermediate stiffeners.

For all segments of nonhybrid girders (except end panels), tension-field action is considered in the design of a web plate if the following conditions are satisfied:

1. The plate girder is nonhybrid.
2. Intermediate transverse stiffeners are provided.
3. $C_\vartheta \leqslant 1.0$.
4. $s/h \leqslant \{260/(h/t_w)\}^2$.
5. $s/h \leqslant 3.0$.

The shear stress and intermediate stiffener spacing constraints may be expressed as follows:

$$\frac{2.89\,V_m}{F_y t_w h C_\vartheta} + \frac{1}{1.15A} - \frac{1}{1.15AC_\vartheta} \leqslant 1$$

and

$$\frac{C_\vartheta}{1.3294A} - \frac{C_\vartheta}{1.156} - \frac{1}{1.3294A} \leqslant 1,$$

where $A = \{1 + (s/h)^2\}^{1/2}$.

For practical considerations related to fabrication, handling, and erection, the spacing of intermediate stiffeners should satisfy the following constraints:

$$hs/(260)^2 t_w^2 \leqslant 1 \qquad \text{and} \qquad s/3h \leqslant 1.$$

Spacing of intermediate stiffeners in each segment is considered to be uniform.

16.6.2.3. Aspect Ratio Constraints. Buckling of the web may occur due to bending in the plane of the web. Such buckling will not occur if

$$h/t_w \leq 760/(F_b)^{1/2}.$$

When higher h/t_w values are used, the web may not have sufficient stiffness to prevent the vertical buckling of the flange.

In the absence of intermediate stiffeners, the h/t_w constraint is written as

$$h\{F_y(F_y + 16.5)\}^{1/2}/14{,}000t_w \leq 1.$$

When transverse stiffeners are provided, the h/t_w constraint is written as

$$(F_y)^{1/2}h/2000t_w \leq 1 \qquad \text{for } s/h < 1.5.$$

The flange and intermediate stiffener plates should satisfy the following constraints in order to prevent local buckling:

$$(F_y)^{1/2}b_f/190t_f \leq 1,$$
$$(F_y)^{1/2}b_s/95t_s \leq 1.$$

16.6.2.4. Constraints on the Size of Intermediate Transverse Stiffeners. Intermediate stiffeners are not required when

$$h/t_w < 260 \qquad \text{and} \qquad f_\vartheta < F_\vartheta.$$

However, when intermediate stiffeners are required, they should satisfy the following constraints in order to provide adequate compression strut capacity during tension-field action:

$$\frac{V_m YD}{2hA_{st}sF_\vartheta} + \frac{V_m YDs^2C_\vartheta}{2h^2A_{st}AF_\vartheta} - \frac{V_m YDs^2}{2h^2A_{st}AF_\vartheta} - \frac{V_m YDsC_\vartheta}{2hA_{st}F_\vartheta} \leq 1,$$

where A_{st} is the gross area of a single or a pair of intermediate stiffeners (in.^{2}), Y is the ratio of yield stress of web steel to yield stress of stiffener steel (F_{yw}/F_{ys}), and D is 1.0 for stiffeners furnished in pairs and 2.4 for single-plate stiffeners. Only rectangular plate stiffeners are considered.

To prevent local buckling of the web between buckled panels, we want the moment of inertia of a pair of intermediate stiffeners or a single stiffener with respect to an axis in the plane of the web to be greater than a specified value. This condition is

$$3h^4/(50)^4 b_s^3 t_s \leqslant 1 \qquad \text{(single stiffener)},$$

$$3h^4/2(50)^4 b_s^3 t_s \leqslant 1 \qquad \text{(pair of stiffeners)}.$$

16.6.2.5. Web Crippling Constraints. Compression stress that results from distributed loads on the compression edge of the web plate shall not exceed a prescribed value. This constraint can be written as

$$\frac{ws^2}{12t_w^3 \times 40{,}000} - \frac{55{,}000s^2}{40{,}000h^2} \leqslant 1,$$

when the compression flange is restricted against rotation, and

$$\frac{ws^2}{12t_w^3 \times 40{,}000} - \frac{20{,}000s^2}{40{,}000h^2} \leqslant 1$$

when the compression flange is not restrained against rotation.

16.6.2.6. Maximum Bending Tensile Stress Constraint. When tension-field action is used in design of the web, the bending tensile stress f_b due to moment in the plane of the web should not exceed a prescribed value. This condition can be written as

$$\frac{0.5M_p h}{0.825F_y I} + \frac{M_p t_f}{0.825F_y I} + \frac{0.375}{0.825t_w h F_y} \leqslant 1,$$

where M_p is the bending moment at points of application of concentrated loads (k-in.).

Note that there are three discontinuities involved in the bending and shear stress constraints. First, the bending stress constraints take into account buckling of the web when the depth-to-thickness ratio of the web exceeds $760/(F_b)^{1/2}$, whereas this instability is not accounted for when the foregoing ratio is less than $760/(F_b)^{1/2}$.

Second, in the shear stress constraint, the coefficient k that relates linear buckling strength of the web plate to its dimensions takes different values, depending on the ratio of s/h. Finally, the coefficient C_ϑ, which is the ratio of the critical web stress to the shear yield stress, is given by two different equations, depending on whether the value of C_ϑ is above or below

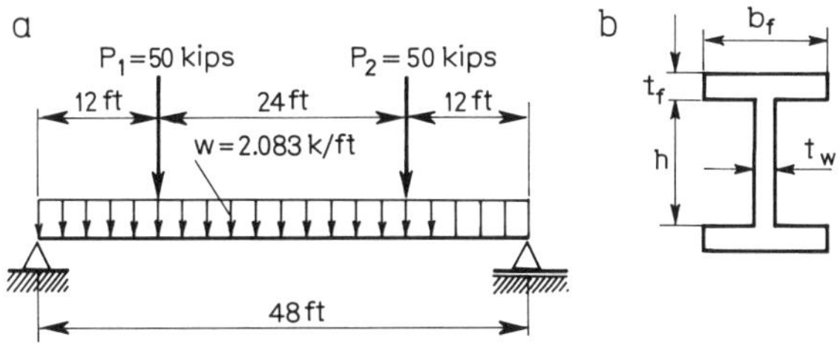

Figure 16.3. Unstiffened Plate Girder of Example 16.2.

the limit of 0.8. Since values of h, t_w, s, and C_ϑ are not known in advance, the three discontinuities, as well as large nonlinearities, may prevent application of most optimization techniques to solve this optimal design problem.

We adopt the following procedure to overcome these difficulties:

1. Design constraints are formulated to accommodate all the cases (i.e., $h/t_w \lesseqgtr 760/(F_b)^{1/2}$ and $s/h <$ or >1).
2. To start the solution, initial values are assumed for the ratio h/t_w. The optimization problem is then solved iteratively.

Note that these constraints are highly nonlinear. Of the many available NLP techniques, very few are suitable for problems with highly nonlinear constraints. The GGP algorithm developed by Avriel, Dembo, and Passy (1975) seems to work effectively on problems with highly nonlinear constraints. In the following we present two examples of simply supported plate girders.

Example 16.2. The girder is unstiffened and subjected to the loading shown in Fig. 16.3a. The compression flange is laterally supported throughout its length, and the girder is restricted against rotation. The type of steel used for the flange and web plates is A36. The minimum-weight design for this unstiffened plate girder is presented in Table 16.1.

Table 16.1. Minimum-Weight Design for Unstiffened Plate Girder (Example 16.2)

Variables[a]	t_w	h	b_f	t_f	W (kips)[b]
Lower bound	0.1	1.0	1.0	0.1	1.0
M.W.D.[c]	0.414	58.917	15.306	0.483	6.4
No. of iterations 2					

[a] All dimensions are in inches.
[b] 1 kip = 4.45 N.
[c] M.W.D. = minimum-weight design.

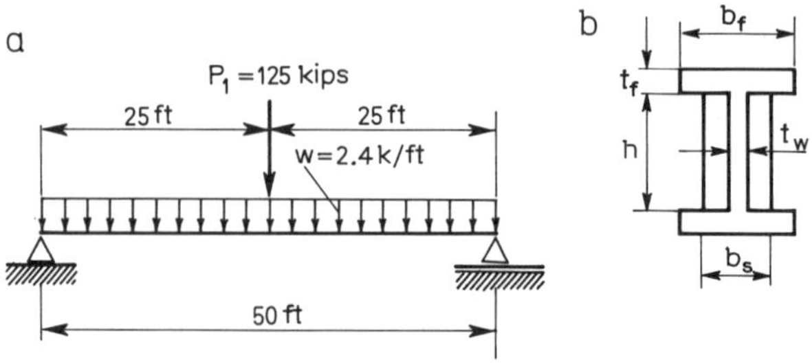

Figure 16.4. Stiffened Plate Girder of Example 16.3.

Table 16.2. Minimum Weight Design for Stiffened Plate Girder (Example 16.3)

Variables[a]	t_w	h	b_f	t_f	t_s	b_s	W (kips)[b]
Lower bound	0.1	1.0	1.0	0.1	1.0	0.1	1.0
M.W.D.[c]	0.222	73.919	22.74	0.718	0.269	2.986	8.56
No. of iterations 3							

[a] All dimensions are in inches.
[b] 1 kip = 4.45 N.
[c] M.W.D. = minimum-weight design.

Example 16.3. The girder is stiffened with double plates and is subjected to the loading shown in Fig. 16.4a. The compression flange is laterally supported throughout its length, and the girder is restrained against rotation. The type of steel used for the flange and web plates and the intermediate stiffeners is A36. The minimum-weight design for this stiffened plate girder is presented in Table 16.2 and Fig. 16.5.

To find the optimal depth-to-span ratio, we carried out a parametric study for nonhybrid stiffened plate girders, subjected to a uniformly distributed loading of intensity w. Results are plotted as optimal L/h ratios versus the span length for different load intensities in Fig. 16.6. These plots are for girders of A36 steel with a yield stress of 36 ksi.

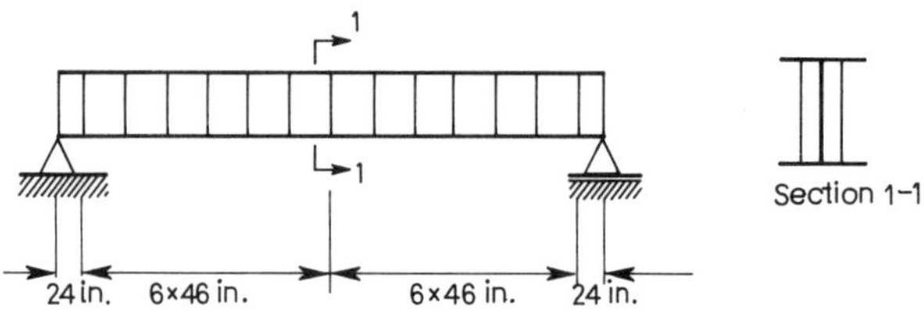

Figure 16.5. Stiffened Plate Girder Elevation of Example 16.3.

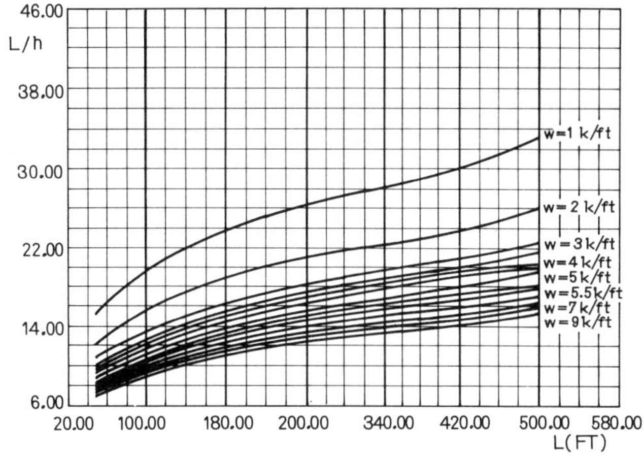

Figure 16.6. Plot of Optimal L/h versus L for $F_y = 36$ ksi [after Abuyounes and Adeli (1986)].

16.6.3. Optimization of a Modular Floor System

Templeman (1972) and Yu, Das Gupta, and Paul (1984, 1987) applied geometric programming to the optimal design of roof and floor systems. Based on the latter papers, we present the results of optimization of a modular floor system consisting of solid and voided precast slabs supported on steel I sections. Both reinforced concrete and prestressed concrete slabs are considered for comparison. Figure 16.7 shows plan and sectional views of such a floor system.

The objective functions of the geometric programming models incorporate the cost of concrete, formwork, reinforcing bars or prestressing, steel beams, and erection of the precast slabs. The constraints are obtained from considerations of both the steel beams and the slabs. For the steel beam, allowable bending, shear, web crippling and web crushing stresses, allowable deflection, minimum requirements of the ratios of flange width to flange thickness and web depth to web thickness are considered. The allowable bending and shear stresses of concrete and reinforcement steel for reinforced concrete and allowable bending stresses of concrete at transfer and service load for prestressed concrete, minimum slab thickness, minimum rib width, flange thickness, and void dimension requirement are considered for the slabs. For details of the objective functions and constraints, refer to Yu, Das Gupta, and Paul (1984), and Das Gupta, Paul, and Yu (1986).

Several dual-based algorithms can be used to solve the aforementioned design problem. SIGNOPT, developed by Templeman (1972), and used in this problem, is one such algorithm. It uses simplex linear programming to

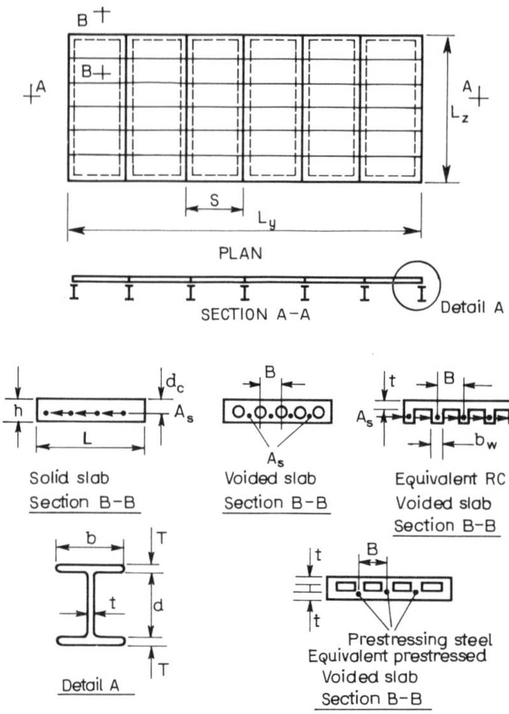

Figure 16.7. Plan and Sectional View of a Modular Floor System.

find feasible starting points and Fletcher-Reeve's conjugate gradient method to determine the approximate location of the optimum. The final optima are obtained by using the quadratic interpolation technique.

Geometric programming models have a maximum of 14 variables and 53 degrees of difficulty, and use an average of 110 s of CPU time on an IBM 3033 mainframe with an average of 14 iterations for the chosen range of imposed loads of 3 to 15 kN/m^2. The variation of the optimal values of a few design parameters is plotted against the imposed loads for the floor system with reinforced concrete solid slabs in Fig. 16.8, and that with prestressed concrete voided slabs in Fig. 16.9. The parameters are as shown in Fig. 16.7, except for Z, the section modulus of the steel beams, and P, the prestressing force. Their nonlinear relationship is clearly indicated.

16.6.4. Optimization of a Prestressed Concrete Box Girder

Based on the papers by Yu, Das Gupta, and Paul (1986, 1987), we present some results of the application of geometric programming to the

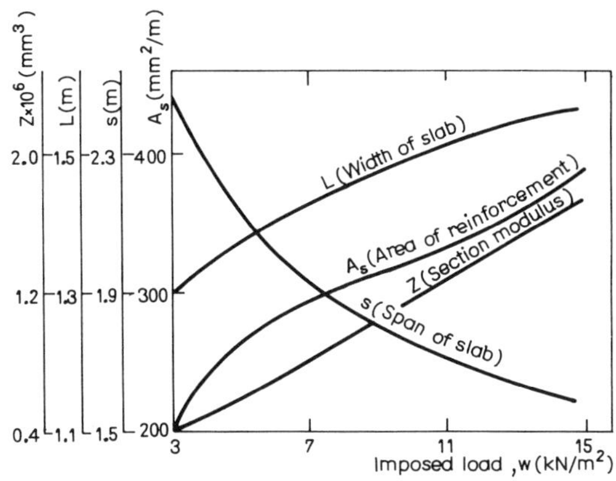

Figure 16.8. Optimal Values of Design Parameters of a Modular Floor System with RC Slabs for Various Imposed Loads [after Das Gupta, Paul, and Yu (1986)].

optimal design of a prestressed concrete box girder for a balanced cantilever bridge. Figure 16.10 shows the side elevation and sectional views of the bridge, which consists of a simply supported main span and two cantilever spans.

The algorithm SIGNOPT was again used to solve the design problem. When the analysis was performed by optimizing both sections for the simple and cantilever spans simultaneously in the original formulation, several difficulties were encountered. The analysis results in very long execution

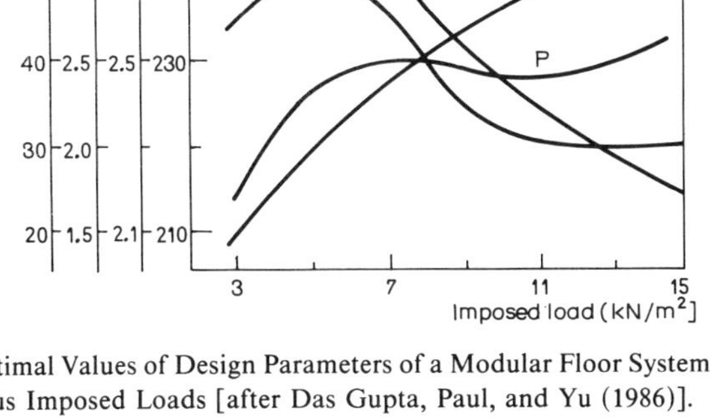

Figure 16.9. Optimal Values of Design Parameters of a Modular Floor System with Prestressed Slabs for Various Imposed Loads [after Das Gupta, Paul, and Yu (1986)].

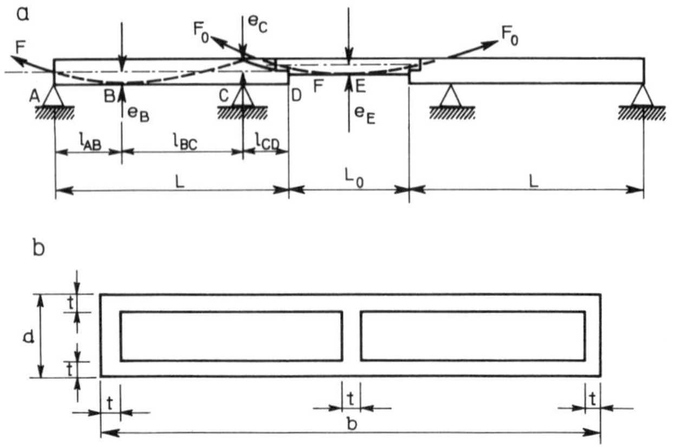

Figure 16.10. (a) Simplified Diagram of a Balanced Cantilever Bridge. (b) Cross section of a Double Box Girder.

time because of the large degree of difficulty (156), which also makes the choice of starting points difficult. Since the solution for the variables of the simple span is governed by the constraint of minimum depth, it is possible to optimize the cantilever span separately.

Thus, the original formulation was modified to that of optimizing the cantilever span only with the depth of the simple span as 1 m. The reduction in the degree of difficulty to 107 with the modification resulted in faster convergence and execution time. With the same starting point, the original formulation used 4.84 min of CPU time, whereas the modified formulation took only 1.062 min on the IBM 3033 computer.

The objective functions incorporate the cost of concrete, prestressing, and formwork. The constraints are obtained from considering the allowable bending stresses at transfer and at service load, section adequacy for bending, allowable shear stresses, and maximum eccentricities in accordance with the British Standard (BS) Codes of Practice for the structural use of concrete. The depth of the girder for the simple span is limited to not less than 1 m for aesthetic reasons. For details of the formulation and results, see Yu, Das Gupta, and Paul (1986).

16.7. Further Applications of Geometric Programming

As mentioned in Section 16.1, the initial stimulus for the development of geometric programming was the work by Zener in engineering design and cost minimization. Next, geometric programming was applied in optimal structural design by Templeman (1970, 1971, 1972, 1976) and Morris (1972,

1974, 1978, 1982). They considered examples concerned with the optimization of tower design, steel floor, and roof systems, reinforced concrete beams, and a vertically corrugated transverse bulkhead of an oil tanker. Another example concerned with optimization of a multistage heat exchanger was considered by Avriel and Williams (1971).

Applications in mechanical engineering are concerned with a machining economics problem [Ermer (1971)], optimal journal bearing design [Unklesbay, Staats, and Creighton (1973)], dry-type natural draft cooling towers [Ecker and Wiebking (1978)], statistical mechanics [Duffin and Zener (1970)], optimal control [Jefferson and Scott (1978b), Salukvadze (1979)], and optimal design of helical springs [Mancini and Piziali (1976)].

Further applications in civil engineering and nuclear engineering are concerned with optimal limit state design of highway bridge superstructures [Edwards (1975)], optimal design of pitched laminated wood beams [Avriel and Barrett (1978)], welded structures [Ragsdell and Phillips (1976)], cofferdam design optimization [Dinkel and Kochenberger (1974), Neghabat and Stark (1973)], optimizing nuclear systems [Bouchey, Beightler, and Koen (1971)], and global optimization for a cooled reactor [Wilde (1976)].

For a bibliography listing many other applications, see Rijckaert and Martens (1978b). Also, Jefferson and Scott (1978a) has an extended bibliography on applications. Fedorowicz and Mazumdar (1968), Wilde (1978), and Zener (1971) contain further engineering applications.

17

Dynamic Programming

17.1. Introduction

The theory of dynamic programming (DP) was developed by Richard Bellman (1957) in the early 1950s. Based on a recursion principle, known as the Bellman principle of optimality, dynamic programming has been quite successful in dealing with optimization problems that can be represented as multistage decision processes—i.e., processes in which a sequence of decisions has to be made, each decision affecting the state of the system and the final result. In structural design this kind of problem is often encountered, for example, when the structure to be optimized consists of a number of components in a serial interaction scheme, as in the case of a bridge whose spans are carried by supports that, in turn, rest on foundations. For a truss or a frame, the components may be individual bars or groups of bars.

The DP technique exploits this sequential feature of a structure to solve the optimization problem in a sequence of steps. In the first step, only the last component of the n-component serial structure is considered, in the second the last two are optimized, and so on. The solution of every successive step uses the result of the preceding step. The optimal solution to the original problem is obtained in the nth step, in which all components are included.

In this way an n-variable design problem can be reduced to a sequence of n single-variable subproblems, which are generally much easier to solve

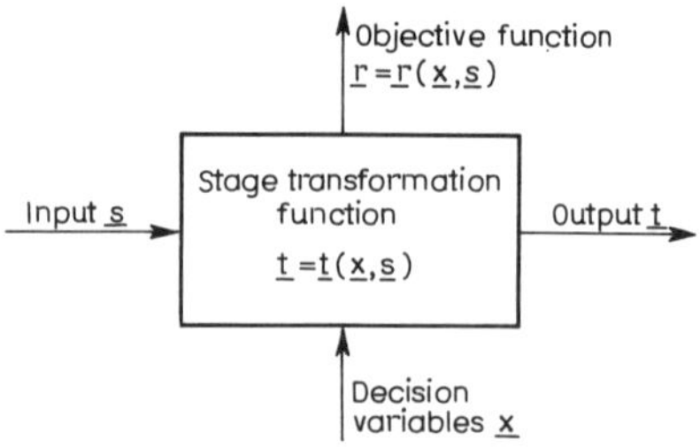

Figure 17.1. Variables in a Single-Stage Decision Process.

than the original problem. The optimization technique to be used in different steps depends on the form of the objective function and constraints. It may be a method of differential calculus, linear or nonlinear programming, or any other technique.

Dynamic programming can handle discrete variables, and it allows objective functions to take complex form, not necessarily convex, differentiable, or continuous. In such cases, we use an integer programming or random search method to solve the successive subproblems.

For a comprehensive exposition of DP theory and examples of its application, see Bellman and Dreyfus (1962), Hadley (1964), Nemhauser (1966), Jacobs (1967), and Larson (1968). Important chapters on this subject are also included in Rao (1978), Kirsch (1981), and Brandt *et al.* (1984).

17.2. Multistage Decision Processes

A typical multistage decision process considered in dynamic programming is a series of single-stage processes representable as shown in Fig. 17.1. A single-stage decision process is characterized by input data, or input state variables, **s**, decision design variables **x**, output state variables **t**, and a return objective function $r = r(\mathbf{x}, \mathbf{s})$ that measures the effectiveness of the decision. The output is related to the input by a stage transformation function (design equation) $\mathbf{t} = \mathbf{t}(\mathbf{x}, \mathbf{s})$.

A scheme of a sequential multistage decision process is shown in Fig. 17.2. The input and the output state vectors in the ith stage are denoted by

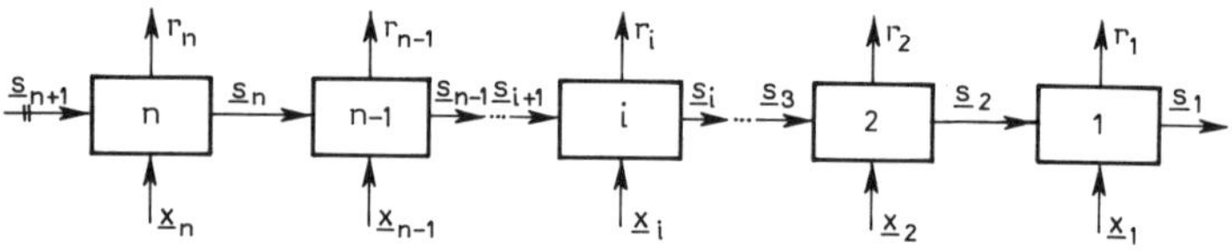

Figure 17.2. Multistage Decision Process.

$\mathbf{s}_{i+1}$ and $\mathbf{s}_i$, respectively (since the arrangement is sequential, the output of stage $i + 1$ is the input of stage i), so the stage transformation and the return function for the ith stage can be written

$$\mathbf{s}_i = \mathbf{t}_i(\mathbf{x}_i, \mathbf{s}_{i+1}), \tag{17.1}$$

$$r_i = r_i(\mathbf{x}_i, \mathbf{s}_{i+1}), \tag{17.2}$$

where $\mathbf{x}_i$ is the decision vector at stage i.

In an optimization problem associated with this process, an overall objective function may be expressed as a function of the individual returns, namely $F(r_1, r_2, \ldots, r_n)$, and the problem is to find the decision vectors $\mathbf{x}_1, \ldots, \mathbf{x}_n$ that minimize F and satisfy Eqs. (17.1) and (17.2). The form of the function F with regard to $r_1, \ldots, r_n$ is crucial for the feasibility of the DP method, whose essential idea is to decompose the original multistage problem into a series of single-variable problems. This will indeed be possible if the objective function F is separable with respect to r_i—for example, when F takes the form of a sum

$$F = \sum_{i=1}^{n} r_i(\mathbf{x}_i, \mathbf{s}_{i+1}) \tag{17.3}$$

or a product

$$F = \prod_{i=1}^{n} r_i(\mathbf{x}_i, \mathbf{s}_{i+1}). \tag{17.4}$$

With regard to the preassigned data, a multistage decision problem may be

(i) an initial value problem if the initial state vector $\mathbf{s}_{n+1}$ is given;
(ii) a final value problem if the final state vector $\mathbf{s}_1$ is given;
(iii) a boundary value problem if both the initial and the final state vectors are prescribed.

17.3. Bellman's Principle of Optimality

Different components in a serial structure are not independent of each other and cannot be optimized individually in an arbitrary order. A classical approach to the problem is to consider all the components jointly and to optimize the objective function with respect to all the variables involved. As mentioned in the introduction, dynamic programming solves the problem in a sequence of steps, beginning from an optimization of the last component of the series, and gradually expanding to involve, one by one, all the components (see Fig. 17.3). At each step, and this is the pivotal point of the method, we consider only those designs that contain optimal substructures found in the preceding steps. In terms of a decision process, this principle of optimality was originally stated by Bellman as follows: "An optimal policy, i.e., an optimal set of decisions, has the property that, whatever the initial state and initial decision are, the remaining decisions must constitute an optimal policy with regard to the state resulting from the first decision."

To show how this principle governs the process of optimization, we consider an initial value multistage problem with the objective function of form (17.3). We shall assume for simplicity that the decision design variables x_i and state variables s_i are all scalars, although the theory is equally applicable when they are vectors. We thus want to find the values of $x_1, \ldots, x_n$ that minimize

$$F = r_n(x_n, s_{n+1}) + r_{n-1}(x_{n-1}, s_n) + \cdots + r_1(x_1, s_2), \qquad (17.5)$$

where the state and the decision variables are related by

$$s_i = t_i(x_i, s_{i+1}), \qquad i = 1, 2, \ldots, n, \qquad (17.6)$$

and the initial input s_{n+1} is given.

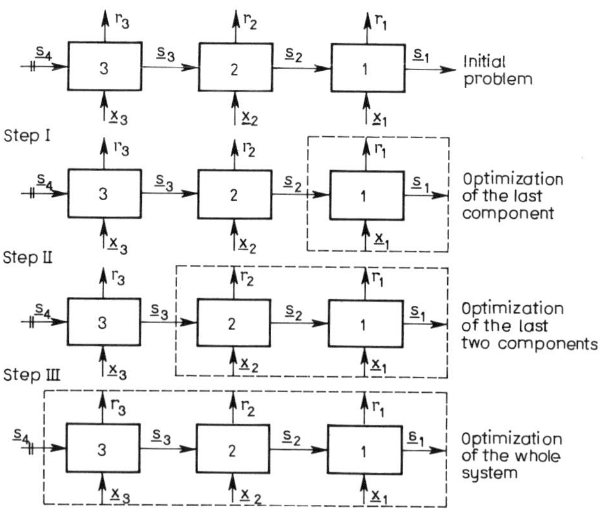

Figure 17.3. Dynamic Programming Method.

We start by minimizing the last component with respect to x_1; i.e., we find

$$\hat{f}_1(s_2) = \min_{x_1}\{r_1(x_1, s_2)\}, \tag{17.7}$$

which is the function of the yet unknown input s_2. Next, we optimize the last two components jointly; i.e., we seek a minimum of the sum $r_2 + r_1$:

$$\hat{f}_2(s_3) = \min_{x_1, x_2}\{r_2(x_2, s_3) + r_1(x_1, s_2)\}. \tag{17.8}$$

By the principle of optimality, x_1 must be chosen to minimize r_1 for a given s_2. Hence, according to Eq. (17.7), we can write

$$\hat{f}_2(s_3) = \min_{x_2}\{r_2(x_2, s_3) + \hat{f}_1(s_2)\}. \tag{17.9}$$

Using the equation of state (17.6), we can rewrite Eq. (17.9) as

$$\hat{f}_2(s_3) = \min_{x_2}\{r_2(x_2, s_3) + \hat{f}_1(t_2(x_2, s_3))\}, \tag{17.10}$$

from which it is apparent that, for a specified input s_3, the second-step optimization is also a single-variable problem.

We continue in a similar manner, involving one by one the successive components and solving a single-variable optimization problem in each step. In the ith step we seek the minimum

$$\hat{f}_i(s_{i+1}) = \min_{x_i, x_{i-1}, \ldots, x_1}\{r_i(x_i, s_{i+1}) + r_{i-1}(x_{i-1}, s_i) + \cdots + r_1(x_1, s_2)\}, \tag{17.11}$$

which, by the Bellman principle, we can write as

$$\hat{f}_i(s_{i+1}) = \min_{x_i}\{r_i(x_i, s_{i+1}) + \hat{f}_{i-1}(s_i)\}, \tag{17.12}$$

where $\hat{f}_{i-1}$ denotes the optimum found in step $i - 1$ and s_i is the input to stage $i - 1$. In the last step, the recursion formula (17.12) gives the optimal value $\hat{f}_n = \min F$ and the optimal decision x_n in terms of the initial input s_{n+1}, which is known (in an initial value problem). By retracing the steps down to the last component, we then find the complete set of optimal decisions, $\hat{x}_n$, $\hat{x}_{n-1}, \ldots, \hat{x}_1$ (see next section).

Let us stress again that, owing to the Bellman principle of optimality, the original problem (17.5)–(17.6), which would normally require a simultaneous variation of n design variables, can be solved in a sequence of n steps, each involving only one decision variable.

17.3.1. Optimal Design of a Truss Structure (Example)

We wish to determine the cross-sectional areas of the bars of the truss in Fig. 17.4, such that the volume of the truss is minimum and the downward displacement δ_A under the load P is 5 mm, when the load P is given as 250 kN, the Young's modulus of the material is $E = 200\ \text{kN/mm}^2$, and the member lengths are $l_1 = l_3 = l_5 = l_7 = 1.0$ m and $l_2 = l_4 = l_6 = 1.2$ m.

The forces induced in the bars of the truss by the load P are readily calculated as

$$N_1 = N_7 = -P/1.6, \qquad N_3 = N_5 = P/1.6,$$
$$N_2 = N_6 = 3P/8, \qquad N_4 = -3P/4.$$

Hence, if x_i $(i = 1, \ldots, 7)$ denotes the cross-sectional area of bar i, the deflection at point A is

$$\delta_A = \sum_{i=1}^{7} \frac{\partial}{\partial P}\left(\frac{N_i^2 l_i}{E x_i}\right) = 2\left(\frac{488}{x_1}\right) + 2\left(\frac{211}{x_2}\right) + 2\left(\frac{488}{x_3}\right) + \frac{844}{x_4}.$$

The optimization problem can now be formulated as follows: find the values of x_1, x_2, x_3, and x_4 that minimize the volume

$$V(x_1, x_2, x_3, x_4) = 1000F(x_1, x_2, x_3, x_4) = 1000(2x_1 + 2.4x_2 + 2x_3 + 1.2x_4)$$

subject to the constraint

$$976/x_1 + 422/x_2 + 976/x_3 + 844/x_4 = 5$$

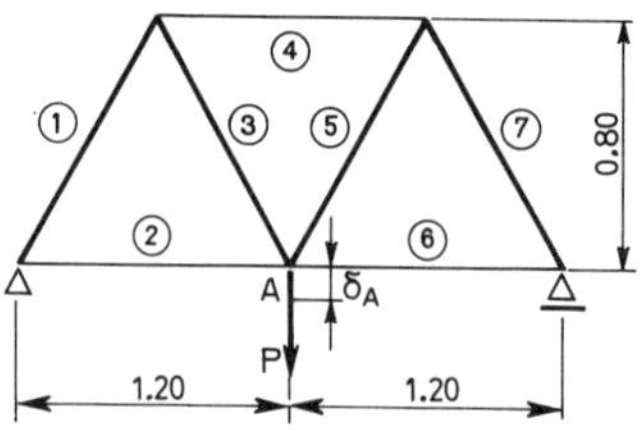

Figure 17.4. Seven-Bar Truss.

and

$$x_i \geqslant 0, \qquad i = 1, 2, 3, 4.$$

Due to the form of the objective function F and the constraint equation, in which the successive terms represent the "contributions" from the pairs of symmetrical members 1 and 7, 2 and 6, 3 and 5, and from member 4, the problem can be regarded as a multistage decision process, as shown in Fig. 17.5, where $s_5 = 5$ mm is the total displacement at point A available for "distribution" among all the components, s_4 is the displacement left after allocation at stage 4, etc.

The solution proceeds as follows. In the first step, we consider component 1 (bars 1 and 7) and find the minimum of the return function r_1:

$$\hat{f}_1(s_2) = \min_{x_1}\{r_1(x_1, s_2)\} = \min_{x_1}\{2x_1(s_2)\},$$

where

$$s_2 = 976/x_1$$

is the unknown displacement at A to be produced by the displacements of bars 1 and 7. We obtain

$$\hat{x}_1 = 976/s_2 \qquad \text{and} \qquad \hat{f}_1 = 1952/s_2.$$

In the second step, we consider components 1 and 2 and seek the minimum of $r_1 + r_2$. According to the optimality principle, the minimum is

$$\hat{f}_2(s_3) = \min_{x_2}\{r_2(x_2, s_3) + \hat{f}_1(s_2)\} = \min_{x_2}\{2.4x_2(s_3) + 1952/s_2\},$$

where s_3, the unknown displacement at A to come from bars 1, 7 (component 1) and 2 and 6 (component 2), is related to s_2 and x_2 by

$$s_3 = s_2 + 422/x_2.$$

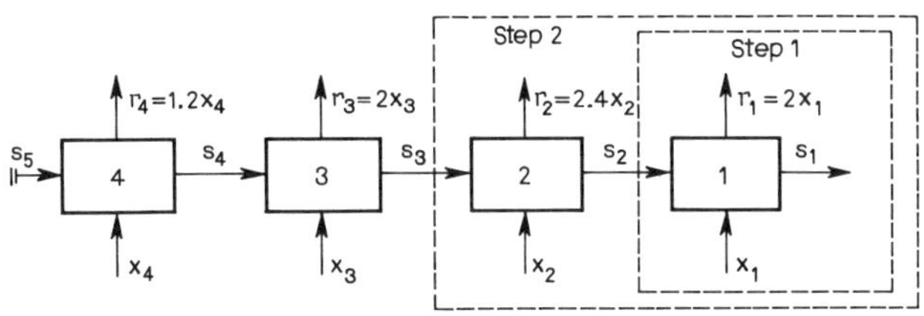

Figure 17.5. Successive Stages of the Truss Problem.

Hence,

$$s_2 = s_3 - 422/x_2,$$

so $\hat{f}_2(s_3)$ can be written as

$$f_2(s_3) = \min_{x_2}\left\{2.4x_2 + \frac{1952}{s_3 - 422/x_2}\right\}.$$

For any fixed s_3, differential calculus gives the minimum point as

$$\hat{x}_2 = 1008/s_3,$$

so

$$\hat{f}_2 = 5777/s_3.$$

A similar procedure at steps 3 and 4 yields

$$\hat{x}_3 = 2655/s_4, \qquad \hat{f}_3 = 14445/s_4,$$

and

$$\hat{x}_4 = 4031/s_5, \qquad \hat{f}_4 = 23107/s_5.$$

Since the initial state variable s_5, or the displacement due to all the members, is 5 mm, we have

$$\hat{x}_4(5) = 806.2 \text{ mm}^2 \qquad \text{and} \qquad \min F = \hat{f}_4(5) = 4622 \text{ mm}^3.$$

We now calculate the optimal values of the remaining design variables:

$$s_4 = 5 - 844/806.2 = 3.95 \text{ mm},$$

$$\hat{x}_3 = 672.2 \text{ mm}^2, \qquad s_3 = 3.95 - 976/672.2 = 2.5 \text{ mm},$$

$$\hat{x}_2 = 403.2 \text{ mm}^2, \qquad s_2 = 2.5 - 422/403.2 = 1.45 \text{ mm},$$

$$\hat{x}_1 = 673.1 \text{ mm}^2.$$

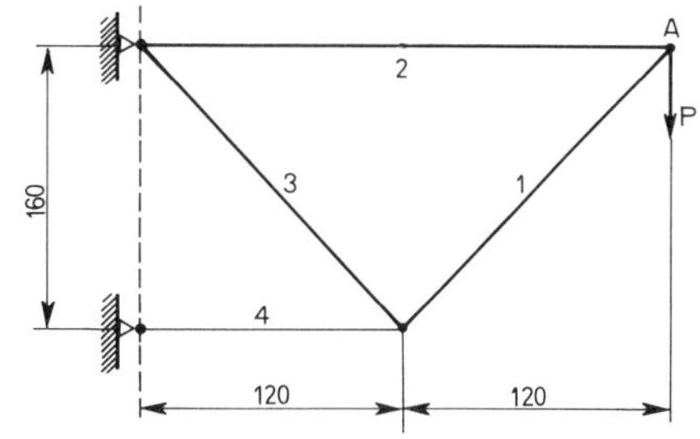

Figure 17.6. Four-Bar Truss.

The minimum volume of the truss is $V = 1000(\min F) = 4.622 \times 10^6\ \text{mm}^3$.

Exercise 17.1

Determine the cross-sectional areas of the bars of the truss in Fig. 17.6 such that the total weight of the truss is a minimum and the vertical deflection at point A is 0.01 m. Assume that the stresses induced are in the elastic range. The material of the truss has a density of $10^4\ \text{kg/m}^3$ and a Young's modulus of 2×10^5 MPa. The load P is 400 kN.

17.4. The Variant Method of Dynamic Programming

Our chief computational problem in implementing the procedure is to construct the successive functions $\hat{f}_i(s_{i+1})$. In particularly simple problems, $\hat{f}_i$ can be represented analytically, but even moderate complexity of the return functions r_i soon leads to extremely complex expressions for $\hat{f}_i$. Besides, the function r_i may be given in tabular form or by graphs, in which case an exact analytical treatment is impossible. A simple alternative is the *variant method.* As we have seen, dynamic programming begins by optimizing the last component, i.e., finding

$$\hat{f}_1(s_2) = \min_{x_1}\{r_1(x_1, s_2)\}. \tag{17.13}$$

Both $\hat{f}_1$ and the value $\hat{x}_1$ that realizes this minimum depend on the input state variable s_2, which is unknown at this stage. In the variant methods, instead of expressing $\hat{f}_1$ and $\hat{x}_1$ as functions of s_2, we limit s_2 to a set of possible values, or variants, and tabulate or plot $\hat{f}_1(s_2)$ and $\hat{x}_1(s_2)$ for each of them. In the second step, we similarly treat s_3 and tabulate or plot $\hat{f}_2(s_3)$ and $\hat{x}_2(s_3)$, using the results of the first step and the optimality principle (17.9). We proceed in this manner until stage n is reached. Since the initial input s_{n+1} is assumed to be given, we obtain a definite optimal value $\hat{f}_n(s_{n+1}) = \min F$ and a definite $\hat{x}_n(s_{n+1})$.

Now we need to return "downstream" and "collect" the optimal decision variables $\hat{x}_i$, as follows. Optimization in the nth step has determined $\hat{x}_n$ and $\hat{f}_n$ for the given s_{n+1}. Using the design equation

$$s_n = t_n(s_{n+1}, \hat{x}_n), \tag{17.14}$$

we can compute the input $\hat{s}_n$ to stage $n-1$. The tabulated values of $x_{n-1}(s_n)$ then show the correct variant $\hat{x}_{n-1}$.

We next use the design equation

$$s_{n-1} = t_{n-1}(s_n, \hat{x}_{n-1}) \tag{17.15}$$

to calculate the input $\hat{s}_{n-1}$ to stage $n-2$, and we again find the decision variant $\hat{x}_{n-2}$ corresponding to $\hat{s}_{n-1}$ from the tabulated values of $x_{n-2}(s_{n-1})$. This procedure is continued until $\hat{s}_2$ and the corresponding $\hat{x}_1$ are determined. The vector $(\hat{x}_1, \hat{x}_2, \ldots, \hat{x}_n)$ so found constitutes the optimal solution of the problem. We illustrate this procedure on a simple example of a continuous beam optimization.

17.4.1. Optimal Design of Beams and Frames (Examples)

Following Cammaert (1971), we consider the optimal design of a multispan beam built from a limited available number of sections for each span. To apply DP formalism, we choose a sequential design for each span. Suppose the cost of each span is related to the bending moments in the ends of that span and that total cost is the sum of the span costs (see Volume 1).

The three-span beam of Fig. 17.7a carries central point loads (whose magnitudes include a suitable load factor). A general moment distribution is shown in Fig. 17.7b. The three support moments M_1, M_2, M_3 are redundant. From the lower-bound theorem of plasticity a safe design is one for which the largest moment in each span does not exceed the full plastic moment. The optimal design problem then consists of finding values for M_1, M_2, M_3 that minimize the total cost. Figure 17.7c shows five economical light sections in the expected range of plastic moment values. In each case the costs of using each section are assumed to be proportional to the product of unit weight and span length; other, more complicated, cost functions can easily be substituted.

Using DP jargon, we call the support moments M_j, which describe the interaction between each span, *state variables.* At any stage of the process, or any span in this case, several decisions about the grid values of the state variables are possible. As shown in Fig. 17.7d, a path linking any number of grid values is called a *subpolicy.* A path linking all state variables, or a *policy*, constitutes an admissible design. As explained in the previous section, the optimal design process must start with the first span. We must choose a section such that the full plastic moment Q_1 is found by

$$Q_1 \geqslant \max(|150 - 0.5M_1|, |M_1|).$$

For $M_1 = 66$ the maximum moment is 117 and the corresponding cost is 700 (product of unit weight 35 and span length 20). When we evaluate the costs for the other values of M_1, we see that the minimum cost of span 1 occurs at $M_1 = 94$ or 109. We now use these results to find the optimal costs for two spans in terms of M_2.

If $M_2 = 66$, for instance, the cost of span 2 for a given value of M_1 is determined by

$$Q_2 \geqslant \max\{|M_1|, |100 - 0.5(M_1 + M_2)|, |M_2|\}.$$

The minimum cost of span 1 for a value of M_1 has already been established. Hence, the minimum cost of spans 1 and 2 is simply the least of the sums of these two costs. From Fig. 17.7d the minimum cost for $M_2 = 66$ is 1200

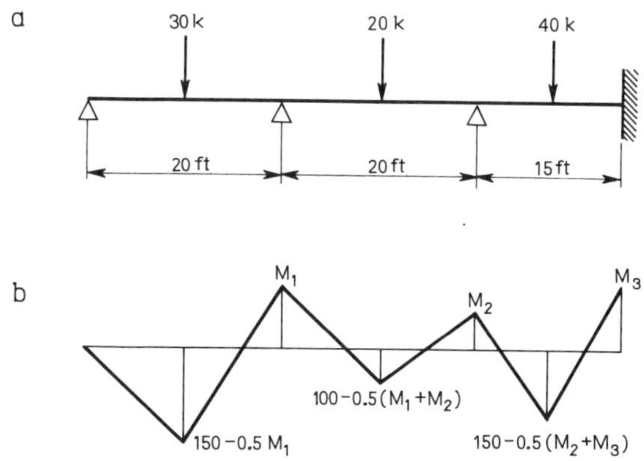

c

UB sections	12×5	12×6.5	14×6.75	14×6.75	15×6
weight/ft, lbs	25	27	30	34	35
Z_p, in^3	32.9	38.0	47.1	54.5	58.5
Q_p, kip-ft	66	76	94	109	117

Figure 17.7. Variant Method for Three-Span Continuous Beam [after Cammaert (1971)].

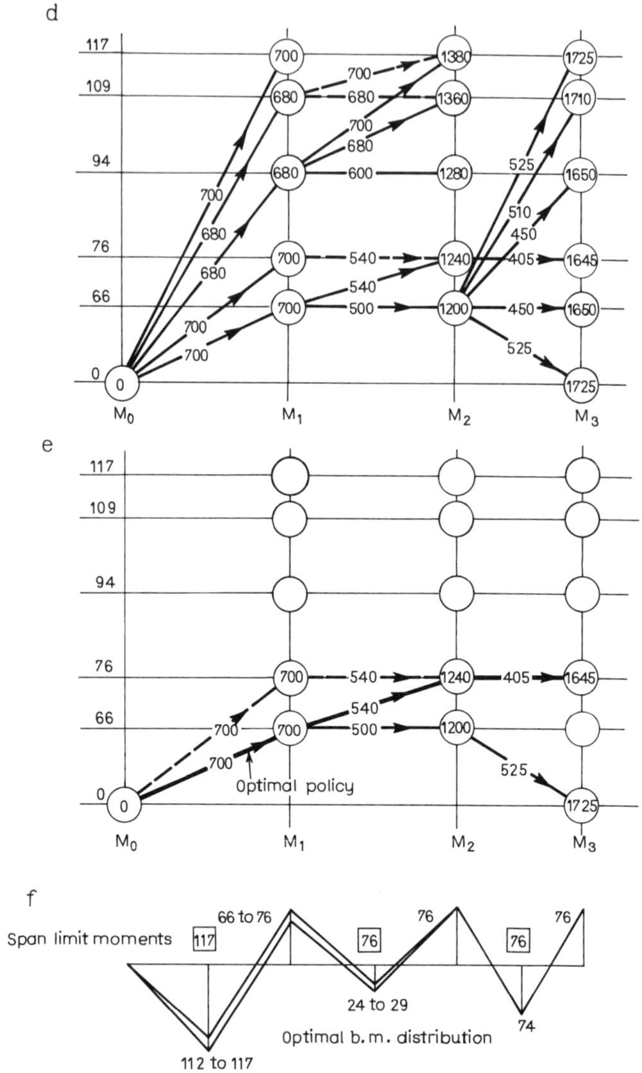

Figure 17.7. Continued.

when $M_1 = 66$. These operations are repeated for all other values of M_2. Occasionally, alternative paths give the same minimum cost, as shown by the broken lines. Now we repeat the whole process for the next stage, giving the minimum total cost of three spans in terms of M_3.

Inspecting Fig. 17.7d, we see that the design defined by $M_3 = 76$ gives the least overall cost of 1645. Retracing the paths or subpolicies to this point, we get $M_2 = 76$, $M_1 = 66$ or 76. These moments complete the optimal policy (Fig. 17.7e). If the beam was simply supported at its right end (i.e., $M_3 = 0$), the corresponding minimum weight design would be 1725. Therefore, we can reduce cost by fixing the right end. The optimal policy for $M_3 = 0$ can also be retraced as shown in Fig. 17.7e. A whole "family" of designs near the optimal design can be investigated by the DP algorithm.

The use of previous results forms the backbone of dynamic programming. It also provides a very efficient search process, since at any stage all unfavorable combinations of moment values are eliminated.

We now consider the general problem of the optimal design of beams and frames, using a recurrence relationship. In the previous example only one state variable was involved. The relationships discussed here extend into multivariable problems. The way in which the subproblems are imbedded in the larger problem must be defined rigorously for dynamic programming to work.

Let us consider span r of a continuous beam (Fig. 17.8). For any loading system, the largest moment in span r can always be found as some function of M_{r-1} and M_r. All other parameters will usually be constant, so the cost of span r can be described as $G_r(M_{r-1}, M_r)$. The total cost of n spans is assumed to be the sum of the individual span costs. The minimum total cost is denoted $T_n(M_n)$. The sum of the costs for n spans is then

$$G_1(0, M_1) + G_2(M_1, M_2) + \cdots + G_n(M_{n-1}, M_n),$$

and the minimum total cost is

$$T_n(M_n) = \min_{\substack{M_i \\ i=1,2,\ldots,n-1}} \{G_1(0, M_1) + G_2(M_1, M_2) + \cdots + G_n(M_{n-1}, M_n)\}. \tag{17.16}$$

If the cost of the last span is distinguished from the rest, the summation can be rewritten as

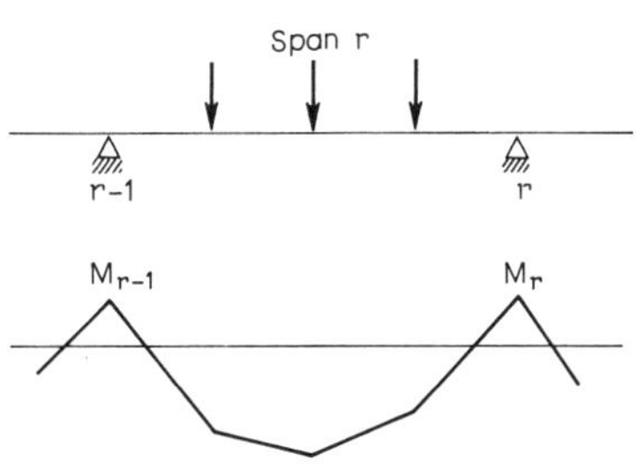

Figure 17.8. Typical Span of a Continuous Beam.

$$T_n(M_n) = \min_{M_{n-1}}\{ \min_{\substack{M_i \\ i=1,\dots,n-2}} (G_1(0, M_1) + G_2(M_1, M_2) + \cdots + G_n(M_{n-1}, M_n))\}. \quad (17.17)$$

The cost G_n does not depend on $M_1, M_2, \ldots, M_{n-2}$, so it can be removed from the inner summation:

$$T_n(M_n) = \min_{M_{n-1}}\{G_n(M_{n-1}, M_n) + \min_{\substack{M_i \\ i=1,\dots,n-2}} (G_1(0, M_1) + G_2(M_1, M_2) + \cdots + G_{n-1}(M_{n-2}, M_{n-1}))\}. \quad (17.18)$$

Comparing the minimization of $n - 2$ spans with Eq. (17.5), we rewrite Eq. (17.7) as

$$T_n(M_n) = \min_{M_{n-1}}\{G_r(M_{n-1}, M_n) + T_{n-1}(M_{n-1})\}. \quad (17.19)$$

For span 1 this expression yields

$$T_1(M_1) = G_1(0, M_1),$$

since $M_0 = 0$; for the second span,

$$T_2(M_2) = \min_{M_1}\{G_2(M_1, M_2) + T_1(M_1)\}.$$

These relationships outline exactly the steps taken in the previous example.

When the ideas of dynamic programming are extended to the design of multistory frames, the recurrence relationship again applies. Figure 17.9

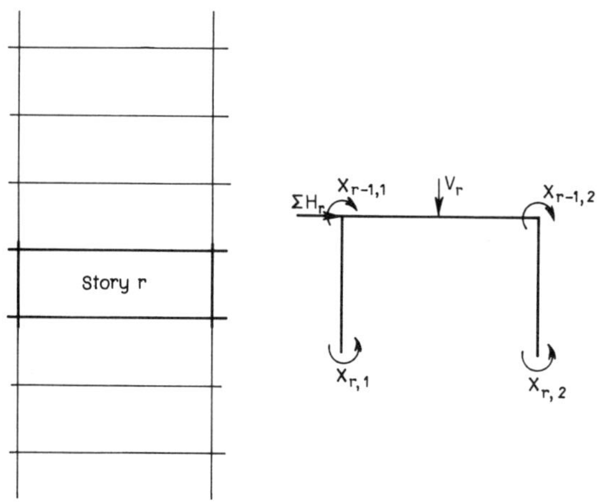

Figure 17.9. Typical Story of Multistory Frame.

shows story r of a multistory frame; the logical choice for the subdivision of the overall problem is the individual story. We assume that simple plastic theory still holds, and we ignore instability and axial force effects. The interaction of story r and story $r-1$ is then completely described by the vector bending moments $X_{r-1,j}$ at the bottom of the columns of story $r-1$ (where j is the column number). The horizontal and vertical moments on each story are assumed fixed. A major difference from the continuous-beam problem is that the moments in story r are not determined uniquely by the interaction moments $X_{r-1,j}$ and $X_{r,j}$. Determining the cost of story r poses a minimization problem within a minimization problem. The redundants for a story have to be chosen such that the cost of story r is minimized. A simplified problem is given when the redundants are assigned arbitrary values; the cost of story r can then be determined uniquely.

Assuming that $G_r(X_{r-1,j}, X_{r,j})$ can be found either by an "interior minimization" process or by some rule, we can write the recurrence relationship for multistory frames as

$$T_r(X_{r,j}) = \min_{X_{r-1,j}} \{G_r(X_{r-1,j}, X_{r,j}) + T_{r-1}(X_{r-1,j})\}, \qquad j = 1, 2, \ldots, m. \tag{17.20}$$

Beginning at the top story, we find the costs $T_1(X_{1,j})$ directly, and use them to find $T_2(X_{2,j})$, and so on, until we obtain the optimal cost of n stories in terms of $X_{n,j}$, the moments at the base of the columns of the bottom story.

17.5. Final to Initial Value Problem Conversion

The theory presented in previous sections dealt with an initial value problem, i.e., one in which the input s_{n+1} to stage n is known. In a final value problem, on the other hand, we are given the final output s_1. To solve such a problem, we convert it to an equivalent initial value problem by interchanging the roles of input and output state variables. This can be done if the design equation of the original problem,

$$s_i = t_i(x_i, s_{i+1}), \qquad i = 1, 2, \ldots, n, \tag{17.21}$$

has an inverse relation

$$s_{i+1} = \bar{t}_i(x_i, s_i), \qquad i = 1, 2, \ldots, n, \tag{17.22}$$

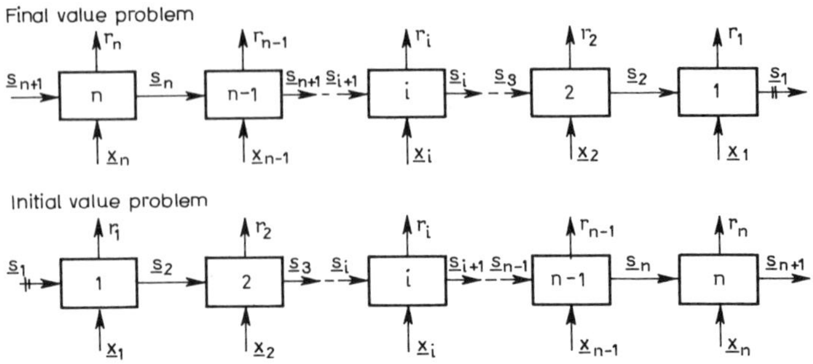

Figure 17.10. Reversing the Flow of Information.

that expresses the input to stage i as a function of its output and decision variable. Replacing Eq. (17.21) by Eq. (17.22) is called *state inversion.* After inversion, the return function of stage i,

$$r_i = r_i(x_i, s_{i+1}), \tag{17.23}$$

takes the form

$$r_i = r_i(x_i, \bar{t}_i(x_i, s_i)) = \bar{r}_i(x_i, s_i). \tag{17.24}$$

The optimization problem can now be formulated as follows: find the values of $x_1, x_2, \ldots, x_n$ that minimize

$$F(x_1, \ldots, x_n) = \sum_{i=1}^{n} r_i(x_i, s_i), \tag{17.25}$$

where the state and decision variables are related by Eq. (17.22). The use of Eq. (17.22) amounts to reversing the flow of information through the state variables (Fig. 17.10). The optimization process can therefore start at stage n and proceed through stages $n-1$, $n-2$, etc., until the last step stage 1 is reached, where the known value s_1 will permit us to determine the required optimum. On changing the stage numbers $1, 2, \ldots, n$ to n, $n-1, \ldots, 1$, respectively, we obtain the standard initial value procedure.

17.6. Further Applications of Dynamic Programming in Optimal Structural Design

The first applications of dynamic programming in structural design involving continuous beams and plate girders are due to Kalaba (1962),

Razani and Goble (1966), and Goble and DeSantis (1966). Porter Goff (1966, 1969) used dynamic programming to optimize the shape of truss structures. Further applications in optimal truss design were given by Palmer (1968, 1969, 1973), Sheppard (1970), Palmer and Sheppard (1970), and Sheppard and Palmer (1972). Palmer (1968) first suggested the extension of the theory to multistory frames. Further investigations in this area are due to Reitman (1968), Swiszczowski (1971, 1974), DiStefano (1972a, b), DiStefano and Rath (1975), Twisdale and Khachaturian (1973, 1975a, b), Kirsch (1981), Ho (1975), and Jendo and Marks (1983, 1984). Pochtman and Baranenko (1975) wrote the first account of applying dynamic programming to structural mechanics, specifically to optimal design of bar structures, membranes, and plates. Raj and Durrant (1976) demonstrated the usefulness of dynamic programming in the very practical problem of optimizing transmission towers that support overhead electric lines. Similar problems had earlier been investigated by Sheppard and Palmer (1972) and were later considered by Howell and Doyle (1978). Blachut (1977) used dynamic programming to optimize a flexible bar in compression, and Vitiello (1977) determined in this way the optimal number of unisection bar families in a large structural system. New possibilities of application of dynamic programming were shown by DiStefano and Samartin (1975), who used this approach to formulate and solve finite element equations. The work of DiStefano (1975) and DiStefano and Bellows (1977) on optimization of two-point boundary value systems has opened yet another area of application of the DP method.

18

Multiobjective Optimization

18.1. Introduction

The first work on multiobjective optimization was by Pareto (1896). After at least 50 years, problems of multiobjective optimization were again considered by von Neumann and Morgenstern (1947) and by Debreu (1959). A relatively modern formulation of the multiobjective optimization problem was presented by Zadeh (1963). But wider interest in optimization theory, operations research, and control theory did not occur until the late 1960s, and since then numerous papers of the subject have been published. Most are concerned with the theory and applications of multiobjective decision making from a general viewpoint, and few applications to engineering design can be found. Comprehensive bibliographies on multicriteria decision making and related areas are in Lin (1976), Hwang and Masud (1979), Stadler (1980), (1984a, b), Chankong and Haimes (1982), and Dauer and Stadler (1986). Cohon (1978), Homenjuk (1983), Yu (1985), Gerasimov, Potchman, and Skalozub (1985), and Dubov, Travkin, and Jakimec (1986) present the foundations of modern multiobjective optimization.

18.2. Problem Formulation

The problem of multicriteria optimization is to generalize a single-criterion optimization, which enables us to more closely approximate real conditions that determine the selection of the final solution. The multicriteria optimization problem can be formulated as follows:

$$\min_{\mathbf{x}\in\Omega} \mathbf{f}(\mathbf{x}), \tag{18.1}$$

where $\mathbf{f}:\Omega \to R^k$ is a vector objective function given by

$$\mathbf{f}(\mathbf{x}) = \{\mathbf{f}_1(\mathbf{x}), \mathbf{f}_2(\mathbf{x}), \ldots, f_k(\mathbf{x})\}. \tag{18.2}$$

The components $f_i:\Omega \to R$, $i = 1, 2, \ldots, k$, are called the *criteria of optimization*, and they represent those conflicting objectives by which the performance of a system is measured. The design variable vector $\mathbf{x}$ belongs to the *feasible set* $\Omega \subset R^n$, defined by equality and inequality constraints as

$$\Omega = \{\mathbf{x} \in R^n: \mathbf{h}(\mathbf{x}) = 0, \mathbf{g}(\mathbf{x}) \leqslant 0\}, \tag{18.3}$$

where $\mathbf{h}: R^n \to R^s$ and $\mathbf{g}: R^n \to R^r$. The notation $\mathbf{z} = \mathbf{f}(\mathbf{x})$ will be used in the sequel for convenience. Furthermore, the image of the feasible set in the criteria space (called here the *attainable* or *consequence* or *image set*) is defined by

$$f(\Omega) = \{\mathbf{z} \in R^k: \mathbf{z} = \mathbf{f}(\mathbf{x}), \mathbf{x} \in \Omega\}.$$

In a multicriteria optimization approach, we usually focus on the attainable set in objective space, whereas single-criterion optimization is concerned with the feasible set in design space. A characteristic feature in multiobjective optimization is that the criteria are in conflict. We can distinguish a local conflict and a global conflict. In the former only the criteria gradients are considered, whereas in the latter a feasible set is also involved. The local conflict can be defined as follows. Criteria f_i and f_j are called *collinear* with no conflict at $\mathbf{x}$ if and only if there exist $\alpha > 0$ such that

$$\nabla f_i(\mathbf{x}) = \alpha \nabla f_j(\mathbf{x}).$$

Otherwise, the criteria are called *locally conflicting at* $\mathbf{x}$. In other words, two criteria are locally conflicting at a point $\mathbf{x}$ if their maximum improvements are achieved in different directions. Consequently, the angle between the gradients of the criteria can be used as a measure of the conflict. In optimization problems, however, the global conflict that is introduced next obviously has greater importance. The global conflict can be defined in the following way. Criteria f_i and f_j are called *globally conflicting in* Ω if and only if the optimization problems

$$\min_{\mathbf{x}\in\Omega} f_i(\mathbf{x}) \qquad \text{and} \qquad \min_{\mathbf{x}\in\Omega} f_j(\mathbf{x})$$

have different solutions.

18.3. Pareto Optimal and Efficient Solutions

Usually there exists no unique solution that would give an optimum for all k criteria simultaneously. Thus a new concept of optimal solution, different from that used in scalar optimization, must be introduced in multicriteria optimization. First a partial order in R^k is generated by the negative cone

$$C = \{\mathbf{z} \in R^k\colon z_i \leqslant 0, i = 1, 2, \ldots, k\} \tag{18.4}$$

as follows:

$$\mathbf{z} \leqslant \mathbf{y} \quad \Leftrightarrow \quad \mathbf{z} - \mathbf{y} \in C. \tag{18.5}$$

Now it is possible to define an *efficient* (minimal, nondominated, noninferior) *solution* in the criterion space in the following way: $\hat{\mathbf{z}} \in f(\Omega)$ is an efficient solution in $f(\Omega) \subset R^k$ if and only if $\mathbf{z} \leqslant \hat{\mathbf{z}}$ and $\hat{\mathbf{z}} \in f(\Omega) \Rightarrow \mathbf{z} = \hat{\mathbf{z}}$.

The corresponding optimal solution in the design space is introduced as follows: a solution $\hat{\mathbf{x}} \in \Omega$ is *Pareto optimal* for problem (18.1) if and only if there exists no $\mathbf{x} \in \Omega$ such that $f_i(\mathbf{x}) \leqslant f_i(\hat{\mathbf{x}})$ for $i = 1, 2, \ldots, k$, with $f_j(\mathbf{x}) < f_j(\hat{\mathbf{x}})$ for at least one j. In other words, $\hat{\mathbf{x}}$ is a Pareto optimal solution if there is no feasible solution $\mathbf{x}$ that would decrease some criterion without causing a simultaneous increase in at least one criterion. The existence of a Pareto optimal solution is guaranteed in the case of continuous criteria and a compact feasible set, as might be expected in most structural design problems. In effect, the problem usually is not the nonexistence of Pareto optimal solutions but the great number of them. This may cause difficulties in generating the solution set and in handling the results. The basic property, following from the definition, is illustrated in Fig. 18.1, where the Pareto optimal curve in the feasible set and the corresponding efficient curve in the attainable set are schematically shown for two design variables and two criteria of optimization.

In some applications weak Pareto optimal solutions may also appear. They can be defined as follows: a solution $\hat{\mathbf{x}}$ is *weakly Pareto optimal* for problem (18.1) if and only if there exists no $\mathbf{x}$ such that $f_i(\mathbf{x}) < f_i(\hat{\mathbf{x}})$ for $i = 1, 2, \ldots, k$. Every Pareto optimal solution is also weakly Pareto optimal, but the converse is not true. This concept is illustrated in Fig. 18.2, where the flat part of the boundary of $f(\Omega)$, which is parallel to the coordinate axis, consists of merely weakly efficient solutions in the criterion space.

As with scalar optimization, the necessary conditions for an optimum for multicriteria optimization can be established. To write the necessary

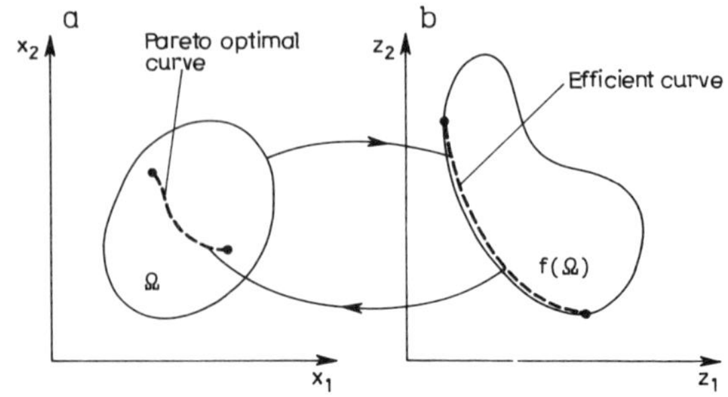

Figure 18.1. (a) Feasible Set and Pareto Optimal Curve. (b) Attainable Set and Corresponding Efficient Curve.

conditions for an optimum, we introduce the concept of a regular point (solution). A vector $\hat{\mathbf{x}}$ is called *regular* with respect to functions $\mathbf{h}$ and $\mathbf{g}$ if the set of gradients

$$\{\nabla h_i(\hat{\mathbf{x}}), \nabla g_j(\hat{\mathbf{x}}): i = 1, 2, \ldots, s, j = 1, 2, \ldots, r, j \in G(\hat{\mathbf{x}})\}, \qquad (18.6)$$

is linearly independent, where $G(\hat{\mathbf{x}})$ is the index set of active inequality constraints at $\hat{\mathbf{x}}$. The letters s and r are used for the number of equality and inequality constraints, respectively. The set of those functions defined in Ω that have continuous first partial derivatives is denoted by $C^1(\Omega)$. Now the necessary conditions for weak Pareto optimality are given as follows:

Theorem. Let $\mathbf{h}$, $\mathbf{g}$, and $\mathbf{f} \in C^1(\Omega)$. Suppose $\hat{\mathbf{x}} \in \Omega$ is a regular point with respect to the constraint functions $\mathbf{h}$ and $\mathbf{g}$. If $\hat{\mathbf{x}}$ is a local weak Pareto

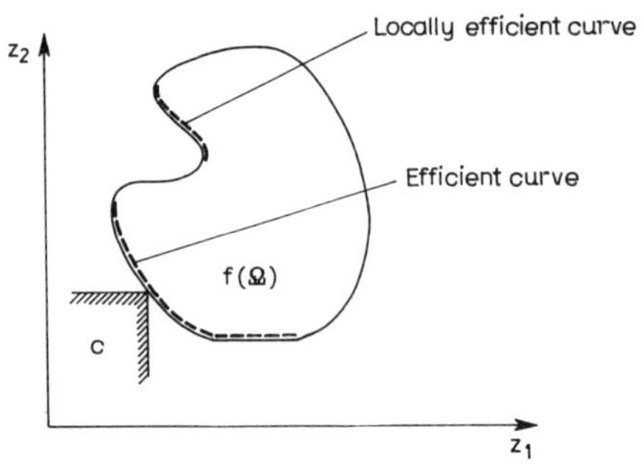

Figure 18.2. Basic Concepts of Multiobjective Optimization in Objective Space.

optimum for problem (18.1), there exist $\boldsymbol{\xi} \in R^k$, $\xi_i \geqslant 0$, $\mathbf{u} \in R^r$, $u_i \geqslant 0$, $\mathbf{v} \in R^s$, such that

$$\sum_{i=1}^{k} \xi_i \nabla f_i(\hat{\mathbf{x}}) + \sum_{i=1}^{r} u_i \nabla g_i(\hat{\mathbf{x}}) + \sum_{i=1}^{s} v_i \nabla h_i(\hat{\mathbf{x}}) = 0, \tag{18.7}$$

$$\mathbf{u}^T \mathbf{g}(\hat{\mathbf{x}}) = 0. \tag{18.8}$$

For convenience the conditions are given for weak solutions only, but they are valid for Pareto optima as well.

The geometrical interpretation of the necessary conditions is presented in Fig. 18.3, where a typical Pareto optimal point and the associated gradients are shown. Conditions (18.7) and (18.8) can be reduced to the well-known Kuhn–Tucker optimality conditions for scalar optimization by simply setting $k = 1$ and $\xi_1 = 1$. We stress that the necessary conditions given here guarantee only local Pareto optimality, as for scalar optimization as well, except for convex problems where the local and global optima agree.

Theoretically, a multiobjective optimization problem is regarded as solved as soon as the Pareto optimal set has been determined. In practical applications, however, we must order this set further, because usually the designer wants only one final solution. In such cases, additional information or a criteria are needed to choose a final (preferable) solution among Pareto optimal solutions.

Several methods are for solving nonlinear multiobjective optimization problems occur in the literature. Usually they turn the original vector optimization problem into a sequence of scalar optimization problems,

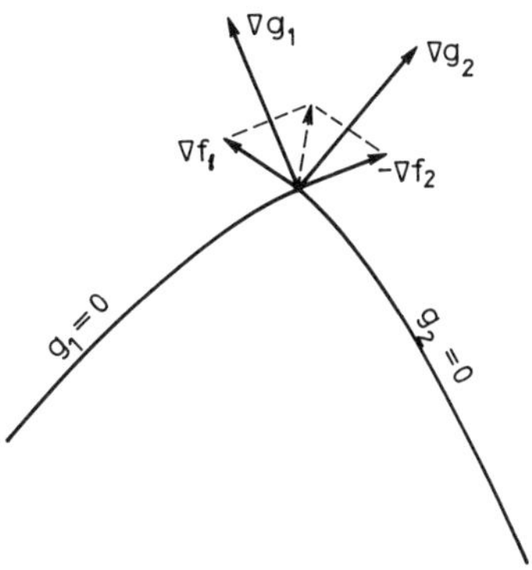

Figure 18.3. Geometrical Interpretation of Necessary Conditions for a Local Pareto Optimum in the Bicriteria Case with Two Design Variables and No Equality Constraints.

which can then be solved numerically by NLP methods. Following Koski (1984a) and Duckstein (1984), we discuss several basic techniques, namely, the *sequential* (hierarchic, lexicographic) *optimization method*, the *constraint method*, the *weighting method*, the *minimax approach*, *compromise programming*, *goal programming*, the *multiattribute utility method*, the *method of pairwise comparisons*, the *trade-off method*, and the *surrogate-worth trade-off methods*.

18.4. Methods for Generating Pareto Optimal Solutions

18.4.1. The Sequential Optimization Method

The sequential optimization method is also sometimes called *hierarchic* or *lexicographic*. It requires that we order the components of the vector objective function according to their importance. The final (preferable) solution is one that first minimizes the objective function having the most important rank, the one that has the next most important rank, and so on. If we rank importance in the hierarchy from 1 to k, then the optimization problem can be formulated as follows:

$$\min_{\mathbf{x}\in\Omega} f_1(\mathbf{x}), \tag{18.9}$$

where $\Omega = \{\mathbf{x} \in R^n: \mathbf{h}(\mathbf{x}) = 0, \mathbf{g}(x) \leqslant 0\}$.

We denote the solution of this problem as $\hat{\mathbf{x}}_1$ and $\hat{f}_1$. If this solution is unique, then $\hat{\mathbf{x}}_1$ is treated as the final (preferable) solution of the original problem. Otherwise, it is possible to solve the second optimization problem as follows:

$$\min_{\mathbf{x}\in\Omega} f_2(\mathbf{x}) \tag{18.10}$$

subject to the additional constraint

$$f_1(\mathbf{x}) = \hat{f}_1.$$

If $\hat{f}_2$ denotes the optimal value of $f_2(\mathbf{x})$ and the solution is not unique, then it is possible to apply the preceding procedure to find

$$\min_{\mathbf{x}\in\Omega} f_j(\mathbf{x}) \tag{18.11}$$

subject to additional constraints

$$f_l(\mathbf{x}) = \hat{f}_l, \qquad l = 1, 2, \ldots, j-1.$$

The procedure ends when we reach a unique solution. Objectives of the lower important ranks in the hierarchy are disregarded. This method often may not be appropriate for structural optimization problems where a unique solution frequently occurs in the first stage.

18.4.2. The Constraint Method

The constraint method is applicable provided that it is possible to choose one criterion as a scalar objective function $f_r(\mathbf{x})$ among all the possible objectives and to establish upper bounds on the rest of the objective functions. The original vector optimization problem (18.1) can then be replaced by

$$\min_{\mathbf{x}\in\Omega\cap\Omega_r(\boldsymbol{\varepsilon})} f_r(\mathbf{x}),$$

where

$$\Omega_r(\boldsymbol{\varepsilon}) = \{\mathbf{x}: f_i(\mathbf{x}) \leqslant \varepsilon_i,\ i \neq r\}$$

and

$$\boldsymbol{\varepsilon} \in E_r = \{\lfloor \varepsilon_1, \varepsilon_2, \ldots, \varepsilon_{r-1}, \varepsilon_{r+1}, \ldots, \varepsilon_k \rfloor : \Omega_r(\boldsymbol{\varepsilon}) \neq \varnothing\}.$$

The main difficulty of this method is in finding upper bounds ε_i that ensure the existence of a nonempty attainable set. By systematically varying these upper bounds on constraints, we can obtain the entire Pareto optimal set in nonconvex cases also. The constraint method may be considered as a basic numerical technique in vector optimization, and it is widely discussed in the literature.

18.4.3. Weighting Method

A commonly used method in vector optimization problems is to form one scalar objective function as a weighted sum of the criteria. One disadvantage of this approach is the difficulty in choosing the weights for the criteria. And in cases where the optimal solution is very sensitive to the variation of weights, this approach may be quite unsatisfactory. In convex multicriteria problems, however, it is possible to use the weighting method in a parametric form to determine the Pareto optimal set. If $\mathbf{w}^T = \{w_1, w_2, \ldots, w_k\}$ denotes the vector of weighting coefficients, the vector optimization problem can be formulated as follows:

$$\min_{\mathbf{x}\in\Omega} \mathbf{w}^T\mathbf{f}(\mathbf{x}),$$

where $\mathbf{w}$ can be normalized so that the sum of its components, which are nonnegative and not all zero, is equal to 1. Now Pareto optimal solutions can be generated by parametrically varying the weights w_i in the objective function. The main drawback of this method is its inability to produce the whole Pareto optimal set for nonconvex problems. This can be seen by a geometrical interpretation of the weighting method in the criteria space, where the problem may be stated as

$$\min_{\mathbf{z}\in f(\Omega)} \mathbf{w}^T\mathbf{z}.$$

Thus, with fixed $\mathbf{w}$, a linear function of the variables $\mathbf{z}$ is minimized in the attainable set. As is shown in Fig. 18.4, even a great majority of Pareto optima may not be reached by this method, since that part of the attainable set associated with the efficient (nondominated) set is not convex. But the weighting method is quite efficient computationally and can also be applied to generate the nondominated solutions to be used as initial solutions in other techniques.

18.4.4. Minimax Approach

The minimax method is based on a certain distance function introduced into the criteria space. This approach leads to a minimax problem where the largest deflection from the so-called ideal solution is minimized. The

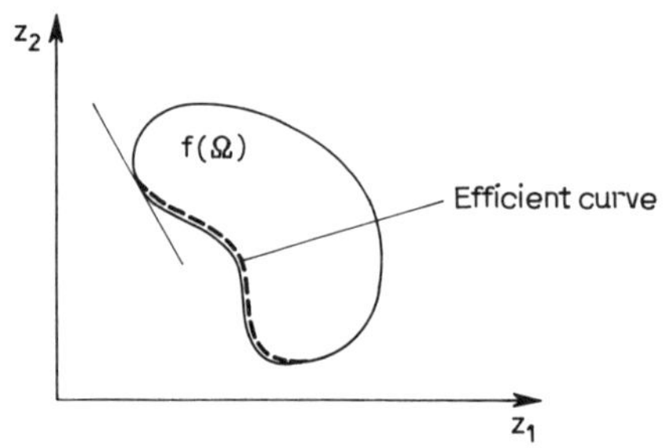

Figure 18.4. Geometrical Interpretation of the Weighting Method in the Bicriteria Case.

first step consists in finding the ideal solution, which is used as a reference point. The ideal solution is defined as follows:

$$\mathbf{z}^{\mathrm{id}} = \{\min_{\mathbf{x}\in\Omega} f_1(\mathbf{x}), \min_{\mathbf{x}\in\Omega} f_2(\mathbf{x}), \ldots, \min_{\mathbf{x}\in\Omega} f_k(\mathbf{x})\}^{\mathrm{T}}. \tag{18.12}$$

To determine this point in the criterion space, we must minimize every criterion f_i separately in the feasible set, which calls for solving k scalar optimization problems. Generally, the ideal point is not feasible; i.e., $\mathbf{z}^{\mathrm{id}} \notin f(\Omega)$. The distance between $\mathbf{z} \in f(\Omega)$ and the ideal point $\mathbf{z}^{\mathrm{id}} \in R^k$ can be measured by the metric function

$$d_\infty(\mathbf{z}, \mathbf{z}^{\mathrm{id}}) = \max_{i=1,\ldots,k} |z_i - z_i^{\mathrm{id}}|. \tag{18.13}$$

In numerical computation, however, it is convenient to deal with all criteria in a nondimensional form. Thus, a scale-invariant solution to the problem is obtained, and the numerical difficulties involved with different scales are eliminated. In the sequel a normalized vector objective function has the form

$$\tilde{\mathbf{f}}(\mathbf{x}) = \{\tilde{\mathbf{f}}_1(\mathbf{x}), \tilde{\mathbf{f}}_2(\mathbf{x}), \ldots, \tilde{\mathbf{f}}_k(\mathbf{x})\}, \tag{18.14}$$

where the components are defined by

$$\tilde{f}_i(\mathbf{x}) = \frac{f_i(\mathbf{x}) - \min f_i(\mathbf{x})}{\max f_i(\mathbf{x}) - \min f_i(\mathbf{x})}. \tag{18.15}$$

Now the values of every normalized criterion are limited to the same range; i.e., $\tilde{f}_i(\mathbf{x}) \in \{0, 1\}$. If we use the notation $\tilde{\mathbf{z}} = \tilde{\mathbf{f}}(\mathbf{x})$, we can restate the problem as looking for a vector $\tilde{\mathbf{z}} \in \tilde{f}(\Omega)$ that, in the sense of the d_∞ metric, is nearest to the ideal $\tilde{\mathbf{z}}^{\mathrm{id}} = \mathbf{0}$; it can be formulated as

$$\min_{\tilde{\mathbf{z}}\in\tilde{f}(\Omega)} d_\infty(\tilde{\mathbf{z}}, \mathbf{0}). \tag{18.16}$$

One Pareto optimal point is usually obtained as a solution to this scalar optimization problem, which is next modified to solve the multicriteria problem (18.1). If certain parameters w_i, $i = 1, 2, \ldots, k$, are used as weights for the criteria in the same way as for the weighting method, the problem can be written as

$$\min_{\mathbf{x}\in\Omega} \max_i \{w_i \tilde{f}_i(\mathbf{x})\}. \tag{18.17}$$

Now the whole Pareto optimal set may be generated by parametrically varying the weights in the objective function. This method is illustrated in

Fig. 18.5, which shows that the minimax approach can be interpreted as finding that point $\hat{\mathbf{z}} \in f(\Omega)$ where the rectangular level set just touches $f(\Omega)$. The corresponding vector $\hat{\mathbf{x}} \in \Omega$ is the Pareto optimum actually needed by the designer. The geometrical interpretation of the weight variation is that of changing the ratio of sides of the rectangle, which in turn may be viewed as an ideal central ball generated by the d_∞ metric in the criteria space. From this discussion it should be clear that, by using the minimax approach, we can obtain all Pareto optima regardless of the nonconvexity of the problem.

One possibility of solving problem (18.17) with fixed weights numerically is to consider the following scalar formula, typical for the minimax approach:

$$\min_{\gamma, \mathbf{x} \in \Omega} \gamma, \tag{18.18}$$

subject to constraints

$$w_i \tilde{f}_i(\mathbf{x}) \leq \gamma, \qquad i = 1, 2, \ldots, k.$$

18.4.5. Compromise Programming

The weighting method, which is a suitable solution technique for convex problems, and the minimax approach, applicable also to nonconvex problems, may both be considered as members of a larger algorithm family called l_p *norm methods*. This is a collection of methods that can be obtained by using the distance function d_p induced by the l_p norm:

$$l_p(\mathbf{x}) = \left[\sum_{i=1}^{k} w_i^p \left| \frac{f_i(\mathbf{x}) - \min f_i}{\max f_i - \min f_i} \right|^p \right]^{1/p}, \tag{18.19}$$

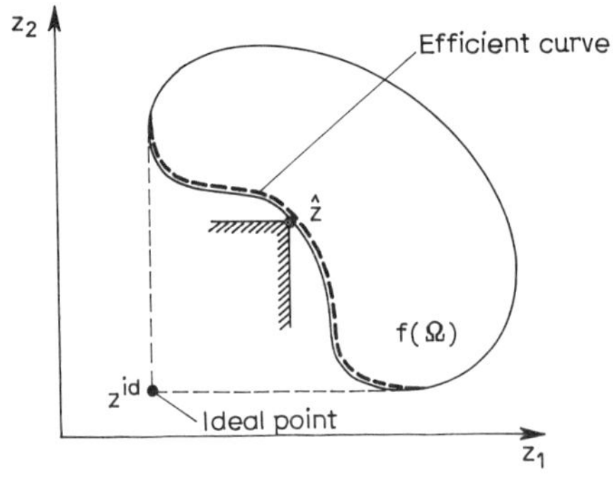

Figure 18.5. Geometrical Interpretation of the Minimax Approach in the Bicriteria Case.

where w_i are the weights, $\min f_i$ is the minimal value, and $\max f_i$ is the worst value obtainable for criterion i; $f_i(\mathbf{x})$ is the result of using the vector of design variables $\mathbf{x}$ in evaluating the ith criterion function. The value of p indicates the type of distance: for $p = 1$, all deviations from $\min f_i$ are taken into account in direct proportion to their magnitudes (weighting method). For $2 \leqslant p < \infty$, the larger deviations carry greater weight in d_∞; for $p \to \infty$, the largest deviation is the only one taken into consideration (minimax approach).

18.4.6. Goal Programming

Goal programming was originally proposed by Charnes and Cooper (1961) for a linear problem. It was further developed by Ijiri (1965), Lee (1972), and Ignizio (1976), among others [see Charnes and Cooper's (1977) review of this method]. This method requires the designer to set goals for each objective that one wishes to obtain. A preferred (final) solution is the one that minimizes the deviations from the set goals. The simplest goal programming problem can be defined as

$$\min_{\mathbf{x} \in \Omega} |f_i(\mathbf{x}) - b_i|,$$

where b_i denotes the target or goal set for the objective function $f_i(\mathbf{x})$, and Ω represents the feasible set. A more general formulation of the goal programming problem is a weighted sum of the pth power of the deviation [Haimes *et al.* (1975)]; it can be stated as follows:

$$\min \left\{ \sum_{j=1}^{k} (d_j^- + d_j^+)^p \right\}^{1/p}, \qquad p > 1,$$

subject to

$$g_j(\mathbf{x}) \leqslant 0, \qquad j = 1, 2, \ldots, m,$$
$$f_i(\mathbf{x}) + d_i^- - d_i^+ = b_i, \qquad i = 1, 2, \ldots, k,$$
$$d_i^- d_i^+ = 0 \qquad \text{for each } i,$$
$$d_i^-, d_i^+ \geqslant 0 \qquad \text{for each } i,$$

where b_i, $i = 1, 2, \ldots, k$, are the goals set by the designer for the objectives, and d_i^- and d_i^+ are the underachievement and overachievement of the ith goal. The value of p is the subjective decision of designer. Such a formulation has been called *generalized goal programming* [Ignizio (1976, 1981)].

The goal programming method credited generally to Charnes and Cooper has been extended into many engineering areas, although few, if any, structural optimization applications can be found. The reasons for this may be the inherently nonlinear nature of structural problems, whereas goal programming works best in linear or, perhaps, quadratic cases.

18.4.7. Multiattribute Utility Method

Von Neumann and Morgenstern (1947) presented an axiomatic utility theory to measure individual or group preferences. Multiattribute utility functions were studied by Keeney and Raiffa (1976), Farquhar (1977), Fishburn (1974), Huber (1974), and Berger (1980). These functions are used to integrate the objective functions into the structure of measuring preferences. The highest degree of utility with respect to all of the objectives is obtained by maximizing the utility function; i.e.,

$$\max_{\mathbf{x}\in\Omega} U(f_1, f_2, \ldots, f_k),$$

where

$$\Omega = \{\mathbf{x} \in R^n : \mathbf{h}(\mathbf{x}) = 0, \mathbf{g}(\mathbf{x}) \leq 0\}.$$

The utility function U may have different forms. It is frequently additive and disjunctive with respect to the objectives f_i: i.e.,

$$U(\mathbf{f}) = U_1(f_1) + U_2(f_2) + \cdots + U_k(f_k).$$

In one case, wieghting factors for individual objectives are given; i.e.,

$$U(\mathbf{f}) = \sum_{i=1}^{k} w_i f_i(\mathbf{x}).$$

A multiplicative form of the utility function is also possible:

$$U(\mathbf{f}) = \prod_{i=1}^{k} U_i(f_i).$$

The contact point in objective space between the efficient (Pareto optimal) set and indifference lines of the utility function U becomes the solution of the optimization problem. The advantage of this method is its simplicity; it replaces the vector optimization problem by a scalar one. The main difficulty lies in finding the appropriate utility function.

18.4.8. Method of Pairwise Comparisons

Sometimes it is enough to generate only a certain subset of the Pareto optimal set. One possibility for getting a subset of the Pareto set is to develop interactive procedures where the designer participates in the process at every Pareto optimum achieved in order to determine the direction and extent of the next step. A method of pairwise comparisons developed and applied successfully to several decision-making problems by Saaty (1979) seems to be one promising alternative for structural design as well.

Choosing the proper weights for the criteria and then applying some scalar optimization method, we can reach directly a Pareto optimal solution that is close to the final (preferable) solution. Difficulties often arise, however, in determining suitable weights for the criteria. Considerable advantage can be gained by applying the present method, which brings out the weights by using pairwise comparisons of the criteria. This approach undoubtedly helps the designer to judge the importance of competing, usually even noncommensurable, objectives. To describe the method, we define a reciprocal matrix **A**, which includes the relative weights of the criteria, by

$$\mathbf{A} = \begin{bmatrix} \dfrac{w_1}{w_1} & \dfrac{w_1}{w_2} & \cdots & \dfrac{w_1}{w_k} \\ \dfrac{w_2}{w_1} & \dfrac{w_2}{w_2} & \cdots & \dfrac{w_2}{w_k} \\ \dfrac{w_k}{w_1} & \dfrac{w_k}{w_2} & \cdots & \dfrac{w_k}{w_k} \end{bmatrix}. \tag{18.20}$$

All its elements are positive and satisfy the reciprocal rule $a_{ij} = 1/a_{ji}$. The matrix has the property that if the vector $\mathbf{w} = \{w_1, w_2, \ldots, w_k\}$ is multiplied by it, then the vector kw is obtained; i.e.,

$$\mathbf{Aw} = k\mathbf{w}. \tag{18.21}$$

This equation has a nonzero solution if and only if k is an eigenvalue of **A**. All the eigenvalues μ_i, $i = 1, 2, \ldots, k$, are zero except one, denoted here by $\mu_{\max}$, which, according to Eq. (18.21), is equal to k. This can be seen immediately from (18.20), where every row is a constant multiple of the first row; i.e., **A** has unit rank. The eigenvector **w** of the problem is any column of **A**, and any two of these solutions differ by a multiplicative constant. If **w** is normalized so that its components sum to unity, the resulting vector is a unique solution of Eq. (18.21). Furthermore, the matrix **A** satisfies the cardinal consistency property $a_{ij}a_{jk} = a_{ik}$ and is called *consistent.*

When applying the method of pairwise comparisons to a specific problem, we only have estimates of the ratios w_i/w_j in the matrix and we want to determine the corresponding weights. The notations **A**′ and **w**′ are used here for the estimated matrix and its eigenvector, respectively. The elements of **A**′ are obtained by pairwise comparisons of the criteria, and the weights $w_1', w_2', \ldots, w_k'$ can be calculated by solving the eigenvalue problem for this matrix.

Now the cardinal consistency relation, however, need not hold. It turns out that **A**′ is consistent if and only if $\mu'_{\max} = k$. With inconsistency the inequality $\mu'_{\max} > k$ always holds, and the departure of $\mu'_{\max}$ from k can be shown to be a convenient measure for that undesirable property. In some cases considerable inconsistency may call for revising the judgments used in forming the matrix **A**′. Also, another consistency criterion, where the computation of a maximum eigenvalue is not needed at all, has been reported by Vargas (1980).

Various applications of this method, only briefly discussed here, to several decision-making problems can be found in Saaty (1979) and Wind and Saaty (1980), and encouraging results have been obtained by using a scale ranging from 1 to 9 for the relative weights. In engineering applications an additional examination of different scales could be advisable, but the method itself seems to have a good deal of unexploited potential in this field.

18.4.9. Trade-off Method

In using an iterative approach for solving the vector optimization problem, the optimization process usually proceeds from one Pareto optimum to another, and each choice of the next step is made by the decision maker. The trade-off approach is probably the most widely accepted one for generating Pareto optimal solutions in different fields of application. The term has been used quite freely in the literature and seems to take different forms, depending on the context. A common feature in most trade-off analyses is, however, that they are concerned with the advantage gained for one criterion by making concessions in another criterion. A comprehensive trade-off study connected with the constraint method has been given by Haimes and Chankong (1979), who also define separately the concepts of partial and total trade-off rates. We briefly discuss the results associated with the partial trade-off rate.

Algorithms used in vector optimization deal usually with local Pareto optima only, and thus it is natural to restrict the survey to some local properties of the efficient surface in this context. By fixing the parameters ε_i in the constraint method, we obtain a Pareto optimal vector $\hat{\mathbf{x}}$ as a solution to the problem. At this point a trade-off number, denoted by $\partial f_k/\partial f_i$, is defined as the rate of change of f_k per unit change in f_i on the efficient surface when all other criteria remain unchanged. Haimes and Chankong (1979) show that under certain regularity conditions the trade-off numbers can be attained as the negatives of the Kuhn–Tucker multipliers of the corresponding constraint problem at the Pareto optimum $\hat{\mathbf{x}}$; i.e.,

$$\frac{\partial f_k}{\partial f_i}(\boldsymbol{\varepsilon}) = -\hat{u}_{ki},$$

where $\hat{u}_{ki}$ is the Kuhn–Tucker multiplier associated with the criterion f_i, which in this scalar approach is one of the constraint functions. Geometrically, useful information about the gradient or the tangent plane of the efficient surface in the criteria space at $\hat{\mathbf{x}}$ can be obtained by computing the optimal multipliers $\hat{u}_{ki}$. In addition, considerable advantage is gained from the decision maker's point of view by introducing properly defined trade-off numbers into the design process, because he or she can then compare two objectives at a time, thus facilitating also the application of Saaty's method of pairwise comparisons.

18.4.10. Surrogate-Worth Trade-off Method

The surrogate-worth function W_{ij}, $i \neq j$, $j = 1, 2, \ldots, n$, is defined as a function of the desirability of the trade-off λ_{ij} on a scale, for example, ranging from -10 to $+10$. A -10 would indicate that marginal units of objective i are worth very much less than one marginal unit of objective j, a $+10$ means the opposite, and a zero indicates an even trade-off.

The surrogate-worth trade-off method can be used when all objective functions are differentiable. We construct a scalar optimization problem by developing a surrogate-worth function in cooperation of the designer. First, we generate the set of efficient solutions, say by the constraint method. Then we search along the efficient boundary using a surrogate-worth function. The trade-off function for any two objectives evaluated at a given efficient solution $\hat{\mathbf{x}}$ is

$$T_{ij}(\hat{\mathbf{x}}) = \frac{\partial f_i(\hat{\mathbf{x}})}{\partial f_j(\hat{\mathbf{x}})}.$$

A complete description of the surrogate-worth method is given in Haimes *et al.* (1975), and an abbreviated version is given in Goicoechea *et al.* (1982).

18.4.11. Choosing a Method

It would be very difficult to try to select one multiobjective decision-making technique as the best for all structural design problems. In order to lessen the computational and human effort in the design process, it is sometimes preferable, to concentrate only on a certain subset of Pareto optima instead of generating all of them. One possibility, which may prove to be a suitable approach in structural optimization, is to develop interactive procedures where the designer participates in the process at every Pareto optimum achieved in order to determine the direction and extent of the next step.

18.5. Multiobjective Structural Optimization

18.5.1. Criteria Used in Structural Optimization

The optimization of civil and mechanical engineering structures usually involves requirements that should be met at the same time to obtain a fully useful design. In single-criterion optimization, one of the requirements is selected as the criterion of optimization and the remaining ones are met by including them into the constraints of optimization. But this approach requires us to determine *a priori* the bounds that these requirements should fulfill. Multiobjective (multicriteria, vector) optimization enables us to consider effectively all the different mutually conflicting requirements inherent in the design problem. The criteria appearing most frequently in the optimization of structures are discussed next.

18.5.1.1. Minimum Volume, Weight, or Cost Criterion. According to these criteria, the optimal structure is one with the minimum volume, weight, or cost that has been selected from the possible structures subjected to the same loading conditions and satisfying the given constraints. If the structure has been built from different materials, we may have to minimize each of them. For example, in a prestressed concrete beam the minimum volumes of both concrete and the prestressing cables may be required.

18.5.1.2. Minimum Compliance or Maximum Structural Stiffness. For this criterion, among all the possible structures subjected to the same loads and satisfying the given constraints, the optimal structure is one in which the internal work done by the loads due to strains has the minimum value. On the other hand, the work done by the loads on displacements in the equilibrium state is equal to the energy (potential) of elastic deformation; thus, this energy also attains its minimum value.

18.5.1.3. Minimum Displacements at Selected Points or Regions of the Structure. In some cases, we must minimize changes in the shape of the structure under action of different loading conditions. Such a requirement is involved, for example, for the supporting structure of radiotelescopes or for some cable-suspended structures.

18.5.1.4. Maximum Critical Load. It is obvious that buckling is important in very slender compressed elements. Therefore, they should be shaped in such a way to obtain the maximum critical load. In some structures, local buckling is most important, and then it is necessary to ensure the maximum stiffness of particular structural elements.

18.5.1.5. Maximum Natural Frequency of Free Vibration. Structures under dynamic loading conditions should be shaped in such a way to obtain extreme values of the natural frequencies of free vibration. In some cases it is sufficient to take into account only the first frequency of free vibration (unimodal optimization), but, in others, we must carry out an analysis for frequencies of the higher vibration modes (multimodal optimization). This criterion can be modified to require maximum detuning of the vibrating system—i.e., maximum separation between the frequencies of natural and forced vibrations.

18.5.1.6. Maximum Reliability or Safety. This criterion is always required, but it is especially important in aerospace engineering and in civil engineering structures built in seismic zones.

18.5.1.7. Maximum Moment of Inertia. It may be required that the cross section moment of inertia of bent or compressed structural elements be maximized. The same requirement may hold for vibrating systems. For twisted elements, it is necessary to maximize the polar moment of inertia of the cross section. The criterion of maximum moment of inertia is consistent with many of the criteria mentioned.

To satisfy any of the criteria already mentioned, we must find the extremum of an objective function, taking into account the related constraints. In structural optimization the most important constraints involve:

allowable stresses or the safety factors of the structure under all possible loading conditions;

allowable displacements in a given structure;

lower and upper bounds for design variables based on serviceability and technological reasons (e.g., on the minimum thickness of metal plates or maximum height of a bridge structure because of the requirements of a transportation system).

18.5.2. Optimization of a Prestressed Concrete I-beam

The optimal design problem for a prestressed concrete I-beam can be formulated as follows. Find the dimensions of the cross sectional area of a prestressed concrete I-beam as well as the eccentricity of prestressing force according to two criteria: minimum volumes of the concrete and the prestressing steel. Two extremal loading conditions represented by the bending moments M_1 and M_2 are considered. The vector of design variables $\mathbf{x}^T = \{x_i\}$, $i = 1, 2, \ldots, 6$, contains the following quantities: beam depth, x_1; flange width, x_2; flange thickness, x_3; web thickness, x_4; values of the prestressing force and its eccentricity, x_5 and x_6, respectively (Fig. 18.6).

The feasible domain Ω is defined by the following constraints:

(a) stress constraints in two loading conditions:

$$x_5/A - x_5x_6x_1/2J + M_1x_1/2J \leqslant \tilde{\sigma}_1,$$
$$x_5/A + x_5x_6x_1/2J - M_1x_1/2J \geqslant \underset{\sim}{\sigma}_1,$$
$$x_5/A - x_5x_6x_1/2J + M_2x_1/2J \geqslant \underset{\sim}{\sigma}_2,$$
$$x_5/A + x_5x_6x_1/2J - M_2x_1/2J \leqslant \tilde{\sigma}_2;$$

(b) side constraints:

$$x_1 \leqslant \tilde{x}_1, \quad x_2 \leqslant \tilde{x}_2, \quad x_3 \geqslant \underset{\sim}{x}_3, \quad x_4 \geqslant \underset{\sim}{x}_4, \quad x_6 \leqslant x_1/2 - d.$$

The optimization criteria can be written as follows:

$$\min_{\mathbf{x}\in\Omega} A = 2x_2x_3 + x_4(x_1 - 2x_3),$$
$$\min_{\mathbf{x}\in\Omega} S = x_5.$$

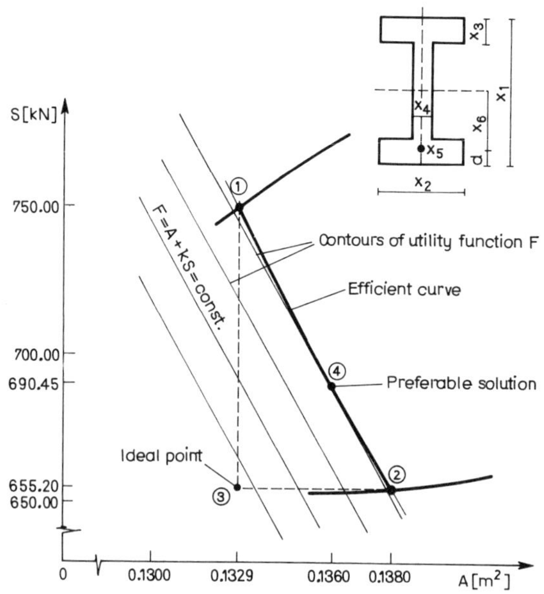

Figure 18.6. Geometrical Interpretation of the Solution for the I-Beam Using a Utility Function.

The curve segment given by

$$x_5 = \frac{\tilde{\sigma}_1 - M_1\tilde{x}_1/2J}{1/A + (\tilde{x}_1/2 - d)\tilde{x}_1/2J}$$

and contained between the curves

$$x_5 = \frac{\tilde{\sigma}_2 + M_2\tilde{x}_1/2J}{1/A + (\tilde{x}_1/2 - d)\tilde{x}_1/2J}$$

and

$$x_5 = \frac{\underset{\sim}{\sigma}_1 + M_1\tilde{x}_1/2J}{1/A + (\tilde{x}_1/2 - d)\tilde{x}_1/2J}$$

is the set of compromise solutions [for details see Jendo and Marks (1986b) and Marks (1978)].

The preferable solution is selected on the basis of the utility function

$$F = A + kS, \qquad k = c_s\gamma/c_c\tilde{\sigma}_s \; [m^2/\mathrm{kN}],$$

where k is the ratio of the unit cost of the prestressing steel over the concrete unit cost. The following notation is used: c_c is the unit cost of 1 m^3 of concrete, c_s is the unit cost of 1 kg of prestressing steel, γ is the bulk density of the prestressing steel, and $\tilde{\sigma}_s$ is the allowable tension stress for the prestressing steel. The preferable solution occurs at one of the end points of the compromise set or at the tangent point between the compromise set and the indifference lines of the utility function $F = A + kS =$ const, which depends on the value of k (Fig. 18.6).

Numerical example: The following data are used: $M_1 = 320$ kN m, $M_2 = 100$ kN m, $\tilde{x}_1 = 0.70$ m, $\tilde{x}_2 = 0.40$ m, $\underset{\sim}{x}_3 = 0.10$ m, $\underset{\sim}{x}_4 = 0.08$ m, $d = 0.07$ m, $\tilde{\sigma}_1 = \tilde{\sigma}_2 = 10{,}000$ kN/m^2, $\underset{\sim}{\sigma}_1 = \underset{\sim}{\sigma}_2 = -500$ kN/m^2, and $k = 0.000054$ m^2/kN.

The solution is presented in Figure 18.6 and contains the following information:

1. The coordinates of the end points of the compromise set:
   point 1 ($A = 0.13286$ m^2, $S = 750.0$ kN);
   point 2 ($A = 0.13797$ m^2, $S = 655.2$ kN);
2. The ideal solution: point 3 ($A^{\mathrm{id}} = 0.13286$ m^2, $S^{\mathrm{id}} = 655.2$ kN);

3. The preferable solution: point 4 ($A^{pr} = 0.13600\ \mathrm{m}^2$, $S^{pr} = 690.45$ kN) and the corresponding vector of design variables $x_1^{pr} = 0.70$ m, $x_2^{pr} = 0.40$ m, $x_3^{pr} = 0.125$ m, $x_4^{pr} = 0.08$ m, $x_5^{pr} = 690.45$ kN, $x_6^{pr} = 0.28$ m;
4. The width of the compromise set: $\Delta A = 0.00511\ \mathrm{m}^2$, $\Delta S = 94.80$ kN;
5. The sensitivity vector of the global criterion F at the preferable solution: $\partial F/\partial A = 1$, $\partial F/\partial S = 0.000054$.

18.5.3. Optimization of Isostatic Trusses

The multicriteria optimization problem of isostatic steel truss structures can be formulated as follows. Find the cross-sectional areas of truss bars $\mathbf{x}^T = \{x_i\}$, $i = 1, 2, \ldots, 7$, according to two criteria: minimum volume of truss and minimum vertical displacement at point B (Fig. 18.7). The feasible domain of solutions is defined by stress and side constraints; i.e.,

$$\Omega = \{\mathbf{x} \in R^n : \underset{\sim}{\sigma} \leqslant \sigma_i \leqslant \tilde{\sigma}, x_i \leqslant \tilde{x}_i, i = 1, \ldots, 7\},$$

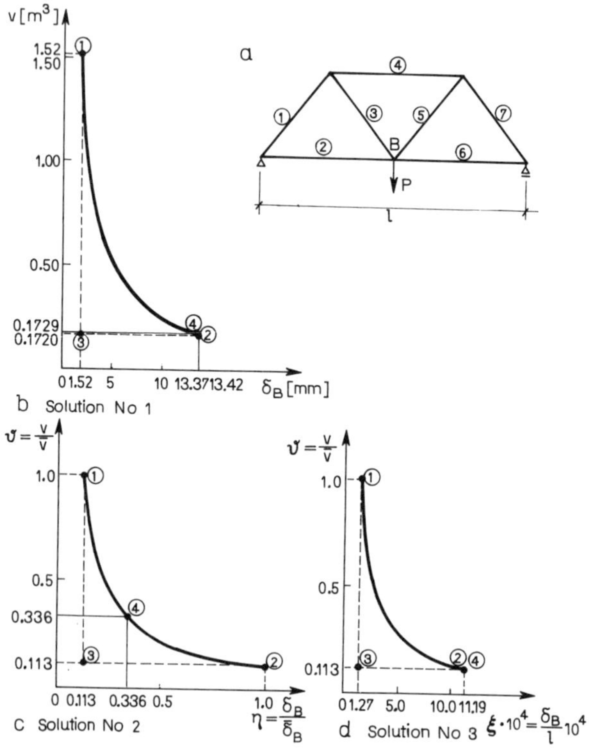

Figure 18.7. Geometrical Interpretation of the Three Solutions for Isostatic Truss Using Distance Functions.

where $\underset{\sim}{\sigma}$ and $\tilde{\sigma}$ denote the allowable compression and tension stresses, respectively, and $\sigma_i = N_i/x_i$.

The optimization criteria for truss structures can be written as

$$\min_{\mathbf{x}\in\Omega} V = \sum_{i=1}^{7} x_i l_i,$$

$$\min_{\mathbf{x}\in\Omega} \delta_B = \sum_{i=1}^{7} \frac{\partial}{\partial P}\left(\frac{N_i^2 l_i}{2Ex_i}\right).$$

We solve the optimization problem by using dynamic programming (see Chapter 17). The compromise set has the form

$$V = P(5l_1 + 3l_2 + 3l_3 + 5l_4)/32E\delta_B, \tag{18.22}$$

and it is contained between $\underset{\sim}{V}$ and $\tilde{V}$ in the coordinate system (V, δ_B). The minimal feasible volume of truss $\underset{\sim}{V}$ is found on the basis of allowable stresses in truss bars:

$$\underset{\sim}{V} = \frac{2l_1 5P}{8\underset{\sim}{\sigma}} + \frac{2l_2 3P}{8\tilde{\sigma}} + \frac{2l_3 5P}{8\tilde{\sigma}} + \frac{l_4 3P}{4\underset{\sim}{\sigma}}.$$

The maximal feasible volume of truss $\tilde{V}$ is found on the basis of side constraints $x_i = \tilde{x}_i = \tilde{x}$; i.e.,

$$\tilde{V} = (2l_1 + 2l_2 + 2l_3 + l_4)\tilde{x}.$$

Since there is no preference between the objective functions V and δ_B, the solution is found based on the minimization of the metric function (see Section 18.4.4). We find three solutions based on different forms of the metric function.

Solution No. 1. The metric function in dimensional form is taken as follows:

$$F^{(2)} = \{(V - V^{\text{id}})^2 + \mu^2(\delta_B - \delta_B^{\text{id}})^2\}^{1/2}, \qquad \mu = 1\ \text{m}^2. \tag{18.23}$$

The ideal solution has coordinates

$$\begin{aligned} V^{\text{id}} &= \frac{P}{4}\left(\frac{5l_1}{\underset{\sim}{\sigma}} + \frac{3l_2}{\tilde{\sigma}} + \frac{5l_3}{\tilde{\sigma}} + \frac{3l_4}{\underset{\sim}{\sigma}}\right), \\ \delta_B^{\text{id}} &= \frac{P(5l_1 + 3l_2 + 5l_3 + 3l_4)^2}{32E\tilde{x}(2l_1 + 2l_2 + 2l_3 + l_4)}. \end{aligned} \tag{18.24}$$

Substituting (18.22) and (18.24) into (18.23), we have

$$F^{(2)} = \left\{ \left\{ V - \frac{P}{4}\left(\frac{5l_1}{\tilde{\sigma}} + \frac{3l_2}{\tilde{\sigma}} + \frac{5l_3}{\tilde{\sigma}} + \frac{3l_4}{\underset{\sim}{\sigma}} \right) \right\}^2 + \mu^2 \left\{ \frac{P^2(5l_1 + 3l_2 + 5l_3 + 3l_4)^4}{1024E^2} \right\} \times \left\{ \frac{1}{V} - \frac{1}{\tilde{x}(2l_1 + 2l_2 + 2l_3 + l_4)} \right\}^2 \right\}^{1/2}.$$

The necessary condition for an extremum gives the equation

$$V^4 - \left\{ \frac{P}{4}\left(\frac{5l_1}{\underset{\sim}{\sigma}} + \frac{3l_2}{\tilde{\sigma}} + \frac{5l_3}{\tilde{\sigma}} + \frac{3l_4}{\underset{\sim}{\sigma}} \right) \right\} V^3 + \mu^2 \left\{ \frac{P^2(5l_1 + 3l_2 + 5l_3 + 3l_4)^4}{1024E^2\tilde{x}(2l_1 + 2l_2 + 2l_3 + l_4)} \right\} V - \mu^2 \left\{ \frac{P^2(5l_1 + 3l_2 + 5l_3 + 3l_4)^4}{1024E^2} \right\} = 0,$$

which allows us to determine the preferable solution $\underset{\sim}{V} \leqslant V^{\text{pr}} \leqslant \tilde{V}$.

Solution No. 2. Taking the metric function in dimensionless form, we have

$$F^{(2)} = \left\{ \left\{ \frac{V - V^{\text{id}}}{\tilde{V}} \right\}^2 + \left\{ \frac{\delta_B - \delta_B^{\text{id}}}{\tilde{\delta}_B} \right\}^2 \right\}^{1/2}.$$

Introducing the notation

$$\nu = V/\tilde{V}, \quad \eta = \delta_B/\tilde{\delta}_B, \quad \beta = \underset{\sim}{V}/\tilde{V}$$

and using an analogous procedure to the foregoing, we find

$$\nu^4 - \beta\nu^3 + \beta^2\nu - \beta^2 = 0,$$

which allows us to determine the preferable solution ν^{pr}.

Solution No. 3. Taking the metric function in another dimensionless form

$$F^{(2)} = \left\{ \left\{ \frac{V - V^{\text{id}}}{\tilde{V}} \right\}^2 + \left\{ \frac{\delta_B - \delta_B^{\text{id}}}{l} \right\}^2 \right\}^{1/2}$$

and introducing notation analogous to what we had before,

$$\nu = \frac{V}{\tilde{V}}, \quad \zeta = \frac{\delta_B}{l}, \quad \alpha = \frac{P(5l_1 + 3l_2 + 5l_3 + 3l_4)^2}{32E},$$

we find

$$\tilde{V}^2\nu^4 - \tilde{V}\underset{\sim}{V}\nu^3 + (\alpha^2/l^2)\nu - \alpha^2/l^2 = 0,$$

which allows us to determine the preferable solution ν^{pr}.

Numerical example. The following data are used: $P = 2000\,\text{kN}$, $E = 200\,\text{kN/mm}^2$, $\underset{\sim}{\sigma} = \tilde{\sigma} = 250\,\text{MPa}$, $l_1 = l_3 = l_5 = l_7 = 5.00\,\text{m}$, $l_2 = l_4 = l_6 = 6.00\,\text{m}$, $\tilde{x} = 40{,}000\,\text{mm}^2$. Three solutions for an optimal truss are presented in Table 18.1 and shown on Fig. 18.7. The compromise sets and ideal solutions are independent of the form of the metric function. Unfortunately, the preferable solution and sensitivity vector essentially depend on the form of the metric function.

18.5.4. Optimization of Cable Structures

18.5.4.1. Characteristics of Cable Structures. This section deals with an optimization problem of single-layer cable systems, which are often used as carrying elements in electric power lines, hanging ropeways, or large-span roofing structures, as shown in Fig. 18.8. Cable structures are substantially different from other types of structures because they can assume various shapes under the action of different loadings. That is why static and dynamic analyses of cable structures are different from those commonly used for stiff structures displacing only slightly from an unloaded reference configuration. The first difference occurs because of the flexible shape of the structure so that the principle of structural rigidity cannot be used here. The second difference consists is that the principle of superposition is inapplicable. This follows from the geometric nonlinearity of cables caused by large changes in cable shapes due to varying loadings. Thus, to write the equilibrium and deformation equations, we must take into account all the loads acting on the structure, which has no *a priori* determined shape.

18.5.4.2. Basic Relationships for Flat Sag Cables. A nomogram for determination of the lowest frequency for symmetric in-plane free vibrations of the extensible cable is shown in Fig. 18.9. It can be seen that maximal values of the lowest natural frequencies occur for very flat sag cables; i.e., $\eta = f/l = 0.03$–0.05 corresponds to the interval (20–200) of the parameter $\xi = ql/A$. The following notation is used: f, l, A, and q denote sag, span, cross-sectional area, and uniform load of cable, respectively. On the other

Table 18.1. List of Solutions for Multicriteria Optimization of an Isostatic Truss

	Solution 1	Solution 2		Solution 3	
Equation of compromise set	$V\,[\mathrm{m}^3] = \dfrac{0.002311}{\delta_B\,[\mathrm{m}]}$	$\nu = \dfrac{0.113137}{\eta}$		$\nu = \dfrac{0.0001267}{\xi}$	
Coordinates of end points of compromise set					
point 1	$V = 1.52\ \mathrm{m}^3$	$\nu = 1$	$V = 1.52\ \mathrm{m}^3$	$\nu = 1$	$V = 1.52\ \mathrm{m}^3$
	$\delta_B = 0.00152\ \mathrm{m}$	$\eta = 0.113$	$\delta_B = 0.00152\ \mathrm{m}$	$\xi = 0.000127$	$\delta_B = 0.00152\ \mathrm{m}$
point 2	$V = 0.172\ \mathrm{m}^3$	$\nu = 0.113$	$V = 0.172\ \mathrm{m}^3$	$\nu = 0.1132$	$V = 0.172\ \mathrm{m}^3$
	$\delta_B = 0.0134\ \mathrm{m}$	$\eta = 1$	$\delta_B = 0.0134\ \mathrm{m}$	$\xi = 0.001119$	$\delta_B = 0.0134\ \mathrm{m}$
Ideal solution					
point 3	$V^{id} = 0.172\ \mathrm{m}^3$	$\nu^{id} = 0.113$	$V^{id} = 0.172\ \mathrm{m}^3$	$\nu^{id} = 0.113$	$V^{id} = 0.172\ \mathrm{m}^3$
	$\delta_B^{id} = 0.00152\ \mathrm{m}$	$\eta^{id} = 0.113$	$\delta_B^{id} = 0.00152\ \mathrm{m}$	$\xi^{id} = 0.000127$	$\delta_B^{id} = 0.00152\ \mathrm{m}$
Preferable solution					
point 4	$V^{pr} = 0.1729\ \mathrm{m}^3$	$\nu_{pr} = 0.336$	$V^{pr} = 0.5107\ \mathrm{m}^3$	$\nu^{pr} = 0.1132$	$V^{pr} = 0.172\ \mathrm{m}^3$
	$\delta_B^{pr} = 0.0134\ \mathrm{m}$	$\eta^{pr} = 0.336$	$\delta_B^{pr} = 0.00453\ \mathrm{m}$	$\xi^{pr} = 0.001119$	$\delta_B^{pr} = 0.0134\ \mathrm{m}$
Design variables $x_1\,[\mathrm{m}^2]$	0.005034	0.014846		0.0050	
$x_2\,[\mathrm{m}^2]$	0.003017	0.008907		0.0030	
$x_3\,[\mathrm{m}^3]$	0.005026	0.014846		0.0050	
$x_4\,[\mathrm{m}^2]$	0.006030	0.017814		0.0060	
Width of compromise set	$\Delta V = 1.348\ \mathrm{m}^3$	$\Delta\nu = 0.887$	$\Delta V = 1.348\ \mathrm{m}^3$	$\Delta\nu = 0.887$	$\Delta V = 1.348\ \mathrm{m}^3$
	$\Delta\delta_B = 0.0119\ \mathrm{m}$	$\Delta\eta = 0.887$	$\Delta\delta_B = 0.0119\ \mathrm{m}$	$\Delta\xi = 0.000993$	$\Delta\delta_B = 0.0119\ \mathrm{m}$
Sensitivity vector	$\partial F/\partial V = 0.0757$	$\partial F/\partial\nu = 0.7071$		$\partial F/\partial\nu = 0$	
	$\partial F/\partial\delta_B = 0.9971$	$\partial F/\partial\eta = 0.7071$		$\partial F/\partial\xi = 1$	

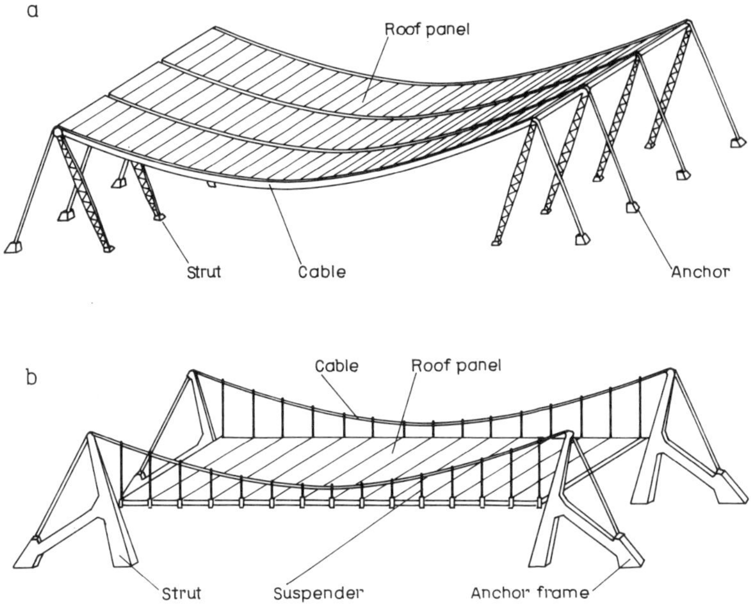

Figure 18.8. Schematic Layouts of Single-Layer Cable Systems.

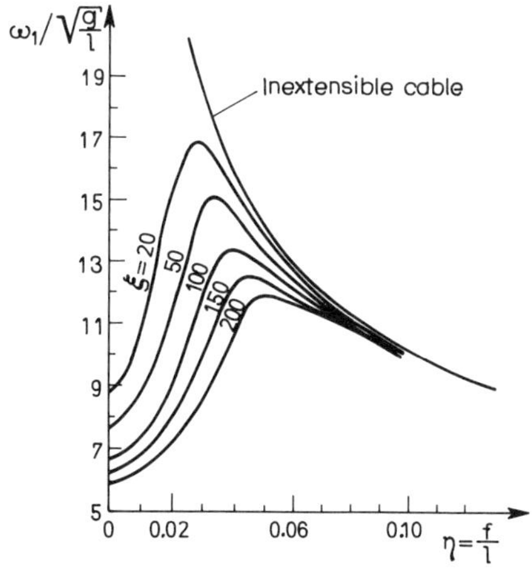

Figure 18.9. Nomogram for Determination of the Lowest Frequency for Symmetric In-Plane Free Vibration.

hand, the optimal cable sags determined according to minimum-weight criterion always occur on the boundary of the feasible domain (Fig. 18.10) given by side, stress, and displacement constraints. Comparing these results, we may conclude that we should find the set of compromise solutions by a multiobjective optimization approach.

18.5.4.3. Formulation of the Biobjective Optimization Problem. The biobjective optimization problem for single-layer cable systems with the assumptions of flat sags and Hooke's law for material behavior can be formulated as follows. Find the cable sag $\eta = f/l$ and cross-sectional area A for a given cable span l, static loading $q(x)$, modulus of elasticity E, dynamic loading $p(x, t)$, and allowable stress $\tilde{\sigma}$ such that the weight of a single-layer cable system

$$W = \gamma A s = \gamma A l(1 + 8\eta^2/3) \tag{18.25}$$

attains a minimum and the lowest natural frequency of the in-plane free vibrations

$$\omega_1 = 2a_1(gH_{st}/ql^2)^{1/2} \tag{18.26}$$

attains a maximum. Here H_{st} and g denote the cable force under static loading q and acceleration of gravity, respectively. This corresponds to the symmetric in-plane eigenmode

$$X_1(x) = C_1\{1 - \cos a_1(2x/l) - \tan a_1 \sin a_1(2x/l)\},$$

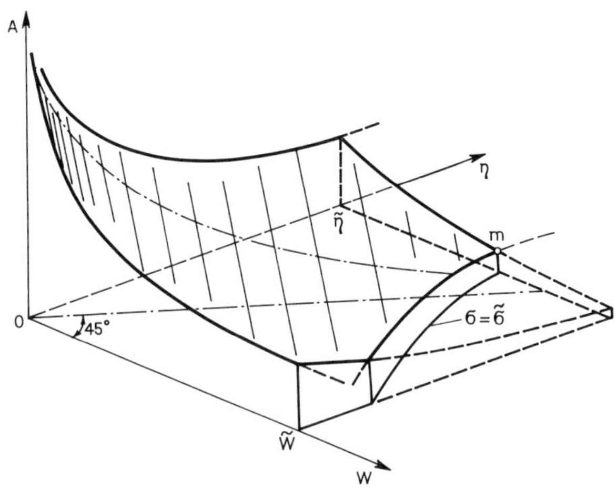

Figure 18.10. Weight of Cable versus Sag (f) and Displacement (w).

where C_1 is a constant and a_1 can be determined from the transcendental equation

$$\tan a_1 - a_1 + a_1^3(H_{st}/16EA\eta^2) = 0. \tag{18.27}$$

We assume that the catenary can be replaced by a parabolic curve of second order

$$y = 4f(x/l - x^2/l^2),$$

because of the flatness. In this case the approximate length of the cable is

$$s = l(1 + 8\eta^2/3),$$

and γ is the bulk density of the cable material.

The optimal solution should satisfy the following system of inequality and equality constraints:

(a) side constraints:

$$0.01 \leqslant \eta \leqslant 0.1, \qquad A > 0; \tag{18.28}$$

(b) static governing equation:

$$\eta = \tfrac{1}{8}\{(ql/H_{st})^2 - 24H_{st}/EA\}^{1/2}; \tag{18.29}$$

(c) stress constraint:

$$\sigma_{\max} = \frac{H_{st} + H_d}{A \cos a_{\max}} = \frac{(H_{st} + H_d)(1 + 16\eta^2)}{A} \leqslant \tilde{\sigma}; \tag{18.30}$$

where

$$H_d(t) = \frac{EA}{l}\int_0^l \left(\frac{\partial w}{\partial x}\right) y'(x)\, dx, \qquad y' = \frac{dy}{dx},$$

with $w(x, t) = \Phi(t)X_1(x)$; and

(d) dynamic displacement constraint:

$$w_{\max} = \Phi(t)X_1(x) \leqslant \tilde{w}, \tag{18.31}$$

where $\Phi(t)$ is determined by the dynamic magnification factor based on only the steady-state response. The last two constraints arise from the dynamic response and can be calculated on the basis of a dynamic loading represented by

$$p(x, t) = p_0 \sin \bar{\omega} t.$$

Substituting Eq. (18.29) into (18.25) gives the cable weight function

$$W = \gamma A l \left\{ 1 + \frac{(ql/H_{st})^2}{24} - \frac{H_{st}}{EA} \right\}. \tag{18.32}$$

18.5.4.4. Solution of the Optimization Problem. To solve the biobjective optimization problem, we must determine the set of feasible solutions in the design space (A, η) and the set of efficient solutions in the objective space (ω_1, W). The sets of feasible and efficient solutions should satisfy the constraints (18.27)-(18.31). To find the sets of feasible and efficient solutions, we check two problem formulations. In the first formulation we maximize the first natural frequency of free vibrations for a given cable weight $W = \text{const.}$ and take the constraints (18.27)-(18.31) into account. The second one deals with the minimization of cable weight for a given frequency for the same group of constraints (18.27)-(18.31). Both formulations give the same sets of feasible and efficient solutions, as shown in Figs. 18.11 and 18.12.

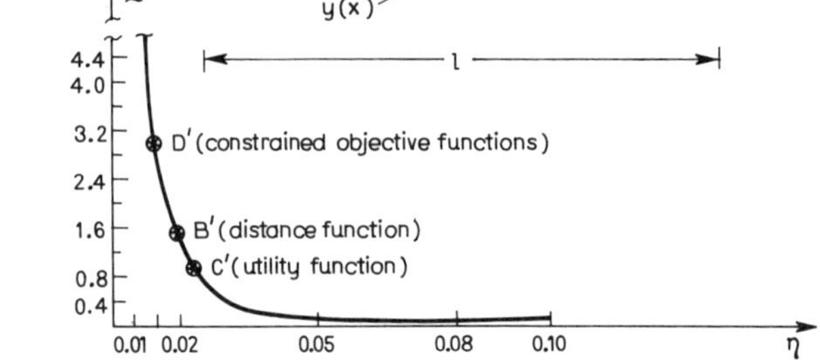

Figure 18.11. Set of Feasible Solutions for Cable.

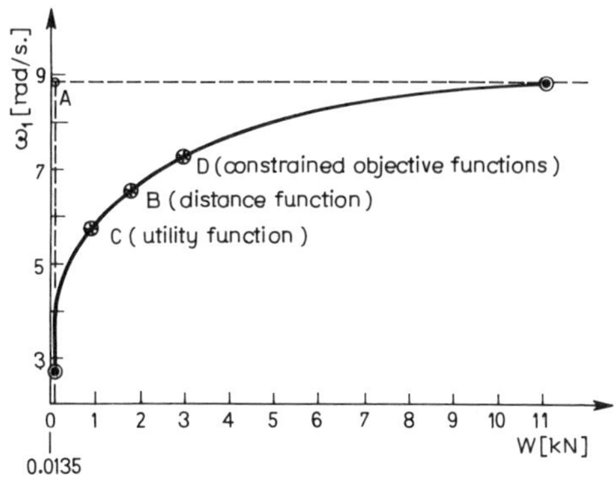

Figure 18.12. Set of Efficient Solutions for Cable.

The optimization problems were solved by two different methods. The first uses Lagrange multipliers and requires the gradients of the objective functions and the constraints. The gradients were evaluated numerically. The second algorithm is based on the evolution strategy, which works with random numbers and can determine by itself the improvements of the objective function. This algorithm does not require any evaluation of gradients.

18.5.4.5. Choosing a Preferable Solution. The set of efficient solutions shown in Fig. 18.12 contains an infinite number of solutions. We must decide which one should be taken as the preferable solution. There are a few methods for choosing a preferable solution from the set of efficient solutions (see Section 18.4).

Using the metric function to find a preferable solution, we have $k = 2$ and we choose $p = 2$. The following data have been used in the numerical example: $q = 1\ \text{kN/m}$, $E = 200\ \text{kN/mm}^2$, $\tilde{\sigma} = 1.2\ \text{kN/mm}^2$, $p_0 = 0.001\ \text{kN/m}$, $\bar{\omega} = 1\ \text{rad/s}$, and $\delta = 0.06$, where δ is the logarithmic decrement of damping. The objective functions are $f_1(\mathbf{x}) = W(\eta, A)$ and $f_2(\mathbf{x}) = \omega_1(\eta, A)$. The ideal point (solution) satisfying the constraints (18.27)-(18.31) was found numerically and is shown in Fig. 18.12 as point A with coordinates $W_{\text{min}} = 0.0135\ \text{kN}$, $\omega_{1\,\text{max}} = 8.91\ \text{rad/s}$. It does not belong to the set of efficient solutions.

The preferable solution was found numerically by minimization of the distance function with $p = 2$; i.e.,

$$\min_{\eta \in \Omega} F^{(2)} = \{(W - W^{\text{id}})^2 - \mu^2(\omega_1 - \omega_1^{\text{id}})^2\}^{1/2},$$

where $\mu = 1\ \text{kN-s/rad}$. The preferable solution obtained by this distance function is shown in Fig. 18.12 as the point $B(W^{pr} = 1.60\,\text{kN},\ \omega_1^{pr} = 6.46\,\text{rad/s})$ corresponding to the point $B'(\eta = 0.0185,\ A = 15{,}990\,\text{mm}^2)$ in Fig. 18.11.

Next, we determine the preferable solution by using the utility function method with weighting factors $w_1 = 0.5$ and $w_2 = 0.5$. To find the preferable solution by this method, we maximize the utility function

$$U(\mathbf{f}) = \sum_{i=1}^{2} w_i f_i(\mathbf{x}) = -\{w_1 W(\eta, A) - w_2 \omega_1(\eta, A)\}$$

subject to the constraints (18.27)-(18.31). The preferable solution found numerically is shown in Fig. 18.12 as the point $C(W^{pr} = 0.97\,\text{kN},\ \omega_1^{pr} = 5.92\,\text{rad/s})$, corresponding to the point $C'(\eta = 0.023,\ A = 9700\,\text{mm}^2)$ in Fig. 18.11.

The preferable solution can also be found by employing the constraint method (see Section 18.4.2). As a primary objective, the first natural frequency of free vibrations has been chosen; i.e.,

$$\max_{\eta \in \Omega} \omega_1 = 2a_1(gH_{st}/ql^2)^{1/2},$$

subject to constraints (18.27)-(18.31) and the additional constraint concerning cable weight

$$\underset{\sim}{W} \leqslant W(\eta, A) \leqslant \tilde{W}.$$

The lower and upper bounds for cable weight can be established on the basis of the allowable interval for cable sags, which was taken as

$$\underset{\sim}{\eta} \leqslant \eta \leqslant \tilde{\eta};$$

i.e., $\underset{\sim}{W} = W(\tilde{\eta})$ and $\tilde{W} = W(\underset{\sim}{\eta})$. We have used $\underset{\sim}{\eta} = 0.015$ and $\tilde{\eta} = 0.1$. The preferable solution is shown in Fig. 18.12 as the point $D(W^{pr} = 3.00\,\text{kN},\ \omega_1^{pr} = 7.176\,\text{rad/s})$, corresponding to the point $D'(\eta = 0.015,\ A = 30{,}000\,\text{mm}^2)$ in Fig. 18.11.

18.5.5. Further Applications of Multiobjective Optimization

Stadler (1975, 1977, 1978) was the first to apply multicriteria optimization in structural mechanics. He considered the bicriteria optimization problem, with weight and stored energy as criteria, and obtained analytical solutions for some structural elements, calling the results "natural structural shapes." Leitmann (1977) and Stadler (1980) used control theory to solve some multicriteria optimization problems in linear viscoelasticity. Baier (1977, 1978) studied multicriteria optimization of structures from a general point of view, taking weight and energy stored under separate loading conditions as optimization criteria. Truss optimization with weight and displacement criteria were considered by Koski (1979, 1984a, b) and Koski and Silvennoinen (1982). Osyczka (1978, 1984) and Rao and Hati (1979) considered some conflicting objectives in solving mechanical optimization problems. Eschenauer (1980, 1981, 1983) and Sattler (1982) applied the multicriteria optimization approach to spatial structures that support radiotelescopes using the following criteria: minimum weight and minimum displacements of the radiotelescope surface from its initial configuration under selected loading conditions. Also, Eschenauer (1984) considered the multiobjective optimization of fiber-reinforced plates and disks, choosing weight and deformation as optimization criteria. Carmichael (1980) solved a multicriteria optimization problem, using the method of constrained objective functions. Jendo and Marks (1986a, b, 1987) applied multicriteria optimization in solving some civil engineering structures like flexural elements and truss structures, using weight and displacements or potential energy as optimization criteria. Jendo, Marks, and Thierauf (1985) considered multicriteria optimization of an isostatic truss using dynamic programming. Jendo (1985, 1986a, b) considered multiobjective optimization of single-layer cable systems, choosing minimum weight and maximum lowest frequency of free vibration as optimization criteria. Some aspects of these problems, including dynamic plasticity of cable systems, were considered also by Eschenauer and Jendo (1986). Eschenauer, Kneppe, and Stenvers (1986) compared deterministic and stochastic multiobjective optimization of beams and shell structures. Watkins and Morris (1987) applied a multicriteria optimization for laminated composites.

19

Other Optimal Design Methods

19.1. Introduction

In this chapter we briefly present a few optimization methods that were not included in previous chapters. Some of these methods are underdeveloped, e.g., optimization with shakedown constraints or multicriteria reliability-based optimization. This chapter contains preliminary information; for more details, see the original source papers. In the following, we will sequentially consider optimal design for plastic shakedown, multilevel optimization and decomposition techniques, optimization with discrete design variables, optimal design with random parameters, and multicriteria reliability-based optimal structural design.

19.2. Optimal Design for Plastic Shakedown

In the introduction to Chapter 13 we quoted the arguments in favor of plastic optimal design, as compared to checking stresses and/or displacements calculated for the elastic response of the structure. A possibility of ensuring the desired safety margin against overloading is offered by the ultimate load theory and should be used whenever the actual properties of the structure and the loading agree with those assumed in that theory.

A basic assumption is that overloading occurs only once and that all components of **p** increase proportionally to a common factor [see Eq. (13.1)].

This assumption might be close to reality when a single load of deadweight type is considered (e.g., the weight of equipment or goods stored on a certain supporting structure). There exist many loads, however, that are of a variable or repetitive nature. Moreover, the values of individual elements of $\mathbf{p}$ can vary in time independently.

A typical example is the wind load. After having converted it into the relevant set of nodal forces, we are left with p_i as unknown functions of time. Nevertheless, it is possible to predict the maximum wind speed that can occur with a certain probability in a given place. This means that after discretization we are able to associate a lower bound p_i^- and an upper bound p_i^+ with each component of the time-dependent $\mathbf{p}$. Hence, the variable loading can be represented by the matrix constraint relation

$$\mathbf{p}^- \leqslant \mathbf{p} \leqslant \mathbf{p}^+, \tag{19.1}$$

which is equivalent to $2n$ ordinary inequalities.

When a structure made of elastic, perfectly plastic material is subjected sufficiently long to such a load, then one of the three following situations can occur:

1. After a few initial load cycles, the plastic strains, if present, stabilize their values and the structure responds purely elastically to the subsequent cycles.
2. After a certain number of cycles, each subsequent one gives rise to identical increments of plastic strain in certain cross sections of the structure.
3. After a certain number of cycles, each subsequent one gives rise to increments of plastic strain that have identical moduli but opposite signs.

The first situation is favorable: we say that the structure shakes down (adapts itself) to the given variable load. The two remaining situations are dangerous. In case 2 the accumulation of plastic strain leads to an increasing distortion of the structure's shape. The repetitive plastic strain increments of opposite sign (case 3) cause fatigue and brittle fracture of the material. In shakedown theory, case 2 is called *incremental* or *progressive collapse*, and case 3 is referred to as *alternating* or *cyclic plasticity*.

Given the reference bounds $\mathbf{p}_0^-$, $\mathbf{p}_0^+$ and the properties of the structure (i.e., the matrices $\mathbf{C}$, $\mathbf{E}$, $\mathbf{N}$, $\mathbf{k}_*$), we can find the ultimate value μ_* of the common factor appearing in the following description of variable load:

$$\mu\mathbf{p}_0^- \leqslant \mathbf{p} \leqslant \mu\mathbf{p}_0^+. \tag{19.2}$$

The value μ_* defines that intensity of loading that corresponds to the exhaustion of the possibility of shakedown. Thus, taking $(\mathbf{p}_0^-, \mathbf{p}_0^+)$ at the level of the service load, we obtain μ_* as the safety factor against plastic nonadaptation. A complete presentation of shakedown theory can be found in König (1987) or Gokhfeld and Cherniavsky (1980).

As in Chapter 13, we can move from analysis to design of the structure. Assuming a certain safety factor μ_*, we then look for the distribution k_* of plastic moduli that is optimal in the sense of a certain cost function (e.g., the linear one given by Eq. 12.34). This leads to modifications of the optimal design problems in Chapter 13 that take into account nonproportional and variable loading.

Optimal design of skeletal structures for shakedown was intensively studied by Čyras and his co-workers [Čyras (1969, 1974, 1975, 1983)]. A few attempts were made to optimize more complex structures; e.g., the numerical solution for plates can be found in Borkauskas and Atkociunas (1975). Generally speaking, the number of papers devoted to optimal design under shakedown constraints is very low when compared with those dealing with the ultimate load factor or with the deflections at shakedown. It seems that the main reason for such a situation is the difficulty introduced by the coupling between elastic and plastic properties of the cross section.

It is clear that varying the dimensions of the cross section changes not only its plastic moduli, such as yield moment or yield axial force, but its elastic stiffness as well. Since the shakedown conditions include the stresses evaluated for elastic response of the structure, the variability of E introduces the same degree of nonlinearity as that encountered in elastic optimal design. Therefore optimization for shakedown cannot be reduced to the most desirable form of dual LP problems. Additional assumptions that keep elastic stiffness constant under variable $\mathbf{k}$ lead to rather academic solutions, whereas sequential linearization requires lengthy calculations. For these reasons, optimal design for shakedown has not been included in this book.

19.3. Multilevel Optimization and Decomposition Techniques

It is well known that optimization of large-scale structures by MP methods can present some difficulties because enormous amounts of computing time and computer memory are needed. We can avoid such difficulties by using multilevel optimization and decomposition methods. In general, an integrated optimization problem involving many subsystems and variables cannot be decomposed into independent subproblems that can be independently optimized. The methods described in this section do permit decomposition of optimization problems into subproblems, each of which, when solved independently, yields the overall system optimum. The interconnection of the subsystems may take many forms, but one of the most common is the hierarchical form in which the second-level unit controls or coordinates the units on the level below, called the *first level* or *optimization level*. The subsystems are solved independently at the first level. Our goal at the second level is to coordinate the action of the first-level units so that we obtain the solution of the original problem.

The first step in the hierarchical approach is to separate the set of design variables Z and the set of constraints H into two classes. The first class contains the coordination variables ZA and the constraints HA, containing only the design variables ZA that are used to create the second, or coordination, level. The second class contains the rest of the design variables $ZB = Z\backslash ZA$ and the constraints $HB = H\backslash HA$, and is used to create the first, or optimization, level. After separating design variables and constraints, we can write the optimization problem as follows:

$$\min_{\mathbf{z}\in Z} C(\mathbf{z}) = \min_{\mathbf{z}\in ZA}\{\min_{\mathbf{z}\in ZB} C(\mathbf{z})\}. \tag{19.3}$$

Now, the basic problem is to decompose the optimization level. Two classes of decomposition methods can be distinguished: intuitive and formal. *Intuitive methods* construct iterative schemes that are reasonable but are not guaranteed to lead to a true optimum. *Formal methods*, on the other hand, converge to a true optimum under well-defined assumptions.

Intuitive optimization methods are based on the use of substructures. Each substructure is optimized under the assumption that loads on the substructure do not change as a result of resizing. The entire structure is analyzed periodically to obtain the new loads acting on each substructure. Giles (1971) applied this approach to aircraft wing design under static loading, where each substructure (cover panel–shear webs) was optimized by using the fully stressed design approach. Sobieszczanski-Sobieski (1979) and Sobieszczanski-Sobieski and Leondorf (1972) treated fuselage and wing structures, using mathematical programming for substructure optimization.

Sobieszczanski-Sobieski, James, and Dovi (1985) described a method for decomposing an optimization problem into a set of subproblems and a coordination problem that preserves coupling between the subproblems. The decomposition is achieved by separating the structural element optimization subproblems from the assembled structural optimization problem. Each element optimization and optimal sensitivity analysis yields the cross-sectional dimensions that minimize the cumulative measure of the element constraint violation as a function of the elemental forces and stiffness. The assembled structural optimization produces the overall mass and stiffness distributions optimized for minimum total mass subject to constraints that include the cumulative measures of the element constraint violations extrapolated linearly with respect to the element forces and stiffnesses. The method is introduced as a special case of a multilevel, multidisciplinary system optimization, and its algorithm is fully described for two-level optimization for structures assembled from finite elements of arbitrary type. Numerical results are given for an example of a framework to show that the decomposition method converges and yields results comparable to those obtained without decomposition. They pointed out that optimization by decomposition should reduce the design time by allowing groups of engineers using different computers to work concurrently on the same large problem.

Kirsch, Reiss, and Shamir (1972) and Kirsch (1975) proposed analyzing the total structure after each substructure optimization to improve the convergence of the method. This approach is ideally suited for local constraints such as stress constraints. For global constraints we need to devise *ad hoc* procedures. Hughes, Mistree, and Zanic (1980) described one such approach for ship structures, and Sobieszczanski-Sobieski and Leondorf (1972) described another for aerospace applications. Schmit and Ramanatham (1978) and Schmit and Mehrinfar (1982) used a two-level approach to design trusses and aircraft wing structures. Ho (1974) applied limit design techniques to multistage planar trusses. He took advantage of the staircase structure of the constraint matrix to apply a nested decomposition technique that is guaranteed to converge to the optimal design. Chang (1981) used the Rosen (1964) decomposition algorithm to solve the problem of limit design of a planar truss structure subject to vulnerability constraints. Woo and Schmit (1981) and Kaneko and Cu Duong Ha (1983) applied a decomposition method to large-scale optimal plastic design problems. Recently, Jendo and Stachowicz (1986, 1988) and Stachowicz and Ziobron (1986) applied multilevel optimization to double-layer cable systems and to underground water tank systems.

19.4. Optimization with Discrete Design Variables

Optimization problems in structural design must often be considered to be of discrete type in which the variables can assume only certain predetermined values. Among the problems that commonly arise in structural design is the selection of standard steel sections for continuous or frame structures. Thus, from a practical standpoint, it is more convenient to formulate the design problem with discrete rather than continuous variables.

Many researchers have attempted to develop methods for discrete optimization of structures. Toakley (1968) formulated the optimal plastic design of frame structures as a discrete problem, in which the solution space consisted of the available standard sections, and solved it by using a mixed integer-continuous programming algorithm. Cella and Soosar (1973) presented a state-of-the-art paper on discrete variables in structural optimization. Lai and Achenbach (1973) presented a direct search algorithm to solve nonlinear optimization problems in which the variables were discrete. Glankwahmdee, Liebman, and Hogg (1979) presented a method specifically for nonlinear discrete unconstrained optimization problems. Next, Liebman, Khachaturian, and Chanaratna (1981), using the last paper as a starting point, developed a method of discrete optimization that uses concepts well-known for problems with continuous variables. Bauer, Gutkowski, and Iwanow (1981) presented a semianalytical approach to the discrete optimization of large space-truss structures. Sepulveda and Cassis (1986) developed a method for efficiently solving the problem of minimum-weight design of plane or space trusses with discrete or mixed variables.

Following Grierson and Lee (1986), we present a formulation of a minimum-weight design problem of planar frameworks under service and ultimate performance conditions, using discrete section members. Service load conditions ensure acceptable elastic stresses and displacements, and ultimate load conditions ensure adequate safety against plastic collapse. We assume that all loads are static and that specified service loads are proportional to specified ultimate loads. The framework is discretized into an assemblage of n prismatic members. For each member, the type of cross section is specified *a priori*, and a corresponding set of discrete sections from which its design cross section is to be selected is identified. The design variable for member i is its cross-sectional area a_i. The minimum-weight design problem can be stated as follows:

$$\min W = \sum_{i=1}^{n} w_i a_i, \tag{19.4}$$

subject to

$$\underset{\sim}{\delta}_j \leqslant \delta_j \leqslant \tilde{\delta}_j, \qquad j = 1, 2, \ldots, d, \tag{19.5}$$

$$\underset{\sim}{\sigma}_k \leqslant \sigma_k \leqslant \tilde{\sigma}_k, \qquad k = 1, 2, \ldots, s, \tag{19.6}$$

$$\underset{\sim}{\alpha}_m \leqslant \alpha_m \leqslant \tilde{\alpha}_m, \qquad m = 1, 2, \ldots, p, \tag{19.7}$$

$$a_i \in A_i, \qquad i = 1, 2, \ldots, n. \tag{19.8}$$

Equation (19.4) defines the weight of the structure (w_i is the weight coefficient for member i); inequality (19.5) defines the d service load constraints on displacements (undertilde and overtilde denote specified lower and upper bounds, respectively); inequality (19.6) defines the s service load constraints on stresses σ_k; inequality (19.7) defines the p ultimate load constraints on plastic collapse load factors α_m; and condition (19.8) requires each member cross-sectional area a_i to belong to a specified set of discrete section areas $A_i = \{a_1, a_2, \ldots\}_i$.

This discrete-weight optimization problem was solved by an iterative method, using an efficient generalized optimality criteria technique due to Fleury (1979). The iterative process terminates with the final design when there is no change in the structural weight from one design stage to the next [for details see Grierson and Lee (1986)].

19.5. Optimal Design with Random Parameters

Two approaches in optimal structural design are the *deterministic* and *probabilistic*; the latter depends on the randomness of basic variables such as loads, strengths, and geometrical quantities. Deterministic optimization has been well known for at least five decades. Brandt (1984) has an historical review with an almost complete bibliography in this area. Probabilistic optimization is a much younger discipline, which started in the early 1960s [for a review see Brandt, Jendo, and Marks (1984)]. Following Davidson, Felton, and Hart (1977a, b) and Moses (1977), we state the minimum-weight optimization problem with random parameters as follows:

Minimize the objective function

$$F(\mathbf{x}) \tag{19.9}$$

subject to the constraints

$$P_0(\mathbf{x}) \leq p_0, \tag{19.10}$$

$$P_i(\mathbf{x}) = P\{g_i(\mathbf{x}) > G_i(\mathbf{x})\} \leq p_i, \qquad i = 1, 2, \ldots, M, \tag{19.11}$$

where $\mathbf{x}^{\mathrm{T}} = (x_1, x_2, \ldots, x_n, \ldots, x_N)$ is a vector of N random variables, the first n of which are the design variables; $P_0(\mathbf{x})$ denotes the overall probability of failure, and p_0 is the allowable limit for this quantity; $g_i(\mathbf{x})$ is the ith response of the structure (e.g., displacement or stress), and $G_i(x)$ its allowable limit. Note that, in many cases, the allowable response G_i may be independent of $\mathbf{x}$. The expression $P\{\ldots\} \leq p_i$ in Eq. (19.11) means that the probability $P_i(\mathbf{x})$ of the ith response being greater than the allowable limit must not exceed a specified failure probability p_i. The probabilities p_0

and p_i will generally have to be very small for constraints (19.10) and (19.11) to be active and sensitive to the distributions of the individual variables.

Assuming all random variables to be normally distributed, we may relate p_i to response quantities by

$$p_i = 1 - \Phi(e_i), \tag{19.12}$$

where $\Phi(e_i)$ is the cumulative normal distribution function for which

$$e_i = \frac{|\bar{G}_i(\mathbf{x}) - \bar{g}_i(\mathbf{x})|}{\{\sigma^2_{G_i}(\mathbf{x}) + \sigma^2_{g_i}(\mathbf{x})\}^{1/2}} \tag{19.13}$$

is the coupling equation between response and allowable response [see Haugen (1968)]. In Eq. (19.13), $\bar{G}_i(\mathbf{x})$ and $\bar{g}_i(\mathbf{x})$ denote mean values of $G_i(\mathbf{x})$ and $g_i(\mathbf{x})$, and $\sigma_{Gi}(\mathbf{x})$ and $\sigma_{g_i}(\mathbf{x})$ are the standard deviations of $G_i(\mathbf{x})$ and $g_i(\mathbf{x})$, respectively.

From Eq. (19.12), we see that specification of p_i also specifies a corresponding value of e_i. Thus, for a given p_i, we may find e_i from Φ-tables and replace the original constraint (19.11) by the equivalent condition

$$\bar{g}_i(\mathbf{x}) + e_i\{\sigma^2_{G_i}(\mathbf{x}) + \sigma^2_{g_i}(\mathbf{x})\}^{1/2} - \bar{G}_i(x) \leq 0, \qquad e_i > 0. \tag{19.14}$$

An important factor in the reliability analysis of a structure is the relationship between overall probability of failure and the probabilities of the individual failure modes. For multiply loaded hyperstatic structures, which are assumed to fail when any constraint is violated ("weakest-link" model), failure modes are usually neither completely statistically dependent nor completely statistically independent, and exact correlations are difficult to determine. However, these two extreme cases provide a lower bound and an upper bound on the overall reliability:

$$\max_i P_i(\mathbf{x}) \leq P_0(\mathbf{x}) \leq \sum_{i=1}^{M} P_i(\mathbf{x}). \tag{19.15}$$

If the design involves relatively few active failure modes, which is often the case, the upper bound can reasonably be used for actual evaluation of P_0:

$$P_0(\mathbf{x}) = \sum_{i=1}^{M} P_i(\mathbf{x}). \tag{19.16}$$

In this relation, $P_i(\mathbf{x})$ is obtained from a modified form of Eq. (19.12):

$$P_i(\mathbf{x}) = 1 - \Phi(\bar{e}_i), \tag{19.17}$$

where $\bar{e}_i$ is the value of e satisfying relation (19.14) as an equality.

The foregoing considerations allow the original optimization problem to be restated as

$$\min F(\mathbf{x}) \tag{19.18}$$

subject to

$$\sum_{i=1}^{M} P_i(\mathbf{x}) - p_0 \leqslant 0, \tag{19.19}$$

$$\bar{g}_i(\mathbf{x}) + e_i\{\sigma_{G_i}^2(\mathbf{x}) + \sigma_{g_i}^2(\mathbf{x})\}^{1/2} - \bar{G}_i(\mathbf{x}) \leqslant 0, \qquad i = 1, 2, \ldots, M. \tag{19.20}$$

To find appropriate approximate expressions for the standard deviations of the response quantities, we replace the responses by the first-order terms of their Taylor series expansions about the means:

$$g_i(\mathbf{x}) = \bar{g}_i(\mathbf{x}) + \sum_{j=1}^{N} \left.\frac{\partial g_i}{\partial x_j}\right|_{\bar{\mathbf{x}}} (x_j - \bar{x}_j). \tag{19.21}$$

The variance of the ith response is defined by

$$\sigma_{g_i}^2(\mathbf{x}) = E\{g_i(\mathbf{x}) - \bar{g}_i(\mathbf{x})\}^2, \tag{19.22}$$

where $E\{\ldots\}$ denotes the expected value. Substituting $g_i(\mathbf{x})$ from (19.21) into (19.22), we get

$$\sigma_{g_i}^2(\mathbf{x}) = E\left\{\sum_{j=1}^{N}\sum_{k=1}^{N} \left.\frac{\partial g_i}{\partial x_j}\right|_{\bar{\mathbf{x}}} \left.\frac{\partial g_i}{\partial x_k}\right|_{\bar{\mathbf{x}}} (x_j - \bar{x}_j)(x_k - \bar{x}_k)\right\}; \tag{19.23}$$

i.e.,

$$\sigma_{g_i}^2(\mathbf{x}) = \sum_{j=1}^{N}\sum_{k=1}^{N} \left.\frac{\partial g_i}{\partial x_j}\right|_{\bar{\mathbf{x}}} \left.\frac{\partial g_i}{\partial x_k}\right|_{\bar{\mathbf{x}}} E\{(x_j - \bar{x}_j)(x_k - \bar{x}_k)\}. \tag{19.24}$$

The expected value term in this equation is, by definition, the covariance of x_j and x_k:

$$E\{(x_j - \bar{x}_j)(x_k - \bar{x}_k)\} = \operatorname{cov}(x_j, x_k). \tag{19.25}$$

For numerical purposes, the covariance is conveniently represented in terms of the correlation coefficient ρ_{jk}, $-1 \leqslant \rho_{jk} \leqslant 1$, as

$$\operatorname{cov}(x_j, x_k) = \rho_{jk}\sigma_j\sigma_k, \tag{19.26}$$

where σ_j and σ_k are standard deviations of the random variables x_j and x_k, respectively. A value $\rho_{jk} = \pm 1$ implies complete statistical dependence for x_j and x_k, and $\rho_{jk} = 0$ implies statistical independence.

Furthermore, σ_j and σ_k can be related to $\bar{x}_j$ and $\bar{x}_k$ through the coefficients of variation γ_j and γ_k, defined by

$$\gamma_l = \frac{\sigma_l}{\bar{x}_l}, \qquad l = 1, 2, \ldots, N. \tag{19.27}$$

Substituting Eqs. (19.25)-(19.27) into (19.24) gives

$$\sigma_{g_i}(\mathbf{x}) = \left[\sum_{j=1}^{N} \sum_{k=1}^{N} \left.\frac{\partial g_i}{\partial x_j}\right|_{\bar{\mathbf{x}}} \left.\frac{\partial g_i}{\partial x_k}\right|_{\bar{\mathbf{x}}} \rho_{jk}\gamma_j\gamma_k\bar{x}_j\bar{x}_k \right]^{1/2}. \tag{19.28}$$

In this way, the standard deviations of the responses have been expressed in terms of the mean values of the random variables, the correlation coefficients, and the coefficients of variation. The values of ρ_{jk}, γ_j, and γ_k must be specified before beginning the design process.

Recently, a few papers have been published on optimal design with reliability constraints. Murotsu *et al.* (1984b) considered frame structures minimizing the structural cost or weight under reliability constraints. For design variables they took the geometrical dimensions of the structure. Its configuration and loading conditions were specified together with their probabilistic nature. The optimization problem was reduced to a NLP problem, which was solved by sequential linear programming. Frangopol (1984) presented a reliability-based computational technique for designing optimal plastic structures. In his computation, the correlation between loads and the correlation between plastic moments was accounted for by using a technique that incorporated the effect of the statistical dependence between any two collapse mechanisms. Cheng and Chang (1985) employed two models of probability distribution, one dead-load model, four live-load models, one seismic force model, and two different expressions on variance formulation in optimal design of steel frames for a set of specified probabilities of failure. Furukawa, Furuta, and Ichikawa (1985) made an attempt to derive a probabilistic optimality deformation criterion for minimum-weight design. Next, they proposed a new probabilistic optimality criterion that combines Switzky's (1965) probabilistic optimality stress criterion and their probabilistic deformation criterion through the envelope method. They demonstrated with examples the efficiency and convergence of their method for unifying stress and deformation requirements. Feng and Moses (1986, 1987) optimized systems with reliability constraints and showed optimal weight characteristics of statically determinate and statically indeterminate structures.

19.6. Multicriteria Reliability-Based Optimal Structural Design

We consider here a general formulation of multicriteria reliability-based optimization where two mutually conflicting criteria, namely minimum cost (weight) and maximum reliability of the structural system, are used. The bicriteria reliability-based optimization problem can be formulated as follows [see Casciati and Faravelli (1985), Frangopol (1985), or Kishi *et al.* (1985)]:

$$\min_{\mathbf{x}\in\Omega} C(\mathbf{x}) \tag{19.29}$$

and

$$\max_{\mathbf{x}\in\Omega} R(\mathbf{x}, \mathbf{u}) = \max_{\mathbf{x}\in\Omega}\{1 - P_f(\mathbf{x}, \mathbf{u})\} = \min_{\mathbf{x}\in\Omega} P_f(\mathbf{x}, \mathbf{u}) \tag{19.30}$$

where $\Omega = \{\mathbf{x}\colon \mathbf{x} \geqslant 0, \beta_s(\mathbf{x}, \mathbf{u}) - \beta_{sa} \geqslant 0\}$ is the domain of feasible solutions, $C(\mathbf{x})$ is the cost function, $\mathbf{x}$ is the vector of design variables, $R(\mathbf{x}, \mathbf{u})$ is the reliability function, P_f is the probability of failure of structural system, $\mathbf{u}$ is the vector of basic random variables, $\beta_s(\mathbf{x}, \mathbf{u})$ is the system reliability index [see Hohenbichler and Rackwitz (1983) or Lind, Krenk, and Overgaard Madsen (1984)], and β_{se} is some target reliability index [see Sorensen and Thoft-Christensen (1986)].

A few procedures allow us to handle the multicriteria optimization problem. One is transforming the multicriteria problem to a single-criterion optimization problem by using the utility function method (see Section 18.4.7); i.e.,

$$\max_{\mathbf{x}\in\Omega} U = \max_{\mathbf{x}\in\Omega}\{-\{C_I(\mathbf{x}) + P_f(\mathbf{x}, \mathbf{u})C_f\}\}, \tag{19.31}$$

where C_I and C_f are the initial and failure costs of the system, respectively, U is the utility function, and P_f is the probability of failure of the system. This formulation requires the monetary value of C_f, which must take into account cost associated with "intangible" aspects, such as the loss of human life.

To avoid such nontechnical and noneconomical items, we can formulate directly the single-criterion optimization problem with reliability constraints as follows:

$$\min_{\mathbf{x}\in\Omega} C_I(\mathbf{x}) \tag{19.32}$$

subject to

$$P_f(\mathbf{x}, \mathbf{u}) \leqslant P_{fa} \tag{19.33}$$

or

$$P_{f_j}(\mathbf{x}, \mathbf{u}) \leqslant P_{f_{ja}}, \qquad j = 1, 2, \ldots, m, \tag{19.34}$$

where P_{fa} is the allowable overall failure probability of the system and $P_{f_{ja}}$ is the allowable failure probability of the jth failure mode.

Another way to handle the multicriteria optimization problem is to generate the set of compromise solutions and to choose a preferable solution by means of additional nontechnical and noneconomical criteria—e.g., the minimax approach (see Section 18.4.4) or fuzzy set theory approach [see Kickert (1978)]. The first step in this process consists of finding the so-called ideal solution, that is, the vector $f_j(\mathbf{x}^{\text{id}})$, $j = 1, 2, \ldots, k$, that satisfies the extremum condition of each objective function $f_j(\mathbf{x})$ considered independently of the remaining ones. Then the metric (distance) function is formulated as follows:

$$F^{(p)} = \left\{ \sum_{j=1}^{k} |f_j(\mathbf{x}) - f_j(\mathbf{x}^{\text{id}})|^p \right\}^{1/p}, \qquad 1 \leqslant p < \infty, \tag{19.35}$$

where k is the number of objective functions. The preferable solution can be found by requiring that the distance between the compromise set and the ideal solution be a minimum. The optimal solution depends substantially on the parameter p [see Cohon (1978)].

To solve the multicriteria reliability-based optimization problem (19.29)–(19.30) as formulated, we must calculate the system reliability index [see Lind *et al.* (1984) or Thoft-Christensen and Baker (1982)]

$$\beta_s = -\Phi^{-1}\{1 - \Phi_h(\boldsymbol{\beta}, \boldsymbol{\rho})\}, \tag{19.36}$$

where Φ^{-1} is the inverse standard normal distribution function and $\Phi_h(\cdot\,;\boldsymbol{\rho})$ is the standard distribution function for h normal variables with correlation coefficient matrix $\boldsymbol{\rho}$. The reliability index $\boldsymbol{\beta}$ is defined as the shortest distance from the origin in the u-space to the failure surface $g(T(\mathbf{y})) = 0$; i.e.,

$$\beta = \min_{\{\mathbf{u}:g(\mathbf{u})=0\}} \left\{ \sum_{i=1}^{N} u_i^2 \right\}^{1/2}, \tag{19.37}$$

where N is the number of basic random variables that are collected in the vector $\mathbf{Y} = (Y_1, Y_2, \ldots, Y_N)$ with given density function $f(y)$ [see Lind *et al.* (1984)].

The reliability of structural systems can be defined in several ways. Many papers and a few books are devoted to the calculation of structural reliability. The most general approach takes its origin in Cornell (1969a, b). Some authors followed this idea and developed the approach, which is now known as the *advanced first-order, second-moment approximation method.* Among the researchers who contributed are Hasofer and Lind (1974), Rackwitz and Fiessler (1978), and Ditlevsen (1979a, b, 1981). Rackwitz and Fiessler (1978) presented an algorithm for calculating structural reliability under combined loading. Leporati (1979) described the principles and methods that characterize the three design and safety checking levels (referred to as level 3, level 2, and level 1) of civil engineering structures. Okada *et al.* (1983) proposed a method for generating safety margins of general frame structures, taking account of interaction of the applied loads on yielding of the sections and structural failure defined as production of large nodal displacements due to plastic collapse. Murotsu (1983) used the combinatorial properties of the branch and bound approach for the appropriate selection of the dominant failure paths. Dolinski (1983) and Murotsu *et al.* (1984a) discussed the advanced first-order second-moment approximation method in reliability of structural systems.

Thoft-Christensen and Soerensen (1984) modeled frame and truss structures as systems with ductile elements. They assumed that all elements were linearly elastic and perfectly plastic, and they presented the reliability analysis of structures in two stages. In the first stage the significant failure modes were determined. In the second stage they estimated the failure probability of the structure on the basis of the significant failure modes. Sikorski and Borkowski (1985) considered a static approach to the probabilistic analysis and optimization of structures. They assumed a rigid, perfectly plastic material and discussed two types of problems. The first corresponds to the chance-constrained approach in the theory of stochastic programming [see Charnes and Cooper (1959)]. It allows us to evaluate the ultimate load factor or to find the optimal distribution of plastic moduli under the prescribed probability that the stress state remains statically admissible. The second approach gives the lower bound on the reliability index of the structure, taking into account possible correlation between stochastic parameters (resistances or loads). In the discrete model of a structure, all the cases considered turn out to be reducible to appropriate deterministic LP or NLP problems. The book by Augusti, Baratta, and Casciati (1984) on the probabilistic method in structural engineering is suitable for use by engineers with no background in probability and statistics.

20

Solutions to Exercises

Solutions for Chapter 10

10.1. The convexity of Ω_1 and the nonconvexity of Ω_2 is seen from Fig. S.1.

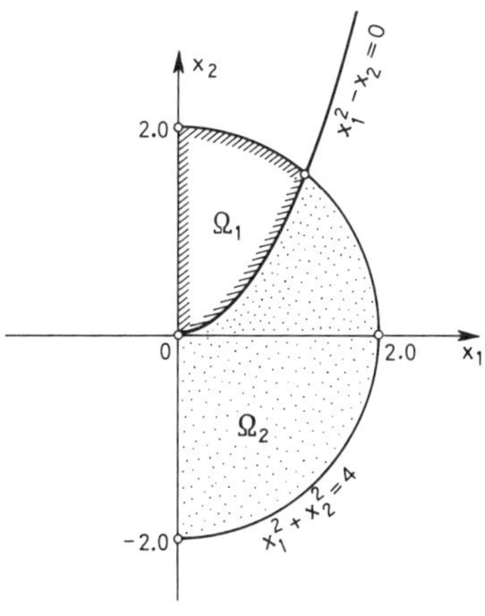

Figure S.1. Domains Ω_1 and Ω_2.

10.2. The Hessian matrix of the function considered is

$$\mathbf{F} = \begin{bmatrix} 4 & 0 \\ 0 & 12 \end{bmatrix}.$$

The product $\mathbf{x}^{\mathrm{T}}\mathbf{F}\mathbf{x} = 4x_1^2 + 12x_2^2$ remains positive for any $x_1 \neq 0$ and $x_2 \neq 0$. Hence, the function is strictly convex.

10.3. Applying formulas (10.2) and (10.3), we obtain $f' = x^2 + y^2$ and $f'' = -x^2 - y^2$.

10.4. The graphical solution of problem (10.8) is shown in Fig. S.2. The cost function attains its constrained minimum value of 2.5 at the point $\mathbf{x}_* = (1.0, 1.5)$. Note that for $f = x_1 + 2x_2$ such a point would not be unique.

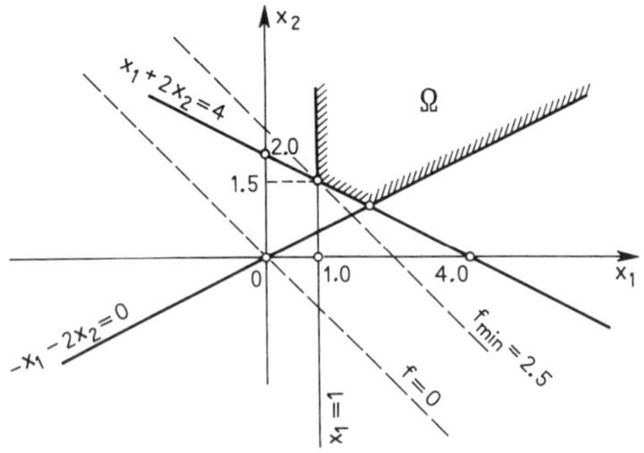

Figure S.2. Graphical Solution of LP Problem (10.8).

10.5. The graphical solution of problem (10.9) is shown in Fig. S.3. The cost function attains its constrained minimum value of 4.5 at the point $\mathbf{x}_* = (1.5, 1.5)$. Note that Ω is closed. Hence, inverting the problem to one of maximization is possible: the same cost function attains its constrained maximum value of 32.0 at the point $\mathbf{x}'_* = (4.0, 4.0)$.

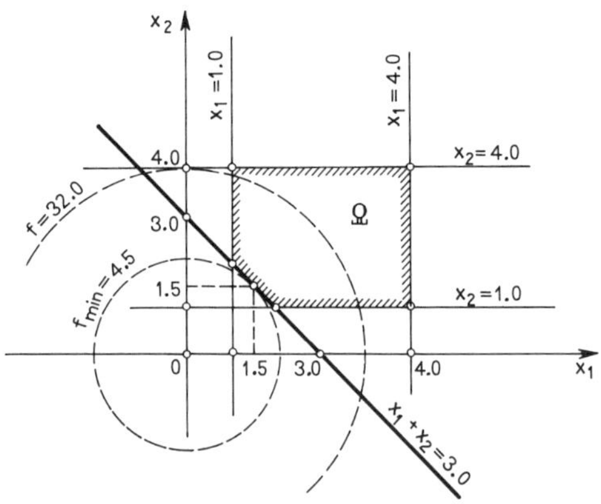

Figure S.3. Graphical Solution of QP Problem (10.9).

10.6. Figure S.4 shows a graphical interpretation of problem (10.10). The admissible domain Ω is formed by the intersection of two convex sets bounded by parabolic curves. The ellipsoidal cost function reaches its constrained minimum value of 4.44 at the vertex $\mathbf{x}_* = (2.0, 4.0)$ of that domain.

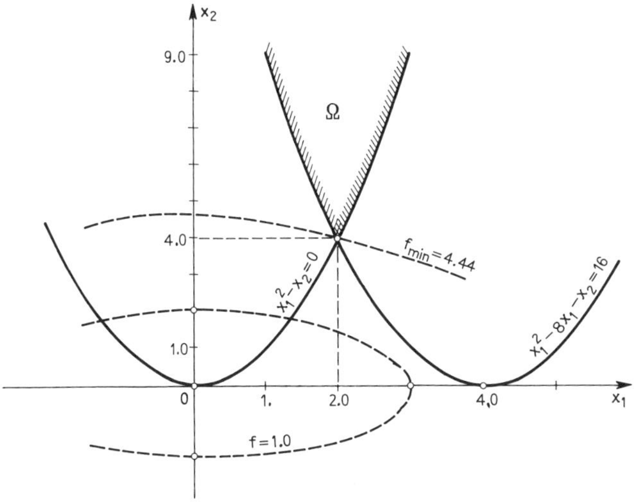

Figure S.4. Graphical Solution of NLP Problem (10.10).

10.7. According to (10.19), the Lagrangian function for problem (10.8) is

$$L(x_1, x_2, y_1, y_2, y_3) = x_1 + x_2 + y_1(4 - x_1 - 2x_2) + y_2(x_1 - 2x_2) + y_3(1 - x_1).$$

Then, according to (10.20), the KKT conditions are

$$\begin{aligned}
&\nabla_{x_1} L = 1 - y_1 + y_2 - y_3 = 0, && y_1 \geqslant 0, y_2 \geqslant 0, y_3 \geqslant 0,\\
&\nabla_{x_2} L = 1 - 2y_1 - 2y_2 = 0, && y_1(4 - x_1 - 2x_2) = 0,\\
&\nabla_{y_1} L = 4 - x_1 - 2x_2 \leqslant 0, && y_2(x_1 - 2x_2) = 0,\\
&\nabla_{y_2} L = x_1 - 2x_2 \leqslant 0, && y_3(1 - x_1) = 0.\\
&\nabla_{y_3} L = 1 - x_1 \leqslant 0, &&
\end{aligned}$$

The solution $\mathbf{x}_* = (1.0, 1.5)$, $\mathbf{y}_* = (0.5, 0.0, 0.5)$ satisfies these conditions.

10.8. According to (10.17), the Lagrangian function for problem (10.9) is

$$L(x_1, x_2, y_1, y_2, y_3, y_4, y_5) = x_1^2 + x_2^2 + y_1(3 - x_1 - x_2) + y_2(-4 - x_1) + y_3(1 - x_1) + y_4(-4 - x_2) + y_5(1 - x_2).$$

Then, according to (10.18), the KKT conditions are

$$\begin{aligned}
&\nabla_{x_1} L = 2x_1 - y_1 - y_2 - y_3 = 0, && y_1 \geq 0,\ y_2 \geq 0, \ldots, y_5 \geq 0,\\
&\nabla_{x_2} L = 2x_2 - y_1 - y_4 - y_5 = 0, && y_1(3 - x_1 - x_2) = 0,\\
&\nabla_{y_1} L = 3 - x_1 - x_2 \leq 0, && y_2(-4 - x_1) = 0,\\
&\nabla_{y_2} L = -4 - x_1 \leq 0, && y_3(1 - x_1) = 0,\\
&\nabla_{y_3} L = 1 - x_1 \leq 0, && y_4(-4 - x_2) = 0,\\
&\nabla_{y_4} L = -4 - x_2 \leq 0, && y_5(1 - x_2) = 0.\\
&\nabla_{y_5} L = 1 - x_2 \leq 0,
\end{aligned}$$

The solution $\mathbf{x}_* = (1.5, 1.5)$, $\mathbf{y}_* = (3.0, 0.0, 0.0, 0.0, 0.0)$ satisfies these conditions.

10.9. According to (10.17), the Lagrangian function for problem (10.10) is

$$L(x_1, x_2, y_1, y_2) = \tfrac{1}{9}x_1^2 + \tfrac{1}{4}x_2^2 + y_1(x_1^2 - x_2) + y_2(x_1^2 - 8x_1 - x_2 + 16).$$

Applying formulas (10.15), we obtain the following KKT conditions:

$$\begin{aligned}
&\nabla_{x_1} L = \tfrac{2}{9}x_1 + 2x_1 y_1 + 2x_1 y_2 - 8y_2 = 0, && y_1 \geq 0,\ y_2 \geq 0,\\
&\nabla_{x_2} L = \tfrac{1}{2}x_2 - y_1 - y_2 = 0, && y_1(-x_1^2 + x_2) = 0,\\
&\nabla_{y_1} L = x_1^2 - x_2 \leq 0, && y_2(x_1^2 - 8x_1 - x_2 + 16) = 0.\\
&\nabla_{y_2} L = x_1^2 - 8x_1 - x_2 + 16 \leq 0,
\end{aligned}$$

The solution is $\mathbf{x}_* = (2.0, 4.0)$, $\mathbf{y}_* = (0.944, 1.056)$.

10.10. According to (10.24), the dual of the LP problem (10.8) is

$$\text{maximize } f = 4y_1 + y_3$$

subject to

$$\begin{aligned}
&y_1 - y_2 = 1,\\
&2y_1 + 2y_2 + y_3 = 1,\\
&y_1 \geq 0,\ y_2 \geq 0,\ y_3 \geq 0.
\end{aligned}$$

Its solution is $\mathbf{y}_* = (0.5, 0.0, 0.5)$ and the value $f_{\max} = 2.5$ coincides with the constrained minimum of the cost function of the primal problem (10.8).

10.11. The dual of type (10.23) for the QP problem (10.9) is

$$\text{maximize } f = -x_1^2 - x_2^2 + 3y_1 - 4y_2 + y_3 - 4y_4 + y_5$$

subject to

$$-x_1 + y_1 - y_2 + y_3 = 0,$$
$$-x_2 + y_1 - y_4 + y_5 = 0,$$
$$y_1 \geqslant 0,\ y_2 \geqslant 0, \ldots, y_5 \geqslant 0.$$

10.12. According to (10.22), the dual of the NLP problem (10.10) is

$$\text{maximize } L = \tfrac{1}{9}x_1 + \tfrac{1}{4}x_2^2 + y_1(x_1^2 - x_2) + y_2(x_1^2 - 8x_1 - x_2 + 16)$$

subject to

$$\tfrac{2}{9}x_1 + 2x_1y_1 + 2x_1y_2 - 8y_2 = 0,$$
$$\tfrac{1}{2}x_2 - y_1 - y_2 = 0,$$
$$y_1 \geqslant 0,\ y_2 \geqslant 0.$$

Its solution is $\mathbf{x}_* = (2.0, 4.0)$, $\mathbf{y}_* = (0.944, 1.056)$.

10.13. Taking

$$\mathbf{A}_{xx} = \begin{bmatrix} 4 & 1 \\ 1 & 4 \end{bmatrix}, \quad \mathbf{A}_{yy} = \begin{bmatrix} -2 & -1 \\ -1 & -1 \end{bmatrix}, \quad \mathbf{A}_{xy} = \begin{bmatrix} -1 & 2 \\ -1 & 0 \end{bmatrix},$$

$$\mathbf{x} = \begin{bmatrix} x_1 \\ x_2 \end{bmatrix}, \quad \mathbf{y} = \begin{bmatrix} x_3 \\ x_4 \end{bmatrix}, \quad \mathbf{b}_x = \begin{bmatrix} -4 \\ -2 \end{bmatrix}, \quad \mathbf{b}_y = \begin{bmatrix} -1 \\ -3 \end{bmatrix},$$

we obtain, according to (10.27), the following saddle potential:

$$f(\mathbf{x}, \mathbf{y}) = 2x_1^2 + 2x_2^2 - x_3^2 - \tfrac{1}{2}x_4^2 - x_1x_3 + 2x_1x_4 - x_2x_3 - 4x_1 - 2x_2 - x_3 - 3x_4.$$

Then the dual QP problems are

$$\text{minimize } 2x_1^2 + 2x_2^2 + x_3^2 + \tfrac{1}{2}x_4^2 - 4x_1 - 2x_2$$

subject to

$$-x_1 - x_2 - 2x_3 - x_4 = 1,$$
$$2x_1 - x_3 - x_4 = 3;$$

$$\text{maximize } -2x_1^2 - 2x_2^2 - x_3^2 - \tfrac{1}{2}x_4^2 - x_3 - 3x_4$$

subject to

$$4x_1 + x_2 - x_3 + 2x_4 = 4,$$
$$x_1 + 4x_2 - x_3 = 2.$$

10.14. For nonnegative variables x_2 and x_3, the QP problems given above are to be changed as follows:

$$\text{minimize } 2x_1^2 + 2x_2^2 + x_3^2 + \tfrac{1}{2}x_4^2 - 4x_1 - 2x_2$$

subject to

$$-x_1 - x_2 - 2x_3 - x_4 \leqslant 1,$$

$$2x_1 - x_3 - x_4 = 3,$$

$$x_2 \geqslant 0;$$

$$\text{maximize } -2x_1^2 - 2x_2^2 - x_3^2 - \tfrac{1}{2}x_4^2 - x_3 - 3x_4$$

subject to

$$4x_1 + x_2 - x_3 + 2x_4 = 4,$$

$$x_1 + 4x_2 - x_3 \geqslant 2,$$

$$x_3 \geqslant 0.$$

Solutions for Chapter 11

11.1. The transposed compatibility matrix is given in Table S.1. It is obtained by considering the equilibrium of each node of the truss (except the fixed node 4).

Table S.1. Matrix $\mathbf{C}^T$ for the Truss in Fig. 11.10

	n_1	n_2	n_3	n_4	n_5	n_6	n_7	n_8	n_9	n_{10}	n_{11}
$p_y^{(1)} =$	0	0	0	0	−1.000	0	0	−0.832	0	0	0
$p_x^{(2)} =$	1.000	−1.000	0	0	0	0	0	0	0.555	−0.555	0
$p_y^{(2)} =$	0	0	0	0	0	−1.000	0	0	−0.832	−0.832	0
$p_x^{(3)} =$	0	1.000	0	0	0	0	0	0	0	0	0.555
$p_y^{(3)} =$	0	0	0	0	0	0	−1.000	0	0	0	−0.832
$p_x^{(5)} =$	0	0	1.000	−1.000	0	0	0	0.555	0	0	−0.555
$p_y^{(5)} =$	0	0	0	0	0	1.000	0	0.832	0	0	0.832
$p_x^{(6)} =$	0	0	0	1.000	0	0	0	0	0	0.555	0
$p_y^{(6)} =$	0	0	0	0	0	0	1.000	0	0	0.832	0

11.2. Table S.2 contains the transposed compatibility matrix of the beam. The corresponding matrices $\mathbf{C}^-$ and $\mathbf{C}^0$ are given in Tables S.3 and S.4, respectively.

Table S.2. Matrix $\mathbf{C}^{\mathrm{T}}$ for the Beam in Fig. 11.11

	$m_a^{(1)}$	$m_b^{(1)}$	$m_a^{(2)}$	$m_b^{(2)}$	$m_a^{(3)}$	$m_b^{(3)}$
$p_y^{(2)} =$	−1.250	−1.250	1.000	1.000	0	0
$p_z^{(2)} =$	0	1.000	1.000	0	0	0
$p_z^{(3)} =$	0	0	0	1.000	1.000	0
$p_z^{(4)} =$	0	0	0	0	0	1.000

Table S.3. Matrix $\mathbf{C}^{-}$ for the Beam in Fig. 11.11

	$\theta_a^{(1)}$	$\theta_b^{(1)}$	$\theta_a^{(2)}$	$\theta_b^{(2)}$	$\theta_a^{(3)}$	$\theta_b^{(3)}$
$w_y^{(2)} =$	−0.80	0	0	0	0	0
$w_z^{(2)} =$	0.80	0	1.0	0	0	0
$w_z^{(3)} =$	0	0	0	0	1.0	0
$w_z^{(4)} =$	0	0	0	0	0	1.0

Table S.4. Matrix $\mathbf{C}^{0}$ for the Beam in Fig. 11.11

	$\theta_a^{(1)}$	$\theta_b^{(1)}$	$\theta_a^{(2)}$	$\theta_b^{(2)}$	$\theta_a^{(3)}$	$\theta_b^{(3)}$
$0 =$	−1.80	1.0	−1.0	0	0	0
$0 =$	0.80	0	0	1.0	−1.0	0

11.3. Table S.5 contains the equilibrium matrix of the frame. The matrices $\mathbf{C}^{-}$ and $\mathbf{C}^{0}$ are given in Tables S.6 and S.7, respectively.

Table S.5. Matrix $\mathbf{C}^{\mathrm{T}}$ for the Frame in Fig. 11.12

	$m_a^{(1)}$	$m_b^{(1)}$	$m_a^{(2)}$	$m_b^{(2)}$	$m_a^{(3)}$	$m_b^{(3)}$
$p_x^{(2)} =$	−1.000	−1.000	0.667	0.667	−1.000	−1.000
$p_z^{(2)} =$	0	1.000	1.000	0	0	0
$p_z^{(3)} =$	0	0	0	1.000	1.000	0

Table S.6. Matrix $\mathbf{C}^{-}$ for the Frame in Fig. 11.12

	$\theta_a^{(1)}$	$\theta_b^{(1)}$	$\theta_a^{(2)}$	$\theta_b^{(2)}$	$\theta_a^{(3)}$	$\theta_b^{(3)}$
$w_x^{(2)} =$	−1.000	0	0	0	0	0
$w_z^{(2)} =$	0.667	0	1.000	0	0	0
$w_z^{(3)} =$	0.667	0	0	1.000	0	0

Table S.7. Matrix $\mathbf{C}^0$ for the Frame in Fig. 11.12

	$\theta_a^{(1)}$	$\theta_b^{(1)}$	$\theta_a^{(2)}$	$\theta_b^{(2)}$	$\theta_a^{(3)}$	$\theta_b^{(3)}$
0 =	−1.667	1.000	−1.000	0	0	0
0 =	−1.667	0	0	−1.000	1.000	0
0 =	−1.000	0	0	0	0	1.000

11.4. If bending moments only are taken into account, then the matrix $\mathbf{C}^T$ for the frame takes the form defined by Table S.8. Note that the absence of axial strain implies a common horizontal displacement of nodes 3 and 6, as well as the zero value of the vertical displacements of nodes 2, 3, 5, and 6. In contrast, for the $M + N$ model each of these nodes possesses three degrees of freedom. The $\mathbf{C}^T$ matrix for the latter case is given in Table S.9.

Table S.8. Matrix $\mathbf{C}^T$ for the Frame in Fig. 11.13 (Model M)

	$m_a^{(1)}$	$m_b^{(1)}$	$m_a^{(2)}$	$m_b^{(2)}$	$m_a^{(3)}$	$m_b^{(3)}$	$m_a^{(4)}$	$m_b^{(4)}$	$m_a^{(5)}$	$m_b^{(5)}$
$p_x^{(2)} =$	−1.667	−1.667	2.500	2.500	0	0	0	0	0	0
$p_z^{(2)} =$	0	1.000	1.000	0	0	0	0	0	0	0
$p_x^{(3)} =$	0	0	−2.500	−2.500	0	0	−2.500	−2.500	0	0
$p_z^{(3)} =$	0	0	0	1.000	0	0	0	0	1.000	0
$p_x^{(5)} =$	0	0	0	0	−1.667	−1.667	2.500	2.500	0	0
$p_z^{(5)} =$	0	0	0	0	0	1.000	1.000	0	0	0
$p_z^{(6)} =$	0	0	0	0	0	0	0	1.000	0	1.000

Table S.9. Matrix $\mathbf{C}^T$ for the Frame in Fig. 11.13 (Model $M + N$)

	$m_a^{(1)}$	$m_b^{(1)}$	n_1	$m_a^{(2)}$	$m_b^{(2)}$	n_2	$m_a^{(3)}$	$m_b^{(3)}$
$p_x^{(2)} =$	−1.667	−1.667	0	2.500	2.500	0	0	0
$p_y^{(2)} =$	0	0	−1.0	0	0	1.0	0	0
$p_z^{(2)} =$	0	1.000	0	1.000	0	0	0	0
$p_x^{(3)} =$	0	0	0	−2.500	−2.500	0	0	0
$p_y^{(3)} =$	0	0	0	0	0	−1.0	0	0
$p_z^{(3)} =$	0	0	0	0	1.000	0	0	0
$p_x^{(5)} =$	0	0	0	0	0	0	−1.667	−1.667
$p_y^{(5)} =$	0	0	0	0	0	0	0	0
$p_z^{(5)} =$	0	0	0	0	0	0	0	1.000
$p_x^{(6)} =$	0	0	0	0	0	0	0	0
$p_y^{(6)} =$	0	0	0	0	0	0	0	0
$p_z^{(6)} =$	0	0	0	0	0	0	0	0

Table S.9. Continued.

	n_3	$m_a^{(4)}$	$m_b^{(4)}$	n_4	$m_a^{(5)}$	$m_b^{(5)}$	n_5
$p_x^{(2)} =$	0	0	0	0	0	0	0
$p_y^{(2)} =$	0	0	0	0	0	0	0
$p_z^{(2)} =$	0	0	0	0	0	0	0
$p_x^{(3)} =$	0	0	0	0	0	0	−1.0
$p_y^{(3)} =$	0	0	0	0	0	0	0
$p_z^{(3)} =$	0	0	0	0	1.000	0	0
$p_x^{(5)} =$	0	2.500	2.500	0	0	0	0
$p_y^{(5)} =$	−1.0	0	0	1.0	0	0	0
$p_z^{(5)} =$	0	1.000	0	0	0	0	0
$p_x^{(6)} =$	0	−2.500	−2.500	0	0	0	1.0
$p_y^{(6)} =$	0	0	0	−1.0	0	0	0
$p_z^{(6)} =$	0	0	1.000	0	0	1.000	0

11.5. The elasticity matrix **E** for the truss is diagonal:

$$\mathbf{E} = \lceil 0.9, 0.9, 0.9, 0.9, 0.6, 0.6, 0.6, 0.499, 0.499, 0.499 \rfloor \times 10^2.$$

11.6. The elasticity matrix for the frame is given in Table S.10.

Table S.10. Matrix **E** for the Frame in Fig. 11.12

	$\theta_a^{(1)}$	$\theta_b^{(1)}$	$\theta_a^{(2)}$	$\theta_b^{(2)}$	$\theta_a^{(3)}$	$\theta_b^{(3)}$
$m_a^{(1)} =$	0.350	0.175	0	0	0	0
$m_b^{(1)} =$	0.175	0.350	0	0	0	0
$m_a^{(2)} =$	0	0	0.467	0.233	0	0
$m_b^{(2)} =$	0	0	0.233	0.467	0	0
$m_a^{(3)} =$	0	0	0	0	0.313	0.156
$m_b^{(3)} =$	0	0	0	0	0.156	0.313

$\times 10^2$

Table S.11. Matrix **N** for the Frame in Fig. 11.12 (Model M)

	Beam 1				Beam 2				Beam 3			
	λ_a^{I}	λ_a^{II}	λ_b^{I}	λ_b^{II}	λ_a^{I}	λ_a^{II}	λ_b^{I}	λ_b^{II}	λ_a^{I}	λ_a^{II}	λ_b^{I}	λ_b^{II}
$\theta_a^{(1)} =$	1	−1	0	0	0	0	0	0	0	0	0	0
$\theta_b^{(1)} =$	0	0	1	−1	0	0	0	0	0	0	0	0
$\theta_a^{(2)} =$	0	0	0	0	1	−1	0	0	0	0	0	0
$\theta_b^{(2)} =$	0	0	0	0	0	0	1	−1	0	0	0	0
$\theta_a^{(3)} =$	0	0	0	0	0	0	0	0	1	−1	0	0
$\theta_b^{(3)} =$	0	0	0	0	0	0	0	0	0	0	1	−1

11.7. If bending moments only are taken into account, then the matrix **N** for the frame takes the form defined by Table S.11. For the rhombic interaction curve, the entries of that matrix are given in Table S.12, with $\xi = 0.894$ and $\eta = 0.447$.

Table S.12. Matrix **N** for the Frame in Fig. 11.12 (Model $M + N$)

	Beam 1								Beam 2			
	λ_a^{I}	λ_a^{II}	λ_a^{III}	λ_a^{IV}	λ_b^{I}	λ_b^{II}	λ_b^{III}	λ_b^{IV}	λ_a^{I}	λ_a^{II}	λ_a^{III}	λ_a^{IV}
$\theta_a^{(1)} =$	ξ	$-\xi$	$-\xi$	ξ	0	0	0	0	0	0	0	0
$\theta_b^{(1)} =$	0	0	0	0	ξ	$-\xi$	$-\xi$	ξ	0	0	0	0
$\varepsilon_1 =$	η	η	$-\eta$	$-\eta$	η	η	$-\eta$	$-\eta$	0	0	0	0
$\theta_a^{(2)} =$	0	0	0	0	0	0	0	0	ξ	$-\xi$	$-\xi$	ξ
$\theta_b^{(2)} =$	0	0	0	0	0	0	0	0	0	0	0	0
$\varepsilon_2 =$	0	0	0	0	0	0	0	0	η	η	$-\eta$	$-\eta$
$\theta_a^{(3)} =$	0	0	0	0	0	0	0	0	0	0	0	0
$\theta_b^{(3)} =$	0	0	0	0	0	0	0	0	0	0	0	0
$\varepsilon_3 =$	0	0	0	0	0	0	0	0	0	0	0	0

	Beam 2				Beam 3							
	λ_b^{I}	λ_b^{II}	λ_b^{III}	λ_b^{IV}	λ_a^{I}	λ_a^{II}	λ_a^{III}	λ_a^{IV}	λ_b^{I}	λ_b^{II}	λ_b^{III}	λ_b^{IV}
$\theta_a^{(1)} =$	0	0	0	0	0	0	0	0	0	0	0	0
$\theta_b^{(1)} =$	0	0	0	0	0	0	0	0	0	0	0	0
$\varepsilon_1 =$	0	0	0	0	0	0	0	0	0	0	0	0
$\theta_a^{(2)} =$	0	0	0	0	0	0	0	0	0	0	0	0
$\theta_b^{(2)} =$	ξ	$-\xi$	$-\xi$	ξ	0	0	0	0	0	0	0	0
$\varepsilon_2 =$	η	η	$-\eta$	$-\eta$	0	0	0	0	0	0	0	0
$\theta_a^{(3)} =$	0	0	0	0	ξ	$-\xi$	$-\xi$	ξ	0	0	0	0
$\theta_b^{(3)} =$	0	0	0	0	0	0	0	0	ξ	$-\xi$	$-\xi$	ξ
$\varepsilon_3 =$	0	0	0	0	η	η	$-\eta$	$-\eta$	η	η	$-\eta$	$-\eta$

Solutions for Chapter 12

12.1. The nondimensional axial forces induced in the isostatic truss by the given load are shown in Fig. S.5. The reference force is $P_0 = 10.0$ kN.

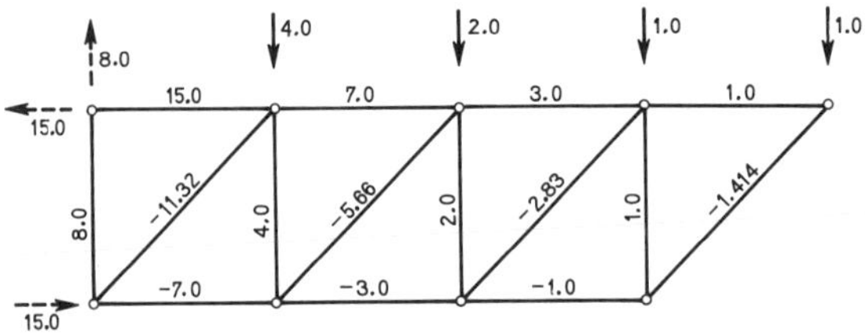

Figure S.5. Axial Forces in an Isostatic Truss.

12.2. Both methods yield the stiffness matrix in Table S.13.

Table S.13. Matrix **K** for the Truss in Fig. 11.10

	$w_y^{(1)}$	$w_x^{(2)}$	$w_y^{(2)}$	$w_x^{(3)}$	$w_y^{(3)}$	$w_x^{(5)}$	$w_y^{(5)}$	$w_x^{(6)}$	$w_y^{(6)}$	
$p_y^{(1)} =$	0.946									
$p_x^{(2)} =$	0	2.108								
$p_y^{(2)} =$	0	0	1.291		Symmetric					
$p_x^{(3)} =$	0	−0.90	0	1.054						
$p_y^{(3)} =$	0	0	0	−0.231	0.946					$\times 10^2$
$p_x^{(5)} =$	−0.231	0	0	−0.154	−0.231	2.108				
$p_y^{(5)} =$	−0.346	0	−0.60	0.231	−0.346	0	1.291			
$p_x^{(6)} =$	0	−0.154	−0.231	0	0	−0.90	0	1.054		
$p_y^{(6)} =$	0	−0.231	−0.346	0	−0.60	0	0	0.231	0.946	

Table S.14. Matrix **K** for the Beam in Fig. 11.11

	$w_y^{(2)}$	$w_z^{(2)}$	$w_z^{(3)}$	$w_z^{(4)}$	
$p_y^{(2)} =$	0.886				
$p_z^{(2)} =$	−0.084	0.225	Symmetric		$\times 10^2$
$p_z^{(3)} =$	0.150	0.050	0.225		
$p_z^{(4)} =$	0	0	0.062	0.125	

12.3. The stiffness matrix of the beam is given in Table S.14. The nondimensional displacements are

$$\mathbf{w}_* = (1.432, 0.831, -1.323, 0.661) \times 10^{-2}.$$

The corresponding bending moments are shown in Fig. S.6.

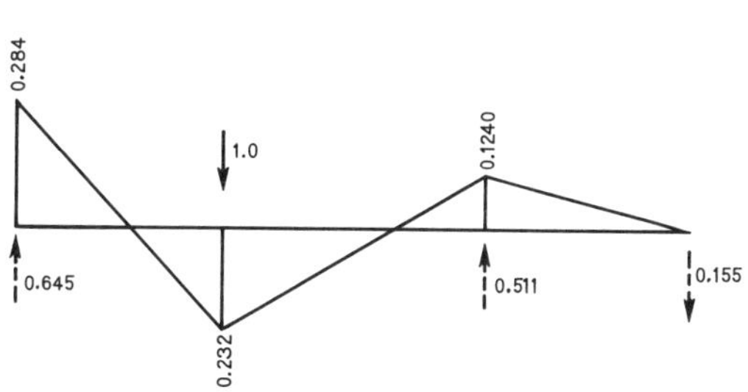

Figure S.6. Bending Moments in a Beam (Elastic Solution).

12.4. The flexibility matrix of the truss is

$$\mathbf{F} = \begin{bmatrix} 0.1011 & 0.0702 \\ 0.0702 & 0.1169 \end{bmatrix}.$$

This matrix relates the redundant unknowns $\mathbf{z} = (n_6, n_9)$ to the free terms $\mathbf{d} = (d_1, d_2)$. The nondimensional displacements are

$$\mathbf{w}_* = (0.0325, 0.0226, 0.0744, 0.0275,$$
$$0.1165, -0.0144, 0.0696, -0.0170, 0.1108).$$

The corresponding axial forces are shown in Fig. S.7.

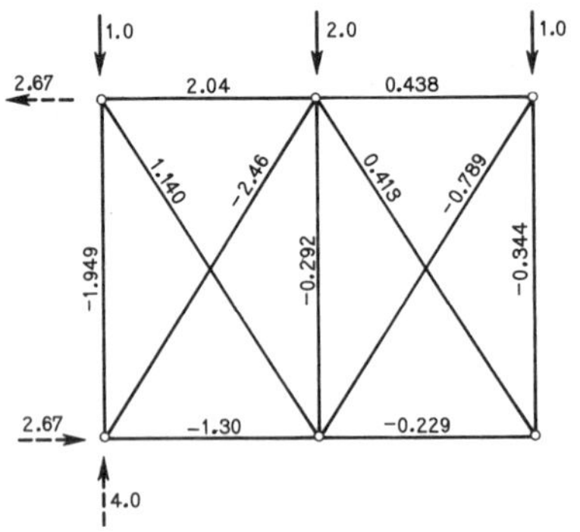

Figure S.7. Axial Forces in a Redundant Truss (Elastic Solution).

12.5. The flexibility matrix of the beam is

$$\mathbf{F} = \begin{bmatrix} 0.778 & -0.130 \\ -0.130 & 0.308 \end{bmatrix} \times 10^{-2}.$$

It relates the redundant unknowns $\mathbf{z} = (m_b^{(1)}, m_b^{(2)})$ to the free terms $\mathbf{d} = (d_1, d_2)$. The nondimensional displacements and bending moments are the same as in Exercise 12.3.

Solutions for Chapter 13

13.1. The ultimate load factor is $\mu_* = 1.374$. The nondimensional axial forces at the incipient plastic collapse are

$$\mathbf{s}_* = (0.861, -0.306, -0.667, -0.611, -1.000,$$
$$-1.000, -0.916, 0.651, -1.000, 1.101, 0.550).$$

These values are not unique, since the right panel of the truss remains rigid at plastic collapse (Fig. S.8).

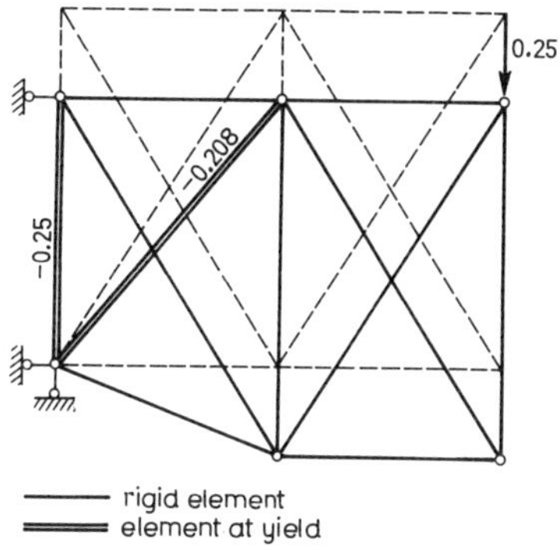

Figure S.8. Collapse Mechanism of a Redundant Truss.

13.2. For the given cross section, $M^0 = 10.0\,\text{kN m}$. Solving the dual problems (13.10), we obtain $\mu_* = 2.250$. The nondimensional bending moments at the incipient plastic collapse are given in Fig. S.9a, whereas the collapse mechanism is shown in Fig. S.9b. The obtained values of the bending moments are unique because the collapse mechanism is complete.

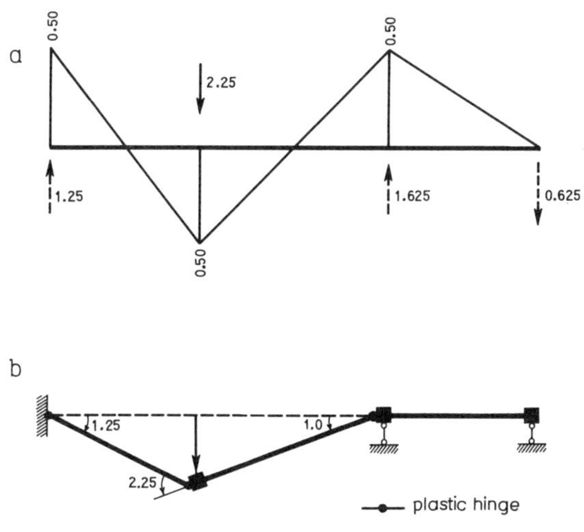

Figure S.9. Solution of the Ultimate Load Problem for a Beam: (a) Bending Moments at Incipient Collapse. (b) Collapse Mechanism.

13.3. The solution of the free design problem is shown in Fig. S.10. It gives the minimum value of the nondimensional volume $v_{\min} = 33.9$. Note that the optimal truss is isostatic and that the optimality criterion (13.33) is fulfilled as an equality only for nonvanishing bars.

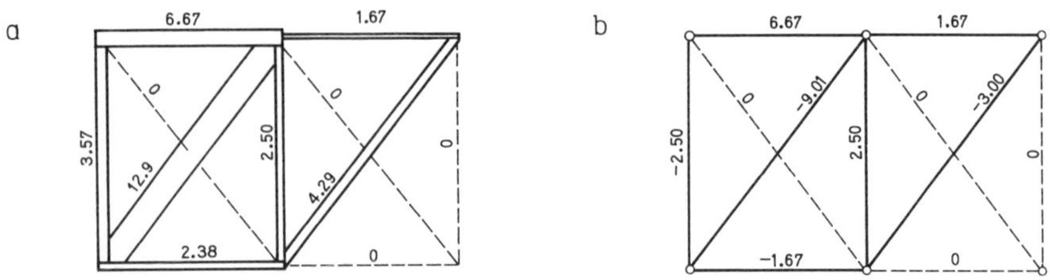

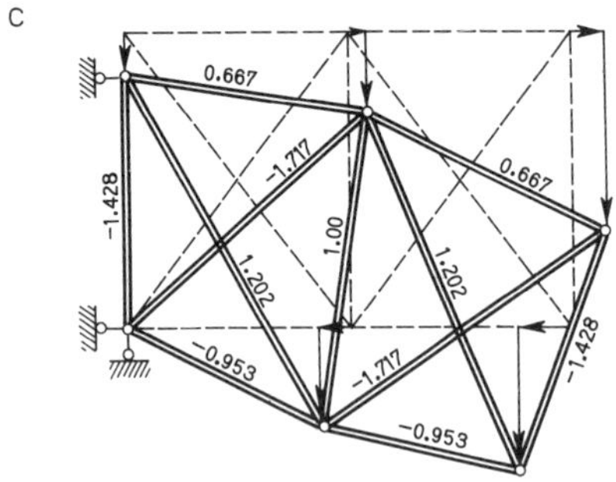

Figure S.10. Solution of the Free Optimal Design Problem for a Redundant Truss: (a) Yield Forces. (b) Axial Forces at Collapse. (c) Collapse Mechanism.

Figure S.11 shows the optimal solution for the regular truss. The volume of such a truss is 72.7, i.e., more than two times greater than that of the previous solution.

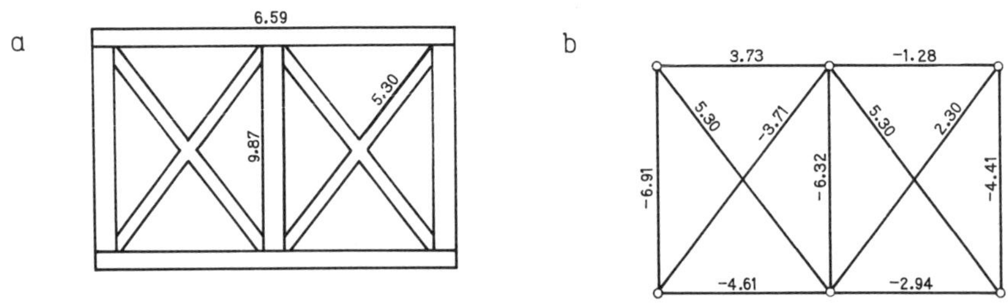

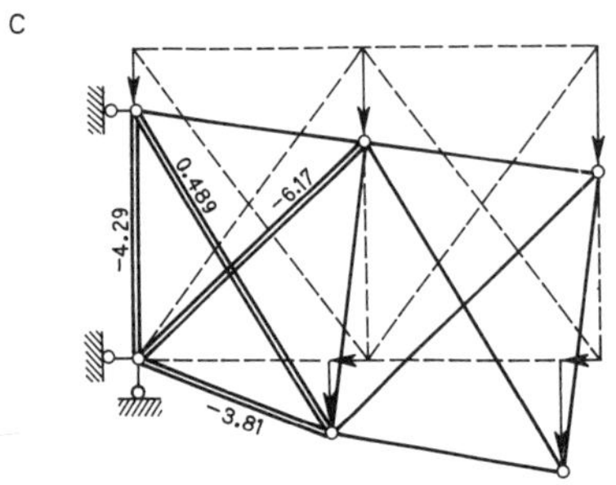

Figure S.11. Solution of the Constrained Optimal Design Problem for a Redundant Truss: (a) Yield Forces. (b) Axial Forces at Collapse. (c) Collapse Mechanism.

13.4. The results of optimization are shown in Fig. S.12. The minimum volume is $v_{min} = 1.388$. Note that the optimal frame has uniform columns even though no such restriction has been imposed *a priori.*

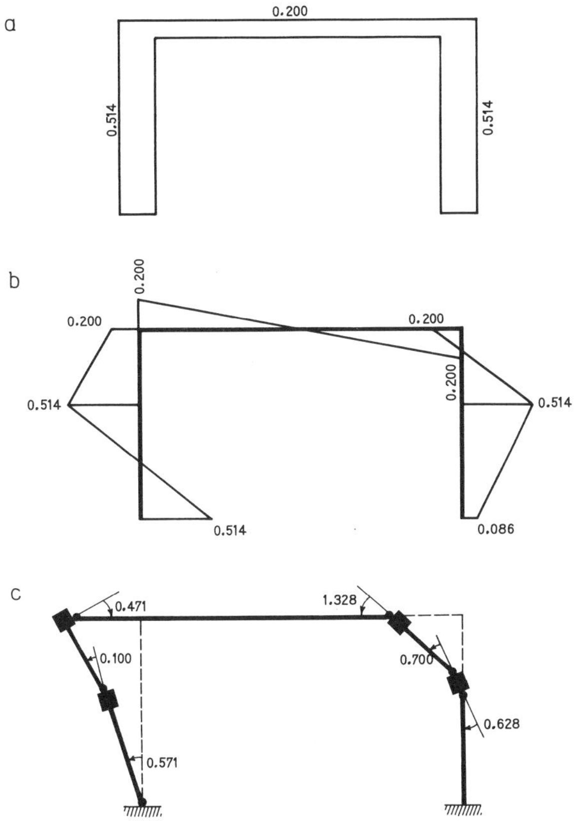

Figure S.12. Solution of the Free Optimal Design Problem for an Orthogonal Frame: (a) Yield Forces. (b) Axial Forces at Collapse. (c) Collapse Mechanism.

13.5. Figure S.13 shows the optimal solution for case (a): two first spans have a common yield moment $m^0 = 0.628$, whereas the beam over the third span vanishes. The volume of such a beam is $v_{min} = 1.131$.

After the constraint $m_i^0 \geqslant 0.30$ is imposed, the collapse mechanism remains unchanged, but the solution changes to $m_1^0 = m_2^0 = 0.543$, $m_3^0 = 0.300$ (Fig. S.14). The volume is now equal to 1.217, which is only about 8% greater than the volume obtained in the course of free optimization.

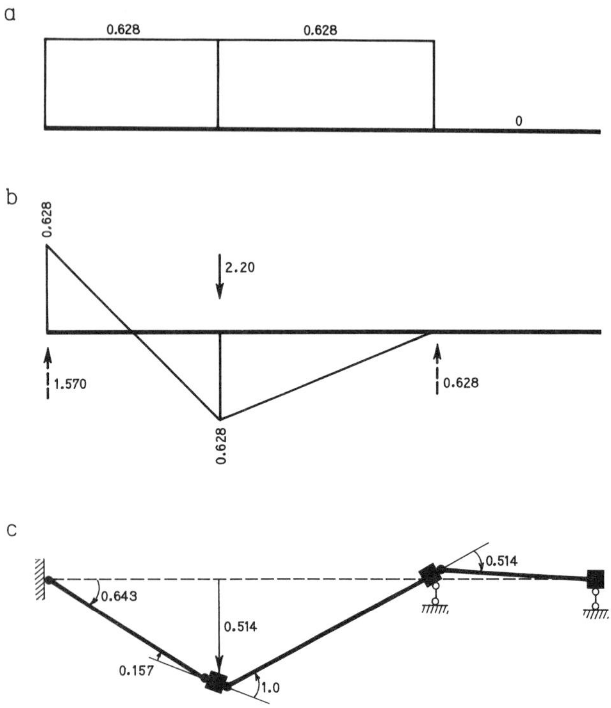

Figure S.13. Solution of the Free Optimal Design Problem for a Beam: (a) Yield Moments. (b) Bending Moments at Collapse. (c) Collapse Mechanism.

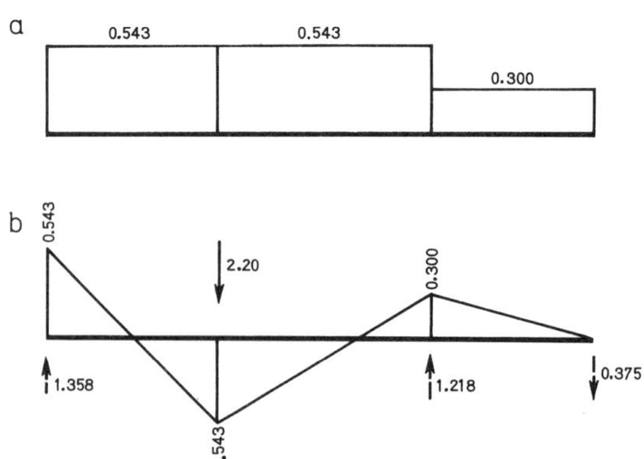

Figure S.14. Solution of the Constrained Optimal Design Problem for a Beam: (a) Yield Moments. (b) Bending Moments at Collapse.

13.6. The yield moment of element 3 vanishes as a result of free optimization (Fig. S.15). After introducing the constraint $m^0 \geqslant 0.40$, we obtain the solution shown in Fig. S.16. Such a constraint increases the volume from 1.250 to 1.497, about 16%.

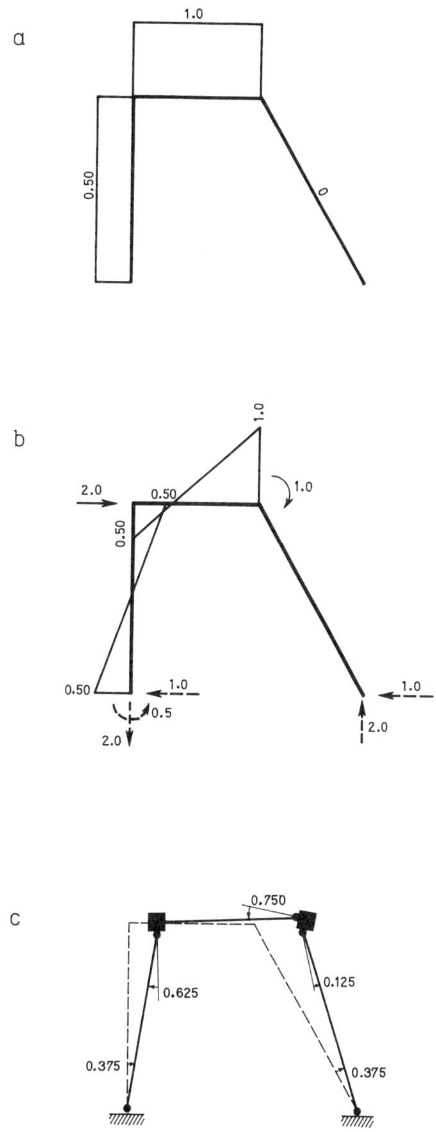

Figure S.15. Solution of the Free Optimal Design Problem for a Nonorthogonal Frame: (a) Yield Moments. (b) Bending Moments at Collapse. (c) Collapse Mechanism.

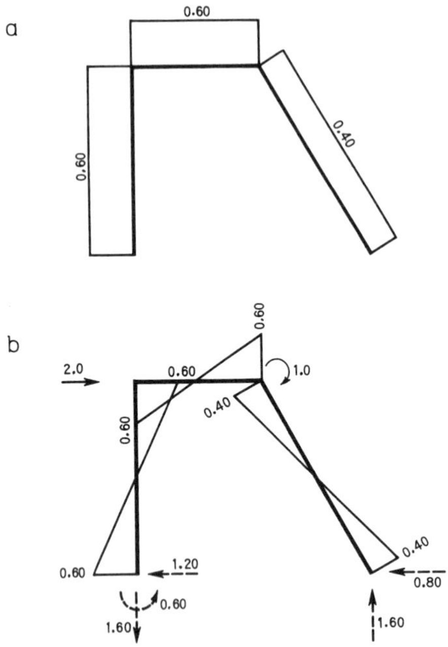

Figure S.16. Solution of the Constrained Optimal Design Problem for a Nonorthogonal Frame: (a) Yield Moments. (b) Bending Moments at Collapse.

13.7. The matrix **Q** in the reduced model (13.46) is given in Table S.15. The right side terms of the inequality constraints are

$$\mathbf{u} = (-1.333, 0.0, 0.0, 1.0, 0.0, 0.0, 1.333, 0.0, 0.0, -1.0, 0.0, 0.0).$$

The solution is

$$\mathbf{z}_* = (m_b^{(1)}, m_a^{(3)}, m_b^{(3)}) = (0.5, 0.0, 0.0),$$
$$\psi_* = (m_1^0, m_2^0, m_3^0) = (0.5, 1.0, 0.0),$$

which coincides with the solution of Exercise 13.6.

Table S.15. Matrix **Q** for the Frame in Fig. 11.12 (Model M)

	Beam 1				Beam 2				Beam 3			
	λ_a^{I}	λ_a^{II}	λ_b^{I}	λ_b^{II}	λ_a^{I}	λ_a^{II}	λ_b^{I}	λ_b^{II}	λ_a^{I}	λ_a^{II}	λ_b^{I}	λ_b^{II}
0 =	−1.67	1.0	−1.0	0.0	0.0	0.0	1.67	−1.0	1.0	0.0	0.0	0.0
0 =	−1.67	0.0	0.0	−1.0	1.0	0.0	1.67	0.0	0.0	1.0	−1.0	0.0
0 =	−1.0	0.0	0.0	0.0	0.0	1.0	1.0	0.0	0.0	0.0	0.0	−1.0

Solutions for Chapter 14

14.1. Using the well-known formula for forces in a spherical shell, we have

$$N_\phi = N_\theta = pr/2.$$

The volume of material is

$$C = 2\pi l^2 H \frac{1 - \cos \phi}{\sin^2 \phi}.$$

Substitution and minimization of C with respect to ϕ gives $2\phi = 120°$. As expected, the optimal shell is shallower than the arch ($120° < 133.5°$), since the significance of the peripheral part of the structure with higher slope is greater than for the arch.

14.2. Many diverse formulations of the problem are possible, permitting us to take into account different factors: thermal and chemical reinforcement, drawing, radioactive processing, etc. [See, for example, Mikeladze (1969), Reitman and Shapiro (1976).] For instance, we may use Hill's plasticity criterion for membrane shells:

$$\frac{N_x^2}{T_{0x}^2} - \frac{N_x N_y}{T_{0x} T_{0y}} + \frac{N_y^2}{T_{0y}^2} + 3\frac{N_{xy}^2}{T_{0xy}} \leqslant 1,$$

where T_{0x}, T_{0y}, and T_{0xy} are material constants. If we assume that they all may be achieved, we obtain the criterion

$$C = \int_S (T_{0x} + T_{0y} + T_{0xy})\, dS \to \min.$$

(Compare with the minimization of steel reinforcement.)

14.3. (*Hint*) The required function may be constructed if we assume that two elements are approximately identical when their mechanical properties are almost the same (for instance, when the heights of shell elements in the group differ by not more than 5%).

14.4. The influence of the shear forces on the plastification of plates may be represented by the following extension of von Mises' plasticity criterion,

$$M_x^2 - M_x M_y + M_y^2 + 3\left[M_{xy}^2 + \frac{H^2}{16}(Q_x^2 + Q_y^2)\right] \leqslant \frac{H^4}{16}\sigma_0^2,$$

if we assume that the shear stresses σ_{zx} and σ_{zy} are distributed uniformly over the height of any section of the plate. It cannot prevent degeneration in some cases (for instance, in zones where $Q_x = Q_y = 0$).

14.5. (*Hint*) The optimal height of a stiffener may be found from the equality of the yield load for the plate and the local buckling load for a reinforcing high beam. The buckling constraint is obtained by studying the stability of the plane shape of bending of the stiffener as a plate that is fixed at the upper edge and free at the lower edge [see a similar problem for a simply supported high beam in Reitman and Shapiro (1976)].

Solutions for Chapter 16

16.1. Optimal values of the decision variables are $\hat{x}_1 = 1.0$, $\hat{x}_2 = 0.5$, and $\hat{x}_3 = 2.0$.

16.2. Optimal width and height of the beam are $b = R$ and $h = \sqrt{3}\,R$.

Solutions for Chapter 17

17.1. Cross-sectional areas of the truss elements are

$$x_1 = 21.52\ \text{cm}^2, \quad x_2 = 12.84\ \text{cm}^2, \quad x_3 = 21.46\ \text{cm}^2, \quad x_4 = 25.76\ \text{cm}^2.$$

References

Abrams, R., and Bunting, M., 1974, Reducing posynomial programs, *SIAM J. Appl. Math.* **27**, 629–640.

Abuyounes, S., and Adeli, H., 1986, Optimization of steel plate girders via general geometric programming, *J. Struct. Mech.* **14**, No. 4, 501–524.

Akita, Y., and Kitamura, K., 1970, Studies on optimization of ship structures; nonlinear programming and optimal design of box and I section girders, *J. Soc. Nav. Archit. Jap.* **128**, 101–119.

Anderhegen, E., and Knöpfel, H., 1972, Finite element limit analysis using linear programming, *Int. J. Solids Struct.* **8**, 1413–1431.

Argyris, J. H., and Mlejnek, H. P., 1982, *Einführung in die Methoden der finiten Elemente*, Vieweg, Braunschweig.

Armand, J.-L., 1971, Minimum mass design of a plate-like structure for specified fundamental frequency, *AIAA J.* **9**, 1739–1771.

Atrek, E., 1984, A theory for optimum design of skeletal structures, in *New Directions in Optimum Structural Design* (Atrek, E., Gallagher, R. H., Ragsdell, K. M., and Zienkiewicz, O. C., Eds.), Wiley, New York.

Augusti, G., Baratta, A., and Casciati, F., 1984, *Probabilistic Methods in Structural Engineering*, Chapman and Hall, London.

Avriel, M., 1976, *Nonlinear Programming: Analysis and Methods*, Prentice-Hall, Englewood Cliffs, NJ.

Avriel, M., and Barrett, J. D., 1978, Optimal design of pitched laminated wood beams, *J. Optimiz. Theory Appl.* **26**, 291–304.

Avriel, M., Dembo, R., and Passy, U., 1975, Solution of generalized geometric programs, *Int. J. Num. Meth. Eng.* **9**, 149–169.

Avriel, M., and Williams, A. C., 1970, Complementary geometric programming, *SIAM J. Appl. Math.* **19**, 125–141.

Avriel, M., and Williams, A. C., 1971, An extension of geometric programming with applications in engineering optimization, *J. Eng. Math.* **5**, No. 3, 187–194.

Baier, H., 1977, Über Algorithmen zur Emittlung und Charakterisierung Pareto-Optimaler Lösungen bei Entwurfsaufgaben elastischer Tragwerke, *ZAMM*, **57**, 318–320.

Baier, H., 1978, *Mathematische Programmierung zur Optimierung von Tragwerken inbesondere bei mehrfachen Zielen*, Dissertation D17, Darmstadt.

Bauer, J., Gutkowski, W., and Iwanow, Z., 1981, A discrete method for lattice structures optimization, *Eng. Optimiz.* **5**, 121–128.

Beale, E. M. L., 1955, On minimizing a convex function subject to linear inequalities, *J. R. Stat. Soc.* **17B**, 173.

Beck, P. A., and Ecker, J. G., 1975, A modified concave simplex algorithm for geometric programming, *J. Optimiz. Theory Appl.*, **15**, 189–202.

Beightler, C. S., and Phillips, D. T., 1976, *Applied Geometric Programming*, Wiley, New York.

Bell, D. J., and Jacobson, D. H., 1975, *Singular Optimum Control Problems*, Academic Press, New York.

Bellman, R. E., 1957, *Dynamic Programming*, Princeton University Press, Princeton, NJ.

Bellman, R. E., and Dreyfus, S. E., 1962, *Applied Dynamic Programming*, Princeton University Press, Princeton, NJ.

Belsare, S. V., and Arora, J. S., 1983, An algorithm for engineering design optimization, *Int. J. Num. Meth. Eng.* **19**, 841–858.

Bennett, J. A., 1980, Topological structural synthesis, *Comput. Struct.* **12**, 273–280.

Berger, J. O., 1980, *Statistical Decision Theory: Foundation, Concepts and Methods*, Springer-Verlag, New York.

Blachut, J., 1977, Optimum design of compression elements with large deflection using dynamic programming (in Polish), *Mech. Teoret. Stos.* **15**, No. 1, 125–130.

Blau, G. E., and Wilde, D. J., 1969, Generalized polynomial programming, *Can. J. Chem. Eng.* **47**, 317–326.

Blau, G. E., and Wilde, D. J., 1971, A Lagrangian algorithm for equality constrained generalized polynomial optimization, *AI Ch.E.J.* **17**, 235–240.

Borkauskas, A., and Atkociunas, J., 1975, Optimal design for cycling loading, in *Optimization in Structural Design* (Sawczuk, A., and Mroz, Z. Eds.), Springer-Verlag, Berlin.

Borkowski, A., 1987, *Analysis of Skeletal Structural Systems in Elastic and Elastic-Plastic Range*, Elsevier-PWN, Warsaw.

Bouchey, G. D., Beightler, C. S., and Koen, B. V., 1971, Optimization of nuclear systems by geometric programming, *Nucl. Sci. Eng.* **44**, 267–272.

Bradley, J., 1975, *The development of polynomial programming algorithms with applications*, Ph.D. Thesis, Dept. of Computer Science, Dublin University, Dublin, Ireland.

Bradley, J., 1978, Polynomial programming, *Eng. Optimiz.* **3**, No. 3, 121–146.

Brandt, A. M., Ed., 1984, *Criteria and Methods of Structural Optimization*, PWN, Warszawa and Martinus Nijhoff, Boston.

Brandt, A. M., Ed., 1989, *Foundations of Optimum Design in Civil Engineering*, PWN, Warszawa and Martinus Nijhoff, Boston.

Brandt, A. M., Jendo, S., and Marks, W., 1984, Probabilistic approach to reliability-based optimum structural design, *Rozp. Inzyn.* (Eng. Trans.) **32**, 57–74.

Brown, D. M., and Ang, A. H. S., 1966, Structural optimization by non-linear programming, *J. Struct. Div. ASCE* **92**, No. ST6, 319–340.

Burns, S. A., 1987, Generalized geometric programming with many equality constraints, *Int. J. Num. Meth. Eng.* **24**, 725–741.

Burns, S. A., and Ramamurthy, S., 1983, Mathematical programming in structural design, *J. Struct. Div., ASCE* **109**, No. 7, 1669–1679.

Cammaert, A. B., 1971, *The optimal design of multi-storey frames using mathematical programming*, Dissertation, University of Cambridge.

Carmichael, D. G., 1981, *Structural Modeling and Optimization*, Ellis Horwood, Chichester.

Casciati, F., and Faravelli, L., 1985, Structural reliability and structural design optimization, in *Structural Safety and Reliability*, Vol. III, (Konishi, I., Ang, A. H.-S., and Shinozuka, M., Eds.), IASSAR, Kyoto, pp. 61-70.

Cella, A., and Soosar, K., 1973, Discrete variables in structural optimization, in *Optimum Structural Design; Theory and Applications*, (Gallagher, R. H., and Zienkiewicz, O. C., Eds.), Wiley, New York, pp. 201-222.

Chang, C., 1981, *Minimum weight design of trusses subject to damage tolerance constraints*, M. S. Thesis, Illinois Institute of Technology.

Chankong, V., and Haimes, Y. Y., 1982, *Multiobjective Decision-making Theory and Methodology*, Elsevier, New York.

Charnes, A., and Cooper, W. W., 1959, Chance constrained programming, *Manage. Sci.* **6**, 73-79.

Charnes, A., and Cooper, W. W., 1961, *Management Models and Industrial Applications of Linear Programming*, Vols. I, II, Wiley, New York.

Charnes, A., and Cooper, W. W., 1977, Goal programming and multiple objective optimization. Part I, *Eur. J. Oper. Res.* **1**, No. 1, 39-54.

Chenais, D., 1987, Optimal design of midsurface of shells: differentiability, proof and sensitivity computation, *Appl. Math. Optimiz.* **16**, 93-133.

Cheng, F. Y., and Chang, C. C., 1985, Optimum design of steel buildings with consideration of reliability, in *Structural Safety and Reliability*, Vol. III (Konishi, I., Ang, A. H.-S., and Shinozuka, M., Eds.), IASSAR, Kyoto, pp. 81-90.

Chiu, T. C., 1982, Structural synthesis of skeletal frameworks under service and ultimate performance constraints, MASc. Thesis, University of Waterloo, Canada.

Cohon, J. L., 1978, *Multiobjective Programming and Planning*, Academic Press, New York.

Cornell, C. A., 1969a, A first-order reliability theory for structural design, Study No. 3, *Structural Reliability and Codified Design*, University of Waterloo, Ontario, Canada.

Cornell, C. A., 1969b, Structural safety specifications based on second-moment reliability analysis, *IABSE Symposium on Concepts of Safety of Structures and Methods of Design*, London.

Cornell, C. A., Reinschmidt, D. F., and Brotchie, J. F., 1965, *Structural Optimization*, Report R65-25, Department of Civil Engineering, MIT.

Corradi, L., and Poggi, C., 1984, A refined finite element model for the analysis of elastic-plastic frames, *Int. J. Num. Meth. Eng.* **20**, 2155-2174.

Cottle, R. W., 1963, Symmetric dual quadratic programs, *Q. Appl. Math.* **21**, 237.

Čyras, A. A., 1969, *Analysis and Design of Elastic-Plastic Structures by Linear Programming* (in Russian), Stroiizdat, Leningrad.

Čyras, A. A., 1975, Optimization theory in the design of elasto-plastic structures, in *Structural Optimization*, CISM courses and lectures, No. 237, Springer-Verlag, Berlin.

Čyras, A. A., 1983, *Mathematical Models for the Analysis and Optimization of Elastic-Plastic Systems*, Wiley, Chichester.

Čyras, A. A., Borkauskas, A. E., and Karkauskas, R. P., 1974, *Theory and Methods of Optimization of Elastic-Plastic Structures* (in Russian), Stroiizdat, Leningrad.

Dantzig, G. B., 1963, *Linear Programming and Extensions*, Princeton University Press, Princeton, NJ.

Das Gupta, N. C., Paul, H., and Yu, C. H., 1986, An application of geometric programming to structural design, *Comput. Struct.* **22**, No. 6, 965-971.

Dauer, J. P., and Stadler, W., 1986, A survey of vector optimization in infinite-dimensional spaces, Part 2, *J. Optimiz. Theory Appl.* **51**, No. 2, 205-241.

Davidson, J. W., Felton, L. P., and Hart, G. C., 1977a, Optimum design of structures with random parameters, *Comput. Struct.* **7**, 481-486.

Davidson, J. W., Felton, L. P., and Hart, G. C., 1977b, Reliability-based optimization for dynamic loads, *J. Struct. Div. ASCE* **ST 10**, 2021-2035.

Davies, A. J., 1980, *The Finite Element Method*, Clarendon Press, London.

Debreu, G., 1959, *Theory of Value*, Wiley, New York.

Dembo, R. S., 1972a, *GGP—A computer program for solving generalized geometric programming problems*, Users Manual, Report 72/59, Department of Chemical Engineering, Technion, Israel Institute of Technology, Haifa, Israel.

Dembo, R. S., 1972b, *The solution of complementary geometric programming problems*, M.Sc. Thesis, Technion, Israel Institute of Technology, Haifa, Israel.

Dembo, R. S., 1976, A set of geometric programming test problems and their solutions, *Math. Program.* **10**, 192-214.

Dembo, R. S., 1978a, Current state of the art of algorithms and computer software for geometric programming, *J. Optimiz. Theory Appl.* **26**, 149-184.

Dembo, R. S., 1978b, *Second order algorithms for the geometric programming dual*, Part I, Technical Report 22, Yale School of Organization and Management, New Haven, CT.

Dembo, R. S., and Rijckaert, M. J., 1978, Computational experiments in geometric programming, *Eng. Optimiz.* **3**, No. 3, 161-173.

Diaz, A., Papalambros, P., and Taylor, J. E., 1984, Geometric programming for continuous design problems, in *New Directions in Optimum Structural Design*, (Atrek, E. *et al.*, Eds.), Wiley, New York, pp. 309-325.

Dinkel, J. J., Elliott, W. H., and Kochenberger, G. A., 1977, Computational aspects of cutting plane algorithms for geometric programming problems, *Math. Program.* **13**, 200-220.

Dinkel, J. J., and Kochenberger, G. A., 1974, A cofferdam design optimization, *Math. Program.* **6**, 114-116.

Dinkel, J. J., Kochenberger, G. A., and McCarl, B., 1974, An approach to the numerical solution of geometric programming, *Math. Program.* **7**, 181-190.

Distefano, N., 1972a, Dynamic programming and max-min problem in the theory of structures, *J. Franklin Inst.* **294**, No. 5, 339-350.

Distefano, N., 1972b, Dynamic programming and the optimum design of rotating disks, *J. Optimiz. Theory Appl.* **10**, No. 2, 109-128.

Distefano, N., 1975, Dynamic programming and the optimization of two-point boundary value system, *J. Math. Anal. Appl.* **52**, 142-150.

Distefano, N., and Bellows, G., 1977, Application of coupled dynamic programming to two-point boundary-value systems, *J. Optimiz. Theory Appl.* **23**, 1.

Distefano, N., and Rath, A., 1975, A dynamic programming approach to the optimization of elastic trusses, *J. Optimiz. Theory Appl.* **15**, No. 1, 13-26.

Distefano, N., and Samartin, A., 1975, A dynamic programming approach to the formulation and solution of finite element equations, *Comput. Meth. Appl. Mech. Eng.* **5**, 37-52.

Ditlevsen, O., 1979a, Generalized second-moment reliability index, *J. Struct. Mech.* **7**, No. 4, 435-451.

Ditlevsen, O., 1979b, Narrow reliability bounds for structural systems, *J. Struct. Mech.* **7**, No. 4, 453-472.

Ditlevsen, O., 1981, The principle of normal tail approximation, *J. Eng. Mech., ASCE* **107**, 1191-1208.

Dolinski, K., 1983, First-order second-moment approximation in reliability of structural systems: critical review and alternative approach, *Struct. Saf.* **1**, 211-231.

Dorn, W. S., Gomory, R. B., and Greenberg, H. J., 1964, Automated design of optimal structures, *J. Mech.* **3**, No. 1, 25-52.

Dubov, Ju. A., Travkin, S. I., and Jakimec, V. N., 1986, *Multiobjective Models for Creating and Choosing a System of Variants* (in Russian), Nauka, Moscow.

Duckstein, L., 1984, Multiobjective optimization in structural design, in *New Directions in Optimum Structural Design* (Atrek, E. *et al.*, Eds.), Wiley, New York, pp. 459-481.

Duffin, R. J., 1970, Linearizing geometric programs, *SIAM Rev.* **12**, 211-227.

Duffin, R. J., and Peterson, E. L., 1972a, The proximity of algebraic geometric programming to linear programming, *Math. Program.* **3**, 250-254.

Duffin, R. J., and Peterson, E. L., 1972b, Reversed geometric programming treated by harmonic means, *Indiana Univ. Math. J.* **22**, 531-550.

Duffin, R. J., and Peterson, E. L., 1973, Geometric programming with signomials, *J. Optimiz. Theory Appl.* **11**, 3-35.

Duffin, R. J., Peterson, E. L., and Zener, C., 1967, *Geometric Programming*, Wiley, New York.

Duffin, R. J., and Zener, C., 1970, Geometric programming and the Darwin-Fowler method in statistical mechanics, *J. Phys. Chem.* **74**, 2419-2423.

Dwyer, W. J., 1974, *An improved automated structural optimization program*, Air Force Dynamics Laboratory, TR-74-96.

Eaves, C. B., and Saigal, R., 1972, Homotopies for computation of fixed points on unbounded regions, *Math. Program.* **3**, 225-237.

Ecker, J. G., 1980, Geometric programming methods, computation and applications, *SIAM Rev.* **22**, 3, 338-362.

Ecker, J. G., Gochet, W., and Smeers, Y., 1976, A modified reduced gradient method for dual posynomial programming, *J. Optimiz. Theory Appl.* **26**, 265-275.

Ecker, J. G., Gochet, W., and Smeers, Y., 1978, Some primal and dual algorithms for posynomial and signomial geometric programs, *Eng. Optimiz.* **3**, No. 3, 147-160.

Ecker, J. G., and Wiebking, R. D., 1978, Optimal design of a dry-type natural-draft cooling tower by geometric programming, *J. Optimiz. Theory Appl.* **26**, 305-324.

Ecker, J. G., and Zoracki, M. J., 1976, An easy primal method for geometric programming, *Manage. Sci.* **23**, 71-77.

Edwards, L. S., 1975, Optimum limit state design of highway bridge superstructures using geometric programming, *Eng. Optimiz.* **1**, No. 4, 201-202.

Emtell, H., and Ovrebo, B., 1965, A systematic analysis of weight of the transverse bulkhead in a 120000 T.DW. oil tanker, *Eur. Shipbuild.* **14**, 68-79.

Ermer, D. S., 1971, Optimization of the constrained machining economics problem by geometric programming, *J. Eng. Ind. Trans. ASME* **93**, 1067-1072.

Eschenauer, H. 1980, *Eine Vektoroptimierungsaufgabe der Strukturmechanik*, Karl-Marguerre-Gedachtniskolloquium, THD Schriftenreihe Wissenschaft und Technik, Bd. 16, pp. 89-101.

Eschenauer, H., 1981, Anwendung der Vektoroptimierung bei räumlichen Tragstrukturen, *Der Stahlbau* **4**, 110-115.

Eschenauer, H., 1983, Vector optimization in structural design and its application on antenna, in *Optimization Methods in Structural Design*, (Eschenauer, H., and Olhoff, N., Eds.), Bibliographisches Institut, Mannheim, pp. 146-155.

Eschenauer, H., 1984, Optimierung ebener Flächentragwerke aus Verbundwerkstoff, *Z. Flugwiss. Weltraumforsch.* **8**, No. 6, 367-378.

Eschenauer, H., 1986, *Numerical and Experimental Investigations on Structural Optimization of Engineering Design, A Research Project of the German Research Association* (*GRA*) *1980-1984,* Research Laboratory for Applied Structural Optimization, Institute of Mechanics and Control Engineering, University of Siegen, Siegen, West Germany.

Eschenauer, H., and Jendo, S., 1986, Multiobjective optimization of cable structures, *ZAMM* **66**, No. 5, T342–T343.

Eschenauer, H., Kneppe, G., and Stenvers, K. H., 1986, Deterministic and stochastic multiobjective optimization of beam and shell structures, *J. Mech. Transm. Autom. Des.* **108**, 31–37.

Evans, J. H., 1958, A structural analysis and design integration with application to the midship-section characteristics of transversally framed ships, *Trans. SNAME* **66**, 244–309.

Evans, J. H., Ed., 1974, *Ship structural design concepts*, SDC-Project SR-200, U.S. Coast Guard Headquarters, Washington, D.C.

Evans, J. H., and Khoushy, D., 1963, Optimized design of midship section structure, *Trans. SNAME*, **71**, 144–191.

Falk, J. E., 1973, *Global solutions of signomial problems*, Technical Report T-274, George Washington University, Program in Logistics, Washington D.C.

Farquhar, P. H., 1977, A survey of multiattribute utility theory and applications, in *Multiple Criteria Decision Making* (Starr, M. K., and Zeleny, M., Eds.), North-Holland, Amsterdam.

Federowicz, A. J., and Mazumdar, M., 1968, Use of geometric programming to maximize reliability achieved by redundancy, *Oper. Res.* **16**, 948–954.

Feng, Y. S., and Moses, F., 1986, A method of structural optimization based on structural system reliability, *J. Struct. Mech.* **14**, No. 4, 437–453.

Feng, Y. S., and Moses, F., 1987, Optimum design, redundancy and reliability of structural systems, *Comput. Struct.* **14**, 239–251.

Fiacco, A. V., and Ghaemi, A., 1979, *Optimal treatment levels of a stream pollution abatement system under three environmental control policies*, Report T-38, School of Engineering and Applied Science, George Washington University, Washington, D.C.

Fiacco, A. V., and McCormick, G. P., 1968, *Nonlinear Programming: Sequential Unconstrained Minimization Techniques*, Wiley, New York.

Finifter, D., 1972, *Structural optimization of a web frame*, S. M. and Naval Architect Thesis, Department of Ocean Engineering, MIT.

Finifter, D., and Mansour, A., 1974, Finite element analysis and optimization of a web frame of a tanker with isolated ballast systems, *J. Ship Res.* **18**, No. 2, 85–95.

Fishburn, P. C., 1974, Lexicographic orders: utilities and decision rules (a survey), *Manage. Sci.* **11**, 1442–1471.

Fleury, C., 1979, Structural weight optimization by dual methods of convex programming, *Int. J. Num. Meth. Eng.* **14**, 1761–1783.

Fleury, C., 1984, Large scale structural optimization by finite elements, in *New Directions in Optimum Structural Design* (Atrek, E. *et al.*, Eds.), Wiley, New York.

Flügge, W., 1960, *Stresses in Shells*, Springer-Verlag, Berlin and New York.

Folkers, J. S., 1973, Ship optimization and design, in *Optimization and Design* (Avriel, M., Rijckaert, M. J., and Wilde, D. J., Eds.), Prentice-Hall, Englewood Cliffs, NJ, pp. 221–226.

Fox, R. L., and Miura, H., 1971, An approximate analysis technique for design calculation, *AIAA J.* **9**, No. 1, 177–179.

Frangopol, D. M., 1984, A reliability-based optimization technique for automatic plastic design, *Comput. Meth. Appl. Mech. Eng.* **44**, 105–117.

Frangopol, D. M., 1985, Multicriteria reliability-based structural optimization, *Struct. Saf.* **3**, 23–28.

Frank, C. J., 1966, An algorithm for geometric programming, in *Recent Advances in Optimization Techniques* (Lavi, A., and Vogl, T., Eds.), Wiley, New York, pp. 145–162.

Furukawa, K., Furuta, H., and Ichikawa, H., 1985, Probabilistic optimality criteria method unifying stress and deformation requirements, in *Structural Safety and Reliability*, Vol. III, (Konishi, I., Ang, A. H.-S., and Shinozuka, M., Eds.), IASSAR, Kyoto, pp. 91–100.

Gallagher, R. H., 1975, *Finite-Element-Analysis: Fundamentals*, Prentice-Hall, Englewood Cliffs, NJ.

Gallagher, R. H., and Gellatly, R. A., 1966, Automated minimum weight design of framework structures, *International Symposium on the Use of Computers in Structural Engineering*, University of Newcastle.

Gelfand, I. M., and Fomin, S. V., 1972, *Calculus of Variations*, Prentice-Hall, Englewood Cliffs, NJ.

Geniev, G. A., and Kissyuk, V. N., 1965, On the generation problem of the theory of strength for concrete (in Russian), *Beton i Zhelezobeton*, No. 2, 16–19.

Gerasimov, E. N., Potchman, Ju, M., and Skalozub, V. V., 1985, *Multicriteria Structural Optimization* (in Russian), Vistcha Schola, Kiev.

Giles, G. L., 1971, Procedure for automated aircraft wing structural design, *J. Struct. Div. ASCE* **97**, 99–113.

Gill, P. E., Murray, W., and Saunders, M. A., 1983, *User's Guide for SOL/QPSOL: A Fortran Package for Quadratic Programming*, Report SOL 83-7, Department of Operations Research, Stanford University, Stanford, CA.

Glankwahmdee, A., Liebman, J. S., and Hogg, G. L., 1979, Unconstrained discrete nonlinear programming, *Eng. Optimiz.* **4**, 95–107.

Goble, G. G., and De Santis, P. V., 1966, Optimum design of mixed steel composite girders, *Proc. ASCE, J. Struct. Div.* **92**, No. ST 6, 25–43.

Gochet, W., Loute, E., and Solow, D., 1974, Comparative computer results of three algorithms for solving prototype geometric programming problems, *Cah. Cent. Etud. Rech. Oper.* **16**, 461–486.

Gochet, W. and Smeers, Y., 1974, On the use of linear programs to solve prototype goemetric programs, *Cah. Cent. Etud. Rech. Oper.* **16**, 23–36.

Gochet, W. and Smeers, Y., 1975, *A branch and bound method for reversed geometric programming*, CORE Discussion Paper 7511, Catholic University of Louvain.

Goicoechea, A., Hansen, D. and Duckstein, L., 1982, *Introduction to Multi-Objective Analysis with Engineering and Business Applications*, Wiley, New York.

Gokhfeld, D. A., and Cherniavsky, O. F., 1980, *Limit Analysis of Structures at Thermal Cycling*, Sijthoff-Noordhoff, Alphen aan den Rijn.

Grierson, D. E., and Chiu, T. C. W., 1982, *Synthesis of frameworks under multilevel performance constraints*, Solid Mechanics Division, University of Waterloo, Canada, No. 172.

Grierson, D. E., and Lee, W. H., 1986, Optimal synthesis of frameworks under elastic and plastic performance constraints using discrete sections, *J. Struct. Mech.* **14**, No. 4, 401–420.

Gudzenchuk, Z. V., and Reitman, M. I., 1977, Shape optimization for roofing shells on arbitrary view-plan by mass criterion (in Russian), *Beton i Zhelezobeton*, No. 8, 20–22.

Gvozdev, A. A., 1934, Evaluation of collapse load for redundant structures (in Russian), *Project and Standard* **3**, 10–16.

Gvozdev, A. A., 1936, Determination of the limit load for redundant structures with plastic strains (in Russian), *Trudy Konferentsii po Plasticheskim Deformatsiyam*, Akademiya Nauk SSSR, Moscow.

Gvozdev, A. A., 1949, *Theory of Ultimate Equilibrium* (In Russian), Stroiizdat, Moscow.

Hadley, G., 1964, *Nonlinear and Dynamic Programming*, Addison-Wesley, Reading, MA.

Haftka, R. T., and Grandhi, R. V., 1986, Structural shape optimization—a survey, *Comput. Math. Appl. Mech. Eng.* **57**, 91–106.

Haimes, Y. Y., and Chankong, V., 1979, Kuhn–Tucker multipliers as trade-offs in multiobjective decision-making analysis, *Automatica* **15**, 59–72.

Haimes, Y. Y., Hall, W., and Freedman, H., 1975, *Multi-Objective Optimization in Water Resources Systems: The Surrogate Worth Trade-Off Method*, Elsevier, Amsterdam.

Hasofer, A. M., and Lind, N. C., 1974, Exact and invariant second moment code format, *J. Eng. Mech., ASCE* **100**, No. EM1, 111–121.

Haug, E. J., 1984, *Computer Aided Analysis and Optimization of Structures*, Springer-Verlag, Berlin and New York.

Haug, E. J., and Cea, J., Ed., 1981, *Optimization of Distributed Parameters Structures*, Sijthoff-Nordhoff, Boston.

Haugen, E. B., 1968, *Probabilistic Approaches to Design*, Wiley, London.

Hestenes, M. R., 1966, *Calculus of Variations and Optimal Control Theory*, Wiley, New York.

Hilleary, R. R., 1966, *The tangent search method of constrained minimization*, Technical Report, No. 59, U.S. Naval Postgraduate School, Monterey, CA.

Ho, J. K., 1974, *Optimal design of multi-stage structures—a nested decomposition approach*, Technical Report 72-4, System Optimization Laboratory Department of Operations Research, Stanford University, Stanford, CA.

Ho, J. K., 1975, Optimal design of multi-stage structures: a nested decomposition approach, *Comput. Struct.* **5**, No. 4, 249–255.

Hodge, P. G., 1959, *Plastic Analysis of Structures*, McGraw-Hill, New York.

Hohenbichler, M., and Rackwitz, R., 1983, First-order concepts in system reliability, *Struct. Saf.* **1**, 177–188.

Hook, R., and Jeeves, T. A., 1962, Direct search solution of numerical and statistical problems, *J. Assoc. Comput. Mach.* **8**.

Homenjuk, V. V., 1983, *Elements of Theory of Multiobjective Optimization* (in Russian), Nauka, Moscow.

Howell, G. C., and Doyle, W. S., 1978, Dynamic programming and direct iteration for the optimum design of skeletal towers, *Comput. Struct.* **9**, No. 6, 621–627.

Huber, G. P., 1974, Multi-attribute utility models: a review of field and field-like studies, *Manage. Sci.* **10**, 1393–1402.

Hughes, O. F., Mistree, F., and Zanic, V., 1980, A practical method for the rational design of ship structures, *J. Ship Res.* **24**, 101–113.

Hwang, C. L., and Masud, A. S. M., 1979, *Multiple Objective Decision Making; Methods and Applications*, Lecture Notes in Economics and Mathematical Systems, Vol. 164, Springer-Verlag, New York.

IBM, 1977, *Mathematical Programming System Extended (MPSX/370)*, Program Reference Manual SH19-1095-2, IBM-France.

ICL, 1971, *XDLA Linear Programming Mk3*, Technical Publication Service, London.

Ignizio, J., 1976, *Goal Programming and Extensions*, D. C. Heath, Lexington, MA.

Ignizio, J. P., 1981, The determination of a subset of efficient solutions via goal programming, *Comput. Oper. Res.* **3**, 9–16.

Ijiri, Y., 1965, *Management Goals and Accounting for Control*, North-Holland, Amsterdam.

Jacobs, O. L. R., 1967, *An Introduction to Dynamic Programming*, Chapman and Hall, London.

Jakubowska, B., and Lesniak, Z. K., 1982, Shape optimization of double curved shells, in *Optimization Methods in Structural Design, Proceedings of Euromech Colloquium 164*, Siegen, West Germany, pp. 210–215.

Jefferson, T. R., 1974, *Manual for the Geometric Programming Code GPROG (CDC) VERSION 2*, Report 1974/OR/2, Mechanical and Industrial Engineering, University of New South Wales, Australia.

Jefferson, T. R., and Scott, C. H., 1978a, Avenues of geometric programming, *NZ Oper. Res.* **6**, 109–136.

Jefferson, T. R., and Scott, C. H., 1978b, Generalized geometric programming applied to problems of optimal control. Part I, *J. Optimiz. Theory Appl.* **26**, 117–129.

Jendo, S., 1979, *Optimization of cable and frame structures in elastic range according to minimum volume, weight or cost criteria* (in Polish), IFTR Reports (Nos. 27 and 28), Polish Academy of Sciences, Warsaw.

Jendo, S., 1985, *Some problems of multicriteria optimization of cable structures*, IFTR Reports, Warsaw, 43.

Jendo, S., 1986a, Multicriteria optimization of cable systems, *Proceedings of the IASS Symposium on Membrane Structures and Space Frames*, Vol. 2, September 1986, Osaka, Japan, Elsevier, Amsterdam, pp. 71-78.

Jendo, S., 1986b, Multicriteria optimization of single-layer cable systems, *Arch. Mech.* **38**, No. 3, 219-234.

Jendo, S., Lawo, M., and Thierauf, G., 1984, Optimum structural design by iterative methods, *ZAMM* **64**, T414-T415.

Jendo, S., and Marks, W., 1983, An application of dynamic programming in optimum structural design (in Polish), *Proceedings of the PAS Conference on Optymalizacja wytrzymalosciowa konstrukcji*, Jablonna, 1980, Ossolineum, pp. 295-313.

Jendo, S., and Marks, W., 1984, Dynamic programming, in *Criteria and Methods of Structural Optimization*, (Brandt, A. M., Ed.), PWN, Warsaw, Martinus Nijhoff, The Hague, pp. 287-298.

Jendo, S., and Marks, W., 1986a, Multiobjective structural optimization, in *Lecture Notes in Control and Information Science*, Vol. 84, (Prekopa, A., Szelezsan, J., and Strazicky, B., Eds.), Springer-Verlag, Berlin, pp. 365-374.

Jendo, S., and Marks, W., 1986b, Formulation of structural optimization problems (in Polish), *Rozp. Inzyn.* (*Eng. Trans.*) **34**, No. 4, 457-481.

Jendo, S., and Marks, W., 1987, Multicriteria optimization in civil engineering design, *Proceedings of the International Conference on Optimization Techniques and Applications*, Singapore, pp. 998-1006.

Jendo, S., Marks, W., and Thierauf, G., 1985, Multicriteria optimization in optimum structural design, in *Large Scale Systems*, (Sage, A. P., Ed.), Elsevier, North-Holland, pp. 141-150.

Jendo, S., Nowacki, H., and Yi-Suang Tein, 1974, *Optimum design of tanker transverse web frames*, Final Report, Department of Naval Architecture and Marine Engineering, University of Michigan, Ann Arbor.

Jendo, S., and Stachowicz, A., 1986, Optimum structural design by multilevel optimization, *ZAMM* **66**, No. 5, T347-T349.

Jendo, S., and Stachowicz, A., 1988, Multilevel optimization in civil engineering structural design, *ZAMM* **68**, No. 4, T199-T200.

Johansen, K., 1962, *Yield Line Theory*, Cement and Concrete Association, London (translated from 1943 Danish Edition).

Jorgensen, L. R., 1915, The constant angle arch dam, *AMSE Trans.* **78**, 685-721.

Kalaba, R., 1962, Design of minimal-weight structures for given reliability and cost, *J. Aerosp. Sci.* **29**, No. 3, 355-356.

Kaneko, I., and Cu Duong Ha, 1983, A decomposition procedure for large-scale optimum plastic design problems, *Int. J. Num. Meth. Eng.* **19**, 873-889.

Karpenko, N. I., 1966, On the behaviour of the reinforced concrete slabs with cracks (in Russian), *Proceedings of the VI All-Union Conference on Reinforced Concrete Structures*, Riga.

Kavlie, D., Kowalik, J., Lund, S., and Moe, J., 1966, Design optimization using a general nonlinear programming method, *Eur. Shipbuild.* **1**, No. 4, 57-62.

Kavlie, D., Kowalik, J., and Moe, J., 1966, *Structural optimization by means of nonlinear programming*, Meddelelse SKB II/M4, Trondheim.

Kavlie, D., and Moe, J., 1969, Application of nonlinear programming to optimum grillage design with nonconvex sets of variables, *Int. J. Num. Meth. Eng.* **1**, No. 4, 351-378.

Keeney, R. L., and Raiffa, H., 1976, *Decision with Multiple Objectives: Preferences and Value Trade-offs*, Wiley, New York.

Kelley, J. E., 1960, The cutting-plane method for solving convex programs, *SIAM J. Appl. Math.* **8**, 703-712.

Khuberyan, K. M., 1956, *Rational Forms of the Pipelines, Rezervoirs and Pressure Vessels* (in Russian), Gosstroyizdat, Moscow.

Kickert, W. J. M., 1978, Fuzzy theories on decision-making. A critical review, in *Frontiers in System Research*, Vol. 3, Martinus Nijhoff, Boston.

Kirsch, U., 1975, Multilevel approach to optimum structural design, *Proc. ASCE, J. Struct. Div.* **101**, No. ST 4, 957-974.

Kirsch, U., 1981, *Optimum Structural Design; Concepts, Methods and Applications*, McGraw-Hill, New York.

Kirsch, U., 1986, Optimal topologies of flexural systems, in *Computer Aided Structural Design*, Vol. 1, Troja, Portugal, pp. 371-382.

Kirsch, U., Reiss, M., and Shamir, U., 1972, Optimum design by partitioning into substructures, *J. Struct. Div. ASCE* **98**, 249-261.

Kiselev, V. A., 1953, *Rational Forms of Arches and Suspended Systems* (in Russian), Stroyizdat, Moscow.

Kishi, M., Murotsu, Y., Okada, H., and Taguchi, K., 1985, Probabilistically optimum design of offshore platforms considering maintenance costs, in *Structural Safety and Reliability* Vol. II (Konishi, I., Ang, A. H.-S., and Shinozuka, M., Eds.), IASSAR, Kyoto, pp. 557-563.

Kitamura, K., 1971, Studies of optimization of ship structures; optimum design of grillages, *J. Soc. Nav. Archit. Jap.*, **130**.

König, J. A., 1987, *Shakedown of Elastic-Plastic Structures*, Elsevier-PWN, Warsaw.

Kos, F., 1983, *Vergleich von Algorithmen der Mathematischen Programmierung bei der automatischen Bemessung*, Diplomarbeit 83/1, Fachbereich Bauwesen, Universität Essen, Essen.

Koski, J., 1979, *Truss optimization with vector criterion*, Tampere University of Technology, Tampere/Finland Publications, No. 6.

Koski, J., 1984a, Multicriteria optimization in optimum structural design, in *New Directions in Optimum Structural Design* (Atrek, E. *et al.*, Ed.), Wiley, New York, pp. 483-503.

Koski, J., 1984b, *Bicriterion optimum design method for elastic trusses*, Acta Polytechnica Scandinavica, Mechanical Engineering Series, No. 86, Helsinki.

Koski, J., and Silvennoinen, R., 1982, Pareto optima of isostatic trusses, *Comput. Meth. Appl. Mech. Eng.* **31**, No. 3, 265-279.

Kruzelecki, J., and Zyczkowski, M., 1985, Optimal structural design of shells, a survey, *SM Arch.* **10**, 101-170.

Kuhn, H. W., and Tucker, A. W., 1951, Nonlinear programming, *Proceedings of the Second Berkeley Symposium on Mathematical Statistics and Probability* (Neyman, J., Ed.), University of California Press, Stanford, CA, pp. 481-492.

Kunar, R. R., and Chan, A. S. L., 1976, A method for the configurational optimization of structures, *Comput. Meth. Appl. Mech. Eng.* **7**, 331-350.

Kuznetsov, E. N., Nekrutman, A. B., and Ostrovsky, A. Yu., 1971, Optimal design of momentless shells of revolution (in Russian), *Transactions of National Center for Structural Engineering*, Stroyizdat, Moscow, pp. 149-156.

Lai, Y. S., and Achenbach, J. D., 1973, Direct search optimization method, *J. Struct. Div., ASCE* **98**, No. ST1, 119-131.

Larson, R. E., 1968, *State Increment Dynamic Programming*, Elsevier, New York.

Lawo, M., 1981, *Automatische Bemessung für stochastische dynamische Belastung*, Dissertation, Universitat Essen, Essen.

Lawo, M., and Thierauf, G., 1980, Stabtragwerke. *Matrizenmethoden der Static und Dynamik*, Teil 1, Statik, Vieweg, Braunschweig.

Lee, S. M., 1972, *Goal Programming for Decision Analysis*, Auerbach, Philadelphia.

Lee, B., and Kapton, J., 1974, Optimum cost design of steel-framed building, *Eng. Optimiz.* **1**, 3.

Leimbach, K. R., and Lieweg, H., 1977, Simple weight optimization scheme for steel frames, *J. Struct. Div., ASCE* **103**, No. ST 12.

Leitmann, G., 1977, Some problems of scalar and vector-valued optimization in linear viscoelasticity, *J. Optimiz. Theory Appl.* **23**, No. 1, 93–99.

Lemke, C. E., 1965, Bimatrix equilibrium points and mathematical programming, *Manage. Sci.* **2**, 681.

Lepik, U. R., 1983, *Optimal Design of Inelastic Structures under Dynamic Loading* (in Russian), Valgus, Tallinn.

Leporati, E., 1979, The assessment of structural safety. A comparative statistical study of the evolution and use of level 3, level 2, and level 1, in *Series in Cement and Concrete Research*, Vol. 1, Forest Grove, Oregon-Letchworth, Herts: Research Studies Press, IX.

Liebman, J. S., Khachaturian, N., and Chanaratna, V., 1981, Discrete structural optimization, *J. Struct. Div., ASCE* **107**, No. ST 11, 2177–2197.

Lin, J. G., 1976, Maximal vector and multi-objective optimization, *J. Optimiz. Theory Appl.* **18**, No. 1, 41–63.

Lippmann, H., 1972, *Extremum and Variational Principles in Mechanics*, CISM courses and lectures, No. 54, Springer-Verlag, Wien.

Lipson, S. L., and Gwinn, L. B., 1977, Discrete sizing of trusses for optimal geometry, *Proc. A.S.C.E.* **103**, No. ST 5, 1031–1046.

Lund, S., 1971, Optimum design of transverse frame structures in tankers, *Eur. Shipbuild.* **20**, Nos. 5 and 6, 3–9 and 2–11.

Lurie, K. A., 1985, *Applied Optimal Control of Distributed Systems*, Plenum Press, New York.

Lurie, K. A., and Cherkaev, A. V., 1981, *G-Closure of some particular sets of admissible material characteristics for the problem of bending of thin plates*, DCAMM Report No. 214, The Technical University of Denmark.

Lurie, K. A., Fedorov, A. V., and Cherkaev, A. V., 1984, Regularization of optimal design problems for bars and plates, *J. Optimiz. Theory Appl.* **16**, 499–543.

Machura, M., and Maluwa, A., 1973, Algorithm 450, Rosenbrock function minimization, *Commun. ACM* **16**, No. 8, 482–485.

Madsen, H. O., Krenk, S., and Lind, N. C., 1986, *Safety of Structures*, Prentice-Hall, New York.

Maier, G., 1970, A matrix structural theory of piecewise linear elastoplasticity with interacting yield planes, *Meccanica* **5**, 54–66.

Majid, K. I., 1974, *Optimum Design of Structures*, Butterworth, London.

Majid, K. I., and Anderson, D., 1968, Elastic-plastic design of sway frames by computer, *Proc. Inst. Civ. Eng.* **41**, 705–729.

Majid, K. I., and Anderson, D., 1972, Optimum design of hyperstatic structures, *Int. J. Num. Meth. Eng.* **4**, No. 4, 561–578.

Majid, K. I., and Elliott, D. W. C., 1971, Optimum design of frames with deflection constraints by non-linear programming, *Struct. Eng.* **49**, No. 4, 179–188.

Malkov, V. P., and Ugodchikov, A. G., 1981, *Optimization of Elastic Systems* (in Russian), Nauka, Moscow.

Mancini, L. J., and Piziali, R. A., 1976, Optimal design of helical springs by geometric programming, *J. Eng. Optimiz.* **2**, 83–95.

Mandel, P., and Chryssostomidis, C., 1972, A design methodology for ships and other complex systems, *Philos. Trans. R. Soc., London, A* **273**, 85–98.

Mandel, P., and Leopold, R., 1966, Optimization methods applied to ship design. *Trans. SNAME* **74**, 477–521.

Mangasarian, O. L., 1969, *Nonlinear Programming*, McGraw-Hill, New York.
Marks, W., 1978, *Optimum design of prestressed bending elements* (in Polish), IFTR Reports, 52.
Martin, J. B., 1964, Impulsive loading theorems for rigid-plastic continua, *Eng. Mech. Div., ASCE* **94**, No. EM5, 27–46.
Massonnet, Ch., and Save, M., 1963, *Calcule Plastique de Constructions*, Bruxelles.
Massonet, C. E. and Save, M. A., 1965, *Plastic Analysis and Design*, Vol. 1, Blaisdell, New York.
Maxwell, J. C., 1890, *The Scientific Papers*, Vol. 2, Cambridge University Press.
McGuire, W. and Gallagher, R. H., 1979, *Matrix Structural Analysis*, Wiley, New York.
Megarefs, G. J., 1966, Method of optimal design of axisymmetric plates, *Proc. ASCE* **92**, 79–99.
Michell, A. G. M., 1904, The limits of economy of material in frame structures, *Philos. Mag. Ser. 6*, **8**.
Mikeladze, M. Sh., 1969, *Introduction to the Engineering Theory of Perfectly Plastic Shells* (in Russian), Metsniereba, Tbilisi.
Miura, H., Kavlie, D., and Moe, J., 1975, Interactive and automated design of ship structures, in *Optimization in Structural Design* (Sawczuk, A., and Mroz, Z., Eds.), Springer-Verlag, Berlin.
Moe, J., 1969a, Finite element techniques in ship structures design, in *ISD Finite Element Techniques* (Soerensen, M., Ed.), Stuttgart University, pp. 1–59; also as Meddelelse SKB II/M, 14, Trondheim.
Moe, J., 1969b, *Optimum design of statically indeterminate frames by means of nonlinear programming*, University of Michigan, Reports 23, 24.
Moe, J. and Lund, S., 1968, Cost and weight minimization of structures with special emphasis on longitudinal strength members of tankers, *Trans. RINA* **110**, 43–70.
Morley, C. T., 1966, On the yield criterion of an orthogonally reinforced concrete slab elements, *J. Mech. Phys. Solids* **14**, No. 1, 65–71.
Morris, A. J., 1972, Structural optimization by geometric programming, *Int. J. Solids Struct.* **8**, No. 7, 847–864.
Morris, A. J., 1974, A primal-dual method for minimum weight design of statically determinate structures with several systems of load, *Int. J. Mech. Sci.* **16**, 801–807.
Morris, A. J., 1978, On condensed geometric programming in structural optimization, *Comput. Meth. Appl. Mech. Eng.* **15**, No. 2, 139–148.
Morris, A. J., Ed., 1982, *Foundations of Structural Optimization: A Unified Approach*, Wiley, New York.
Moses, F., 1977, Structural system reliability and optimization, *Comput. Struct.* **7**, 283–290.
Moses, F., and Onoda, S., 1969, Minimum weight design of structures with application to elastic grillages, *Int. J. Num. Meth. Eng.* **1**, 311–331.
Moussouros, M., 1976, Finite element optimal design of transverse web frames, *Comput. Struct.* **6**, No. 3, 253–266.
Murotsu, Y., 1983, Combinatorial properties of identifying dominant failure paths in structural systems, *Bull. Univ. Osaka Prefect. Ser. A* **32**, 107–116.
Murotsu, Y., Kishi, M., Okada, H., Yonezawa, M., and Taguchi, K., 1984a, Probabilistically optimum design of frame structure, in *System Modelling and Optimization* (Thoft-Christensen, P., Ed.), Springer-Verlag, Berlin.
Murotsu, Y., Yonezawa, M., Okada, H., Matsuzaki, S., and Matsumoto, T., 1984, A study on first-order-second-moment method in structural reliability, *Bull. Univ. Osaka Prefect. Ser. A* **33**, 23–36.
Murphy, R. D., Sabat, D. J., and Taylor, R. J., 1965, Least cost ship characteristics by computer techniques, *Mar. Technol.* **2**, No. 2, 174–202.
Murtagh, B. A., and Saunders, M. A., 1983, *MINOS 5.0 User's Guide*, Report SOL 83-20, Department of Operations Research, Stanford University, Stanford, CA.
Nakamura, T. N., 1976, Minimum weight design of multistory, multispan plane frames subject to reaction constraints, *J. Struct. Mech., ASME*, **4**, 3.

Neal, B. G., 1956, *The Plastic Methods of Structural Analysis*, Chapman and Hall, London.

Neghabat, F., and Stark, R. M., 1973, A cofferdam design optimization, *Math. Program.* **3**, 263–276.

Nemhauser, G. L., 1966, *Introduction to Dynamic Programming*, Wiley, New York.

Nemirovsky, Yu. V., 1975, Effect of shear forces on optimal design of plastic beams and plates, in *Optimization in Structural Design*, Warsaw, pp. 484–496.

Von Neumann, J., and Morgenstern, O., 1947, *Theory of Games and Economic Behaviour*, Second Edition, Princeton University Press, Princeton, NJ.

Novozhilov, V. V., 1964, *Thin Shell Theory*, Nordhoff, Groningen.

Nowacki, H. G., Brusis, F., and Swift, P. H., 1970, Tanker preliminary design: an optimization problem with constraints, *Trans. SNAME.*

Okada, H., Matsuzaki, S., and Murotsu, Y., 1983, Safety margins for reliability analysis of frame structures, *Bull. Univ. Osaka Prefect. Ser. A* **32**, 155–163.

Olhoff, N., 1973, Singularities, local optima and formation of stiffeners in optimal design of plates, in *Optimization in Structural Design*, Warsaw, pp. 82–103.

Olhoff, N., 1986, Structural optimization by variational methods, in *Computer Aided Optimal Design*, Vol. 1 (Soares, C. M., Ed.), Troja, Portugal, pp. 7–96.

Osyczka, A., 1978, An approach to multicriterion optimization problems for engineering design, *Comput. Meth. Appl. Mech. Eng.* **15**, 309–333.

Osyczka, A., 1984, *Multicriterion Optimization in Engineering*, Ellis Horwood, Chichester, England.

Palmer, A. C., 1968, Optimal structural design by dynamic programming, *Proc. ASCE J. Struct. Div.* **94**, No. ST 8, 1887–1906.

Palmer, A. C., 1969, Limit analysis of cylindrical shells by dynamic programming, *Int. J. Solid Struct.* **5**, 289–302.

Palmer, A. C., 1973, Dynamic programming and structural optimization, in *Optimum Structural Design* (Zienkiewicz, O. C., and Gallagher, R. H., Eds.), Wiley, New York.

Palmer, A. C., and Sheppard, D. J., 1970, Optimizing the shape of pin joined structures, *Proc. Inst. Civ. Eng. London* **47**, 363–376.

Pareto, V., 1896, *Cours d'Economie Politique*, Vols. I, II, F. Rouge, Lausanne.

Pascual, L. D., and Ben-Israel, A., 1970, Constrained maximization of posynomials by geometric programming, *J. Optimiz. Theory Appl.* **5**, 73–80.

Passy, U., 1971, Generalized weighted mean programming, *SIAM J. Appl. Math.* **20**, 763–778.

Passy, U., 1974, Signomial geometric programming: determining the global minimum, in *Mathematical Programs for Activity Analysis* (Van Moeseke, P., Ed.), North-Holland, Amsterdam.

Passy, U., and Wilde, D. J., 1967, Generalized polynomial optimization, *SIAM J. Appl. Math.* **15**, No. 5, 1344–1356.

Pedersen, P., 1969, On the minimum mass layout of trusses, *Proceedings No. 36, AGARD-CP-36-70 Symposium*, Istanbul.

Pedersen, P., 1972, On the optimal layout of multi-purpose trusses, *Comput. Struct.* **2**, 695–712.

Pedersen, P., 1973, Optimal joint positions for space trusses, *J. Struct. Div., ASCE* **99**, No. ST 12, 2459–2476.

Pedersen, P., 1981, The integrated approach of FEM-SLP for solving problems of optimal design in *Optimization of Distributed Parameter Structures*, Vol. 1 (Haug, E. J., and Cea, J., Eds.), Sijthoff & Noordhoff, Amsterdam, pp. 757–780.

Pedersen, P., 1983, A unified approach to optimal design, in *Optimization Methods in Structural Design* (Eschenauer, H., and Olhoff, N., Eds.), Bibliographisches Institut, Zurich, pp. 182–187.

Pedersen, P., and Jorgensen, L., 1984, Minimum mass design of elastic frames subjected to multiple load cases, *Comput. Struct.* **18**, No. 1, 147–157.

Peterson, E. L., 1970, Symmetric duality for generalised unconstrained geometric programming, *J. Appl. Math.* **19**, 487-526.

Peterson, E. L., 1976, Geometric programming, *SIAM Rev.* **18**, 1-51.

Pettersen, E., 1973, *Automated design program for plane framework structures in tankers*, User's Manual for NV383, Report No. 72-15-S, Norske Veritas.

Plaut, R. H., 1986, Optimal forms of shallow shells, in *Computer Aided Structural Design*, Vol. 1 (Soares, C. M., Ed.), pp. 138-147.

Plaut, R. H., Johnson, L. W., and Parbery, R. 1984, Optimal forms of shallow shells with circular boundary, Parts 1-3, *J. Appl. Mech.* **51**, 526-539.

Pochtman, Yu., and Baranenko, V., 1975, *Dynamic Programming in the Problems of Structural Mechanics* (in Russian), Stroizdat, Moskva.

Polizzotto, C., and Mazzarella, C., 1979, Limit analysis of frames with axial force-bending moment interaction, *J. Struct. Mech.* **7**, 83-106.

Pontryagin, L. S., Gamkrelidze, R. V., and Mischenko, E. F., 1962, *The Mathematical Theory of Optimal Processes*, Interscience, New York.

Porter Goff, R. F. D., 1966, Decision theory and the shape of structures, *J. R. Aeronaut. Soc.* **70**, No. 3, 448.

Porter Goff, R. F. D., 1969, *Dynamic programming for optimal structural design*, Report 69-4, Department of Engineering, University of Leicester, England.

Powell, M. J. D., 1983, Report DAMTP 1983/NA 17, Department of Applied Mathematics, University of Cambridge, Cambridge.

Prager, W., 1956a, Minimum weight design of portal frames, *J. Eng. Mech., ASCE* **82** (October), Paper 1073.

Prager, W., 1956b, General theory of limit design, *Proceedings of the Eighth International Congress on Applied Mechanics*, Istanbul, 1952, Vol. 2, Istanbul, pp. 65-72.

Prager, W., 1959, On a problem of optimal design, *Proceedings of the IUTAM Symposium on Non-Homogeneity in Elasticity and Plasticity*, Pergamon Press, London, pp. 125-152.

Prasad, B., and Haftka, R. T., 1979, Optimal structural design with plate finite elements, *J. Struct. Div., ASCE* **105**, No. ST 11, 2367-2382.

Rackwitz, R., and Fiessler, B., 1978, Structural reliability under combined random load sequences, *Comput. Struct.* **9**, 489-494.

Ragsdell, K., and Phillips, D., 1976, Optimal design of a class of welded structures using geometric programming, *Trans. ASME*, **3**.

Raj, P. P., and Durrant, S. O., 1976, Optimum structural design by dynamic programming, *Proc. ASCE J. Struct. Div.* **102**, No. ST 8, 1575-1589.

Ramaswamy, G. R., 1958, The theory of shells in the form of a Prandtl membrane, *Civ. Eng. Public Works Rev.* **53**, N626.

Rao, S. S., 1978, *Optimization, Theory and Applications*, Wiley, New Delhi.

Rao, S. S., and Hati, S. K., 1979, Game theory approach in multi-criteria optimization of function generating mechanisms, *Trans. ASME, J. Mech. Eng.* **101**, 398-405.

Raskazov, A. O., and Dekhtyar', A. S., 1978, *The Limit Design of Shells* (in Russian), Vyssha Shkola, Kiev.

Ratner, M., Lasdon, L. S., and Jain, A., 1978, Solving geometric programs using GRG: results and comparisons, *J. Optimiz. Theory Appl.* **26**, 253-265.

Razani, R., and Goble, G. G., 1966, Optimum design of constant depth plate girders, *J. Struct. Div., ASCE* **92**, ST 2.

Rechenberg, I., 1973, *Evolutionsstrategie*, Fromann Verlag, Stuttgart Bad Cannstatt.

Reitman, M. I., 1967, Optimum design of the three-dimensional reinforced concrete structures (in Russian), in *Computer Science in Construction and Design*, No. II-3, Giprotis, Moscow.

Reitman, M. I., 1968, Optimum design and unification of structures by means of dynamic programming (in Russian), *Stroit. i Arkhitekt.* **5**, Novosibirsk, 138-142.

Reitman, M. I., 1983, Optimum design of concrete shells, *N.Y. Acad. Sci. Ann.* **410**, 263-269.

Reitman, M. I., 1985, Optimum design of metal shells, *N.Y. Acad. Sci. Ann.* **452**, 54-63.

Reitman, M. I., and Shapiro, G. S., 1976, *Problems and Methods of Optimal Design of Deformable Bodies* (in Russian), Nauka, Moscow.

Reitman, M. I., and Yarin, L. I., 1974, *Optimal design of the reinforced concrete structures using computers* (in Russian), Stroyizdat, Moscow.

Reklaitis, G. V., and Wilde, D. J., 1974, Geometric programming via a primal auxiliary problem, *AIIE Trans.* **6**, 308-317.

Richardson, J. A., and Kuester, J. L., 1973, Algorithm 454, The complex method for constrained optimzation, *Commun. ACM*, **16**, No. 8, 487-489.

Rijckaert, M. J., and Martens, X. M., 1976, A condensation method for generalized geometric programming, *Math. Programm.* **11**, 89-93.

Rijckaert, M. J., and Martens, X. M., 1978a, Comparisons of generalized geometric programming algorithms, *J. Optimiz. Theory Appl.* **26**, 205-242.

Rijckaert, M. J., and Martens, X. M., 1978b, Bibliographical note on geometric programming, *J. Optimiz. Theory Appl.* **26**, 325-337.

Robinson, J., 1973, *Integrated Theory of Finite Element Methods*, Wiley, London.

Rosen, J. B., 1960, The gradient projection method for nonlinear programming, Part I. *J. Soc. Ind. Appl. Math.* **8**, 181-217.

Rosen, J. B., 1961, The gradient projection method for nonlinear programming, Part II. *J. Soc. Ind. Appl. Math.* **9**, 514-532.

Rosen, J. B., 1964, Primal partition programming for block diagonal matrices, *Num. Math.* **6**, 250-260.

Rozvany, G., 1966, Rational approach to the plate design, *ACI J.* **63**, 1077-1089. (See also discussion by D. Clyde, T. Katow, and M. Reitman, 1966, 1551-1559).

Rozvany, G., 1976, Optimal Design of Flexural Systems, Pergamon Press, Oxford.

Rzhanitsyn, A. R., 1960, *Shallow Shells and Wavy Roofs*, Stroyizdat, Moscow.

Saaty, T. L., 1979, Applications of analytical hierarchies, *Math. Comput. Simul*, **21**, 1-20.

Salukvadze, M. E., 1979, *Vector-Valued Optimization Problems in Control Theory*, Academic Press, New York.

Sarma, V. L. N., Martens, X. M., Reklaitis, G. V., and Rijckaert, M. J., 1978, A comparison of computational strategies for geometric programs, *J. Optimiz. Theory Appl.* **26**, 185-204.

Sattler, H. J., 1982, *Ersatzprobleme für Vektoroptimierungsaufgaben und ihre Anwendung in der Strukturmechanik*, VDI-Verlag, GmbH, Düsseldorf, Series 1, No. 88.

Save, M. A., 1981, *Plastic Analysis and Design of Plates, Shells and Disks*, North-Holland.

Save, M., and Prager, W., Eds., 1985, *Structural Optimization*, Vol. 1, Plenum Press, New York.

Schittkowski, K., 1980, *Nonlinear Programming Codes. Information, Tests, Performance*, Lectures Notes in Economics and Mathematical Systems, Vol. 183, Springer-Verlag, Berlin.

Schittkowski, K., 1981, The nonlinear programming methods of Wilson, Han and Powell with an augmented Lagrangian type line search function, *Num. Math.* **38**, 83-114 and 115-127.

Schittkowski, K., 1984, *User's Guide for the Nonlinear Programming Code NLPQL*, Report, Institute fur Informatik, Universitat Stuttgart.

Schittkowski, K., 1985, *Software for Mathematical Programming.* NATO ASI Series, Vol. F15, *Computational Mathematical Programming*, (Schittkowski, K., Ed.), Springer-Verlag, Berlin.

Schmit, L. A., 1960, Structural design by systematic synthesis, in *Proceedings of the Second Conference on Electronic Computations*, ASCE, Pittsburgh, pp. 105-132.

Schmit, L. A., Jr., and Mallet, R. H., 1963, Structural synthesis and design parameter hierarchy, *J. Struct. Div., ASCE* **89**, No. ST 4, 269-299.

Schmit, L. A., Jr., and Mehrinfar, M., 1982, Multilevel optimum design of structures with fiber-composite stiffened panel components, *AIAA J.* **20**, 138-147.

Schmit, L. A., and Miura, H., 1978, An advanced structural analysis/synthesis capability—

ACCESS-2, *Int. J. Num. Meth. Eng.* **12**, 353–377.
Schmit, L. A., Jr., and Ramanatham, R. K., 1978, Multilevel approach to minimum weight design including buckling constraints, *AIAA J.* **16**, 97–104.
Schwefel, H. P., 1977, *Numerische Optimierung von Computer-Modellenmittels der Evolutionsstrategie*, Birkhauser Verlag, Basel.
Scott, C. H., and Jefferson, T. R., 1977, Duality in infinite-dimensional mathematical programming: convex integral functionals, *Pacific J. Math.* **24**, 525–539.
Sepulveda, A., and Cassis, J. H., 1986, An efficient algorithm for the optimum design of trusses with discrete variables, *Int. J. Num. Meth. Eng.* **23**, 1111–1130.
Sheppard, D. J., 1970, *Structural design optimization by dynamic programming*, Ph.D. Dissertation, University of Cambridge.
Sheppard, D. J., and Palmer, A. C., 1972, Optimal design of transmission towers by dynamic programming, *Comp. Struct.* **2**, No. 4, 455–468.
Sheu, C. Y., and Schmit, L. A., Jr., 1972, Minimum weight design of elastic redundant trusses under multiple static load conditions, *AIAA J.* **10**, 155–162.
Shield, R. T., 1960, On the optimum design of shells, *J. Appl. Mech.* **27**, 316–322.
Sikorski, K. A., and Borkowski, A., 1985, On reliability analysis and optimization of plastic plane frames, *Rozpr. Inz.* (*Eng. Trans.*) **33**, 311–322.
Sin, Y. T., and Reklaitis, G. V., 1981, On the computational utility of generalized geometric programming solution methods. Parts I, II, in *Seventh ASME Design Automation Conference, Hartford, Progress in Engineering Optimization*, (Mayne, R. W., and Ragsdell, K. M., Eds.), ASME Design Division, New York.
Smeers, Y., 1972, *Studies in geometric programming with applications to management science*, Ph.D. Thesis, Carnegie-Mellon University, Pittsburgh, PA.
Smith, G. H., and Woodhead, R. G., 1973, A design scheme for ship structures, *Trans. RINA* **115**, 1–12.
Soares, C. M., Ed., 1986, *Computer Aided Optimal Design, Proceedings of the Troja, Portugal, Conference*, Springer, New York.
Sobieszczanski-Sobieski, J., 1979, *An integrated computer procedure for sizing composite airframe structures*, NASA TP-1300.
Sobieszczanski-Sobieski, J., James, B. B., and Dovi, A. R., 1985, Structural optimization by multilevel decomposition, *AIAA J.* **23**, 1775–1782.
Sobieszczanski-Sobieski, J., and Leondorf, D. D., 1972, A mixed optimization method for automated design of fuselage structures, *J. Aircr.* **9**, 805–811.
Soerensen, J. D., and Thoft-Christensen, P., 1986, Structural optimization with reliability constraints, in *Lecture Notes in Control and Information Science*, Vol. 84, *System Modelling and Optimization* (Prekopa, A., Szelezsan, J., and Strazicky, B., Eds.), Springer-Verlag, Berlin, pp. 876–887.
Spillers, W. R., 1966, Artificial intelligence and structural design, *Proc. ASCE* **92**, No. ST 6, 491–497.
Spillers, W. R., 1971, Graph theory, switching theory, and structural design, *Appl. Mech. Rev.* **29**, 501–504.
Spillers, W. R., 1975, *Iterative Structural Design*, North-Holland, Amsterdam.
Spillers, W. R., and Friedland, L., 1972, On adaptive structural design, *Proc. ASCE* **98**, No. ST 10, 2155–2163.
Stachowicz, A., and Ziobron, W., 1986, *Underground Water Tank Systems* (in Polish), Arkady, Warsaw.
Stadler, W., 1975, Preference optimality and applications of Pareto optimality, in *Multi-Criteria Decision Making* (Leitmann, G., and Marzollo, A., Eds.), CISM courses and lectures, No. 211, Springer-Verlag, Wien and New York.
Stadler, W., 1977, Natural structural shapes of shallow arches, *J. Appl. Mech.* **44**, No. 2, 291–298.

Stadler, W., 1978, Natural structural shapes (the static case), *Q. J. Mech. Appl. Math.* **31**, No. 2, 196-217.
Stadler, W., 1980, Preference optimality in multicriteria control and programming problems, *Nonlinear Anal. Theory, Meth. Appl.* **4**, No. 1, 51-65.
Stadler, W., 1984a, A comprehensive bibliography on MCDM, in *MCDM—Past Decade and Future Trends; A Source Book of Multiple Criteria Decision Making,* (Zeleny, M., Ed), JAI Press, Greenwich, CT.
Stadler, W., 1984b, Multicriteria optimization in mechanics (a survey), *AMR* **37**, No. 3, 277-286.
Swiszczowski, S., 1971, Optimum design of continuous beams using dynamic programming (in Polish), *Czasop. Tech.* **75**, No. 9, 9-12.
Swiszczowski, S., 1974, Two cases of the optimum design of beam structures by dynamic programming (in Polish), *Arch. Inz. Lad.* **20**, No. 1, 83-98.
Switzky, H., 1965, Minimum weight design with structural reliability, *J. Aircr.* **2**, 228.
Templeman, A. B., 1970, Structural design for minimum cost using the method of geometric programming, *Proc. Inst. Civ. Eng. London* **46**, 459-472.
Templeman, A. B., 1971, The application of geometric programming to the optimum design of bridge structures, *Proceedings of the Interational Symposium on Optimization in Road Design,* PTRC London, pp. 125-130.
Templeman, A. B., 1972, Geometric programming with examples of the optimum design of floor and roof systems, *Proceedings of the International Symposium on Computer-Aided Structural Design,* Warwick University, Coventry.
Templeman, A. B., 1976, Optimum truss design by sequential geometric programming, *J. Struct. Eng.* **3**, No. 4, 155-163.
Templeman, A. B., Wilson, A. J., and Winterbottom, S. K., 1972, *SIGNOPT—a computer code for solving signomial geometric programming problems,* Research Report, Department of Civil Engineering, University of Liverpool, Liverpool, England.
Thoft-Christensen, P., and Baker, M. J., 1982, *Structural Reliability Theory and Its Applications,* Springer-Verlag, Berlin.
Thoft-Christensen, P., and Soerensen, J. D., 1984, Reliability analysis of elasto-plastic structures, in *System Modelling and Optimization,* Lecture Notes in Control and Information Sciences, Vol. 59, Springer-Verlag, New York, pp. 555-565.
Thomas, H. R., Jr., and Brown, D. M., 1977, Optimum least cost of a truss roof systems, *Comput. Struct.* **7**, 13-22.
Toakley, A. R., 1968, Optimum design using available sections, *J. Struct. Div., ASCE* **94**, No. ST 5, 1219-1241.
Twisdale, L. A., and Khachaturian, N., 1973, Absolute minimum weight structure by dynamic programming, *Proc. ASCE* **99**, No. ST 11, 2339-2344.
Twisdale, L. A., and Khachaturian, N., 1975a, Multistage optimization of structures, *Proc. ASCE* **101**, No. ST 5, 1005-1020.
Twisdale, L. A., and Khachaturian, N., 1975b, Application of dynamic programming to optimization of structures, in *Optimization in Structural Design* (Sawczuk, A., and Mroz, Z., Eds.), Springer-Verlag, Berlin.
Unklesbay, K., Staats, G. E., and Creighton, D. L., 1973, Optimal design of journal bearings, *Int. J. Eng. Sci.* **11**, 973-983.
Vanderplatts, G. N., and Moses, F., 1972, Automated design of trusses for optimum geometry, *J. Struct. Div., ASCE* **98**, No. ST 3, 671-689.
Vargas, L. G., 1980, A note on the eigenvalue consistency index, *Appl. Math. Comput.* **7**, 195-203.
Vitiello, E., 1977, Standardization and optimum structural design by dynamic programming, *J. Optimiz. Theory Appl.* **23**, No. 1, 183-191.
Wang, S. H., and Ragsdell, K. H., 1983, *OPT2 User's Manual,* DOL, University of Arizona.
Wangdahl, G., 1972, *The external penalty function optimization technique and its application to*

ship design, Report No. 129, Department of Naval Architecture and Marine Engineering, University of Michigan, Ann Arbor.

Watkins, R. I., and Morris, A. J., 1987, A multicriteria objective function optimization scheme for laminated composites for use in multilevel structural optimization schemes, *Comput. Meth. Appl. Mech. Eng.* **60**, 233–251.

Wilde, D. J., 1976, Global minimization of a cooled reactor using a posynomial lower bounding function, *AICh.E.J.* **22**, 685–686.

Wilde, D. J., 1978, *Globally Optimal Design*, Wiley, New York.

Wilde, D. J., and Beightler, C. S., 1967, *Foundations of Optimization*, Prentice-Hall, Englewood Cliffs, NJ.

Wind, Y., and Saaty, T. L., 1980, Marketing applications of the analytic hierarchy process, *Manage. Sci.* **26**, No. 7, 641–658.

Wolfe, P., 1959, The simplex method for quadratic program, *Econometrica* **27**, 382.

Wolfe, P., 1967, Methods of nonlinear programming, in *Nonlinear Programming* (Abadie, J., Ed.), Wiley, New York.

Wolfensberger, G., 1964, *Traglast und optimale Bemessung von Platten*, Eidgenoss. Techn. Hochschule, Zurich.

Woo, T. H., and Schmit, L. A., 1981, Decomposition in optimal plastic design of structures, *Int. J. Solids Struct.* **17**, 39–56.

Woodhead, R. G., 1972, The application of optimization methods to structural design, *BSRATM*, 412.

Young, D. T., 1985, *Optimal forms of the rectangular-base, shallow shell with respect to buckling.* Ph.D. Dissertation, Virginia Polytechnic Institute and State University, Blacksburg, VA.

Yu, C. H., Das Gupta, N. C., and Paul, H., 1984, Optimization of modular floor system, *International Conference on Computer Technology and Its Applications*, Bangkok, pp. 44–60.

Yu, C. H., Das Gupta, N. C., and Paul, H., 1986, Optimization of prestressed concrete bridge girders, *Eng. Optimiz.* **10**, 1.

Yu, C. H., Das Gupta, N. C., and Paul, H., 1987, Application of geometric programming to structural design, *International Conference on Optimization Techniques and Applications*, Singapore, pp. 1078–1086.

Yu, P. L., 1985, *Multiple-Criteria Decision Making: Concepts, Techniques, and Extensions*, Plenum Press, New York and London.

Zadeh, L., 1963, Optimality and non-scalar-valued performance criteria, *IEEE Trans., Autom. Control*, **AC-8**.

Zangwill, W. I. 1967a, The convex simplex method, *Manage. Sci.* **14**, 221–238.

Znagwill, W. I., 1967b, Nonlinear programming via penalty function, *Manage. Sci.* **13**, No. 5, 344–358.

Zangwill, W. I., 1969, *Nonlinear Programming: A Unified Approach*, Prentice-Hall, Englewood Cliffs, NJ.

Zener, C. M., 1961, A mathematical aid in optimising engineering designs, *Proc. Natl. Acad. Sci.* **47**, 537–539.

Zener, C. M., 1971, *Engineering Design by Geometric Programming*, Wiley, New York.

Zienkiewicz, O. C., 1977, *The Finite Element Method*, McGraw-Hill, New York.

Zienkiewicz, O. C., and Campbell, J. S., 1973, Shape optimization and sequential linear programming, in *Optimum Structural Design* (Gallagher, R. H., and Zienkiewicz, O. C., Eds.), Wiley, New York.

Zyczkowski, M., and Kruzelecki, J., 1975, Optimal design of shells with respect to their stability, in *Optimization in Structural Design* (Sawczuk, A., and Mroz, Z., Eds.), Springer-Verlag, Berlin.

Index